Molecular Genetics

Part III

Chromosome Structure

MOLECULAR BIOLOGY

An International Series of Monographs and Textbooks

Editors: BERNARD HORECKER, NATHAN O. KAPLAN, JULIUS MARMUR, AND HAROLD A. SCHERAGA

A complete list of titles in this series appears at the end of this volume.

Molecular Genetics

Edited by

J. HERBERT TAYLOR

Institute of Molecular Biophysics
Florida State University
Tallahassee, Florida

Part III

Chromosome Structure

1979

ACADEMIC PRESS · New York San Francisco London
A Subsidiary of Harcourt Brace Jovanovich, Publishers

ACADEMIC PRESS, INC.
111 Fifth Avenue, New York, New York 10003

United Kingdom Edition published by
ACADEMIC PRESS, INC. (LONDON) LTD.
24/28 Oval Road, London NW1 7DX

Library of Congress Cataloging in Publication Data

Taylor, James Herbert,   Date   ed.
    Molecular genetics.

    (Molecular biology; an international series of mono-
graphs and textbooks, v. 4, pt. 1–

    CONTENTS: --v. 3.
Chromosome structure.
    1. Genetics--Collected works. I. Title.
QH426.5.T38   574.8'732   62--22114
ISBN 0–12–684403–8

PRINTED IN THE UNITED STATES OF AMERICA

79 80 81 82   9 8 7 6 5 4 3 2 1

Contents

Chapter I

The Role of Restriction Endonucleases in Molecular Genetics

MARC ZABEAU AND RICHARD J. ROBERTS

Chapter II

DNA Topoisomerases: Enzymes That Catalyze the Concerted Breaking and Rejoining of DNA Backbone Bonds

JAMES C. WANG AND LEROY F. LIU

Chapter III

Enzymatic Methylation of DNA: Patterns and Possible Regulatory Roles

J. HERBERT TAYLOR

Chapter IV

Transcriptional Units in Eukaryotic Chromosomes

ANN L. BEYER, STEVEN L. MCKNIGHT, AND OSCAR L. MILLER, JR.

Chapter V

Recognition and Control Sequences in Nucleic Acids

P. ANDREW BIRO AND SHERMAN M. WEISSMAN

Chapter VI

Nucleosomes: Composition and Substructure

RANDOLPH L. RILL

Chapter VII

Nucleosomes and Higher Levels of Chromosomal Organization

JOHN C. HOZIER

Contributors to Part III

Numbers in parentheses indicate the pages on which the authors' contributions begin.

ANN L. BEYER (117), *Department of Biology, University of Virginia, Charlottesville, Virginia 22901*

P. ANDREW BIRO (177), *Department of Human Genetics, Yale University School of Medicine, New Haven, Connecticut 06510*

JOHN C. HOZIER (315), *Department of Laboratory Medicine and Pathology, University of Minnesota Medical School, Minneapolis, Minnesota 55455*

LEROY F. LIU* (65), *Department of Biochemistry and Molecular Biology, Harvard University, Cambridge, Massachusetts 02138*

STEVEN L. MCKNIGHT (117), *Department of Embryology, Carnegie Institution of Washington, Baltimore, Maryland 21210*

OSCAR L. MILLER, JR. (117), *Department of Biology, University of Virginia, Charlottesville, Virginia 22901*

RANDOLPH L. RILL (247), *Department of Chemistry and Institute of Molecular Biophysics, Florida State University, Tallahassee, Florida 32306*

RICHARD J. ROBERTS (1), *Cold Spring Harbor Laboratory, Cold Spring Harbor, New York 11724*

J. HERBERT TAYLOR (89), *Institute of Molecular Biophysics, Department of Biological Science, Florida State University, Tallahassee, Florida 32306*

JAMES C. WANG (65), *Department of Biochemistry and Molecular Biology, Harvard University, Cambridge, Massachusetts 02138*

SHERMAN M. WEISSMAN (177), *Department of Human Genetics, Yale University School of Medicine, New Haven, Connecticut 06510*

MARC ZABEAU (1), *European Molecular Biology Laboratory, 69 Heidelberg, West Germany*

Present address: Department of Biochemistry and Biophysics, School of Medicine, University of California at San Francisco, San Francisco, California 94143.

Preface

Twelve years have elapsed since "Molecular Genetics," Part II, was published. Although there was no intention of including another volume at that time, new information has accumulated at such an alarming rate that concepts which were widely held must now be highly modified or even abandoned. The first part was written soon after the structure of DNA proposed by Watson and Crick had been accepted and numerous advances were being made based on the proposed structure and relating to replication, chromosome reproduction, mechanisms of mutation, coding, and the molecular control of transcription and the protein synthetic system. Many of the presentations have withstood the test of time and require limited modification, while others are no longer relevant.

The most recent and rapid advances impinge on the problems of chromosome structure and the functional role of the structural organization of the genetic apparatus. Therefore, in preparing to expand this treatise, the decision was made to consider a number of topics related to the structure and modification of DNA, chromatin, and the higher order organization insofar as treatment at the molecular level was possible.

We begin with the restriction enzymes (site-specific endonucleases), which have been so useful in making many of the analyses possible and reveal interesting and still poorly understood roles that these enzymes may serve in the cells which produce them. Another group of nucleases, which have been less useful to the molecular biologist but seem to be so essential to the replication, organization, and function of DNA, are the topoisomerases (nicking and closing enzymes). Many of the earlier speculations on the unwinding of DNA and the problems posed have been solved in evolution by this interesting group of enzymes, which appear to conserve the bond energy for reversible reactions that were not envisioned a few years ago. The role of methylation of DNA has puzzled and intrigued some of us since it was discovered many years ago. Its role in the modification–restriction systems has revived interest, and it is likely that surprises await us in this area. A major role for methylation in eukaryotes has yet to be discovered even though it is of almost universal occurrence in the higher forms. Insects may present an important exception, but some other type of DNA modification may be substituted in these animals.

Transcription was beginning to be understood, we thought, but now cloning and sequence analysis has changed concepts of posttranscriptional modifications of the premessenger RNA that reveal a new dimension in the organization of the whole genome. These developments are so recent that we have not assessed their full impact in this volume. The significance and role in the evolution of the genome will have to wait for later treatment. However, in Chapter IV, Oscar Miller and his associates give us a molecular view of the organization and operation of the genetic apparatus which has both astonished and pleased the chemically oriented molecular biologist as well as the electron microscopist.

The cloning and sequencing of DNA are beyond the scope of our treatment and are not far enough advanced to be covered fully, but Biro and Weissman present a basis for following the new developments and also discuss certain regulatory features of the genetic systems so far examined on the basis of sequence information.

The major changes in concepts of chromatin structure and packaging of DNA that have evolved from studies of nuclease digests and electron micrographs have been traced in the last two chapters. The first, written by Rill, considers the nucleosome and its substructure, with emphasis on histone–DNA interactions and arrangements. The second treats the higher orders of organization and possible subunits of chromosomes based on the knowledge gained from the analysis of the nucleosomes and their components.

Plans for this volume were made with the hope that all chapters could be written simultaneously and that the information in all would be equally up-to-date. Such plans seem never to work in reality and, as in earlier volumes, there was considerable variation in the time the different chapters were finally completed. All authors were given a chance to update their material, but there are limits to how much one can revise manuscript after its initial conception. Thanks are due to those who finished early and had to bear with delays of others, but we trust most of the work is durable enough to withstand the test of time. If the latest references are missing from some chapters it is probably related to this variation in completion time.

We wish to thank all of the contributors and especially those scientists and publishers who generously contributed illustrations, graphs, and other illustrative materials to the volume at the request of the various authors. For expediting the final stages of publication and for managing many of the technical details, the publisher is due much of the credit. We hope you will find the collection timely, informative, and interesting reading.

J. HERBERT TAYLOR

Contents of Part I

Contents of Part II

Chapter I

The Role of Restriction Endonucleases in Molecular Genetics

MARC ZABEAU AND
RICHARD J. ROBERTS

I. INTRODUCTION

The acronyms echoing through the halls of our academic institutions usually reflect the haute couture of scientific research. Although old friends such as DPN, FAD, PEP still visit occasionally, a new circle has developed, and *Eco*RI, *Sal*I, and *Pst*I have moved into the vocabulary of the molecular biologist. With them have come terms such as agarose gels, ligase, maps, vectors, blotting, molecular cloning, and *restriction endonucleases*. The new acronyms serve to identify this

1

latter class of important enzymes, which have revitalized molecular biology and have finally allowed direct access to the mysteries of the eukaryotic chromosome.

The information contained within the genome of every organism provides a precise program upon which its biological processes depend. To comprehend these processes in molecular terms, a detailed analysis of gene organization and structure is essential. Molecular approaches have been hampered by the extreme complexity of the DNA molecules encoding this program. Even a simple bacterial chromosome, such as that of *Escherichia coli*, consists of a single polynucleotide chain containing several million nucleotides. The analysis of such a complex molecule requires that it be dissected into discrete segments, amenable to biochemical analysis. The restriction endonucleases have made this feasible. A new technology is now available to investigate the organization of chromosomes, and to analyze genes at both the functional and structural levels. In addition, restriction enzymes have played a key role in the development of recombinant DNA procedures which permit genes to be isolated and manipulated in a fashion hitherto unimaginable.

The study of restriction enzymes can be traced back to the early 1950s when Luria and his collaborators reported the phenomenon of host-controlled variation (Luria and Human, 1952; Bertani and Weigle, 1953; Luria, 1953). They showed that the ability of bacteriophages to grow on particular strains was dependent upon the specific "modifications" induced by the host in which they had been propagated previously. It was concluded that these bacteria must contain some "specificity systems" able to restrict the host range of phages. Extensive genetic and biochemical studies of these systems in *E. coli* strains led to the identification of the two components involved (Linn and Arber, 1968; Meselson and Yuan, 1968). The first was an endodeoxyribonuclease (restriction endonuclease) which could distinguish between host DNA and foreign DNA. This was made possible by strain-specific modification, accomplished by the second component of the system—a modification enzyme. Usually this is mediated by methylation of specific DNA sequences, within either the host DNA or phage DNA grown on that host, which then prevents their cleavage by the restriction enzyme.

Despite continued interest in the biological role of these enzymes from *E. coli*, work on them has been overshadowed by the events which followed the characterization of a similar enzyme, *Hind*II, from *Haemophilus influenzae* Rd (Kelly and Smith, 1970; Smith and Wilcox, 1970). For this enzyme, unlike the *E. coli* enzymes, cleaved DNA at specific sites.[1] Only 6 years have elapsed since the first report appeared exploiting the specificity of *Hind*II (Danna and Nathans, 1971), and yet the present proliferation of papers is almost overwhelming. More than 140 similar enzymes are now known, and a highly sophisticated technology is being applied to the study of gene structure and function. This chapter will attempt to provide a summary of the general properties of restriction enzymes and to describe their various applications in molecular genetics. Several earlier reviews have appeared (Arber, 1965, 1971, 1974; Arber and Linn, 1969; Boyer, 1971; Meselson *et al.*, 1972; Nathans and Smith, 1975; Roberts, 1976).

II. GENERAL PROPERTIES OF RESTRICTION ENZYMES

A. INTRODUCTION

The term "restriction endonuclease" was originally used to designate an endodeoxyribonuclease that was involved in a genetically defined process of host-controlled restriction. The first enzymes (*Eco*B and *Eco*K) were isolated from *E. coli* strains B and K (Linn and Arber, 1968; Meselson and Yuan, 1968) and were detected by their ability to selectively degrade, *in vitro*, DNA isolated from bacteriophages susceptible to *in vivo* restriction. Only phage DNAs which lacked the proper strain-specific modification were cleaved, whereas modified phage DNA was resistant to *in vitro* degradation. This property of restriction enzymes to degrade unmodified DNA selectively, but not modified DNA, was subsequently exploited to detect site-specific endonucleases in other bacterial strains (Smith and Wilcox, 1970; Yoshimori, 1971; Gromkova and Goodgal, 1972; Middleton *et al.*, 1972). These new enzymes differed significantly from the *Eco*B and *Eco*K

---

[1] The nomenclature used throughout this review is detailed in Smith and Nathans (1973). Restriction endonucleases bear a three-letter system name that abbreviates the genus and species of the organism from which they were isolated. Where necessary, a fourth letter is added to designate the strain. Roman numbers following the system name are assigned to differentiate multiple enzymes from the same source. Where only one enzyme has been isolated, the Roman number I is used to avoid later confusion if a second enzyme should be discovered. The prefix endo R (endonuclease R) is omitted to conserve space, and also because in most cases, the endonucleases have not been shown to form part of a genetic restriction-modification system. Examples are *Hind*II, one of multiple restriction enzymes from *Haemophilus influenzae* serotype d; *Hin*fI, an enzyme from *Haemophilus influenzae* serotype f; *Alu*I, an enzyme from *Arthrobacter luteus*; and *Hph*I, an enzyme from *Haemophilus parahaemolyticus* (in this case, *Hpa* was already used for enzymes from *Haemophilus parainfluenzae*).

endonucleases, and it soon became apparent that at least two different types of restriction endonucleases must exist (Boyer, 1971). The principal difference, which was to have far-reaching consequences, lay in the nature of the degradation products, for while the type I enzymes (EcoB and EcoK) gave a heterogeneous array of products, the type II enzymes gave a specific set of discrete fragments. The development of an agarose gel electrophoresis system to fractionate DNA fragments of different sizes (Aaij and Borst, 1972), gave a simple and rapid assay for the type II enzymes (Sharp et al., 1973). This assay has been used to screen many bacterial strains, and a large number of type II enzymes have been isolated (Roberts, 1976). Activities identified using this gel assay procedure are most properly designated site-specific endonucleases; nevertheless they are often referred to as "restriction enzymes," even though most have not been shown to participate in a restriction–modification system. Throughout this chapter, we will use the terms specific endonuclease, restriction endonuclease, and restriction enzyme interchangeably. After a brief description of the type I enzymes, the rest of this chapter will focus on the type II enzymes, which are rapidly becoming indispensable as the molecular scalpels of the contemporary biologist.

B. TYPE I RESTRICTION ENZYMES

Interest in the type I enzymes, EcoB and EcoK, has centered around their role in the biological process of host-controlled restriction and modification (Arber and Linn, 1969; Arber, 1974). Both restriction enzymes and their companion modification enzymes have been purified to near homogeneity, and their subunit structure, catalytic properties, and cleavage mechanism have been examined (Eskin and Linn, 1972a; Lautenberger and Linn, 1972; Yuan et al., 1975). Although both EcoB and EcoK bind to specific sites on the DNA (Arber and Kuhnlein, 1967; Murray et al., 1973b; Horiuchi et al., 1975; Brack et al., 1976b), they show no cleavage specificity (Horiuchi and Zinder, 1972; Murray et al., 1973a). In addition, their endonucleolytic activity requires the cofactors Mg^{2+}, ATP, and S-adenosylmethionine (Linn and Arber, 1968; Meselson and Yuan, 1968; Roulland-Dussoix and Boyer, 1969). The purified enzymes exist as complexes composed of three nonidentical subunits (Eskin and Linn, 1972a; Meselson et al., 1972), and this complex has been shown to catalyze both endonucleolytic cleavage (restriction) and methylation (modification) (Haberman et al., 1972; Vovis et al., 1974; Vovis and Zinder, 1975). Following cleavage, the restriction endonuclease is converted into a potent

ATPase (Eskin and Linn, 1972b; Yuan *et al.*, 1972). Enzymes exhibiting properties similar to those described for *Eco*K and *Eco*B have been isolated from *Haemophilus influenzae* strains R_d and R_f (Gromkova *et al.*, 1973; Piekarowicz *et al.*, 1976).

The main difference between the type I and the type II enzymes is that the latter recognize a specific sequence and *cleave* at a *specific* site. They require only Mg^{2+} as a cofactor and have a much simpler subunit structure. In particular, the restriction and modification enzymes exist as separate entities. Two enzymes, *Eco*P1 and *Eco*P15, specified by bacteriophage P1 (Meselson and Yuan, 1968; Haberman, 1974) and plasmid P15 (Reiser and Yuan, 1977) share properties with both type I and type II enzymes, but can be distinguished from both. They may be the progenitors of yet another type of restriction enzyme. Although they possess a subunit structure similar to that of the type I enzymes (Arber, 1974) and can catalyze both cleavage and methylation, they show no absolute requirement for S-adenosylmethionine, yet are stimulated by it, and do not catalyze a massive ATP hydrolysis (Haberman, 1974; Reiser and Yuan, 1977). Both enzymes display a cleavage specificity comparable with that of the type II enzymes, although complete digest patterns have not yet been observed (Risser *et al.*, 1974; Reiser and Yuan, 1977).

C. TYPE II RESTRICTION ENZYMES

1. *Detection and Purification Procedures*

The first procedures used to assay restriction enzyme activity were based upon the selective degradation of foreign DNA as opposed to host DNA. Degradation was measured as the loss of biological activity (Takano *et al.*, 1966; Linn and Arber, 1968; Meselson and Yuan, 1968; Gromkova and Goodgal, 1972; Takanami, 1973; Takanami and Kojo, 1973; Bron *et al.*, 1975) or as change in either sedimentation velocity (Meselson and Yuan, 1968; Roulland-Dussoix and Boyer, 1969) or viscosity (Smith and Wilcox, 1970; Middleton *et al.*, 1972). A filter-binding assay has also been described (Reiser and Yuan, 1977). These rather laborious assays, which are still the only ones available to monitor the type I enzymes, have now been superceded by the agarose gel assay (Sharp *et al.*, 1973). This assay takes advantage of the fact that a type II restriction enzyme generates a specific set of fragments upon digestion of a small substrate DNA. When fractionated by agarose gel electrophoresis in the presence of ethidium bromide, these fragments can be visualized directly by their fluorescence upon uv irradiation of the gel. When slab gels with multiple slots are used (Sugden *et al.*,

6   MARC ZABEAU AND RICHARD J. ROBERTS

1975), numerous assays can be performed simultaneously, thus permitting the direct visualization of the results of a chromatographic fractionation. The discrete banding pattern not only shows the elution profile of the enzyme but also reveals the presence of two different restriction enzymes within the same bacterial strain. Moreover, the presence of contaminating nonspecific nucleases in fractions containing a restriction enzyme can be inferred from the sharpness of the bands obtained under various digestion conditions. The simultaneous detection of both the desired specific endonuclease and the undesired nonspecific nucleases illustrates the power of this assay.

Most purification procedures are aimed at quickly obtaining an enzyme preparation which is relatively free of contaminating nonspecific nucleases and the required degree of purity strongly depends upon its particular usage. In comparative restriction enzyme analysis and genome mapping experiments, less pure enzyme preparations suffice, whereas DNA sequencing requires highly purified enzymes devoid of nonspecific contaminants. Only for studies of catalytic properties is homogeneous protein needed or its acquisition attempted. Because restriction enzymes are isolated from widely different bacterial sources each containing a different set of contaminants, it would be naive to suppose that a general purification procedure exists for all restriction enzymes. Nevertheless, many of the schemes reported in the literature (see the reference list in Table I) often represent only minor variations of that used for purifying HindII (Smith and Wilcox, 1970). The first steps in the isolation usually involve the preparation of a high-speed supernatant of the cell lysate and the removal of nucleic acids, by gel filtration or precipitation with either streptomycin sulfate or polyethylene imine. Further purification is achieved by column chromatography and, as with most enzymes involved in nucleic acid metabolism, phosphocellulose has proved immensely useful. Other ion exchangers, such as DEAE-cellulose, QAE-Sephadex, etc., have been used extensively, and, recently, several more exotic adsorbents have become fashionable. Columns of single-stranded DNA agarose (Schaller et al., 1972) have sometimes given dramatic purification (Sack, 1974; P. A. Myers and R. J. Roberts, unpublished results) as have the hydrophobic matrices provided by the ω-aminoalkyl-Sepharose derivatives (Shaltiel and Er-El, 1973; Gelinas et al., 1977b; Mann et al., 1978). Heparin-agarose has been introduced recently and may have general utility (Bickle et al., 1977). Finally, rapid and specialized procedures have been devised for purifying EcoRI (Bingham et al., 1977; Sumegi et al., 1977) and BglII (Bickle et al., 1977).

One of the key factors responsible for the success in purifying restriction enzymes has been their quite remarkable stability. Indeed, many enzymes will continue to digest DNA in a linear fashion for periods in excess of 12 hours. This must reflect both their inherent stability and also the absence of significant amounts of proteases in partially purified enzyme preparations. Since assays of crude cell extracts rarely give distinct fragment patterns, due to the high concentration of nonspecific nucleases, it is difficult to quantitate the amounts of enzyme originally present. For this reason, enzyme yield is usually described in terms of the amount of enzyme finally obtained. Phage λ DNA is a commonly used substrate for monitoring cleavage, and the yield is conveniently expressed in arbitrary units, where one unit is defined as the amount of enzyme necessary to completely digest 1 μg of λ DNA in 1 hour at 37°C. This unit definition must be viewed with caution, because it does not necessarily give an accurate reflection of the total amount of DNA which one might expect to cleave with a given amount of enzyme. Both the degree of purity of the enzyme and the DNA concentration can markedly influence the cleavage efficiency. Although a more rigorous definition of a unit would be desirable, the kinetic parameters necessary to establish an absolute rate are difficult to obtain when homogeneous enzyme preparations are not available. A satisfactory unit has been measured only in the case of *Eco*RI, where one unit is defined as the amount of enzyme that cleaves 1 pmole of phosphodiester bonds per minute (Greene *et al.*, 1975; Modrich and Zabel, 1976).

2. *Characterization*

The key feature which distinguishes one type II enzyme from another lies in the specificity of the double-strand break, so that the most useful characteristic is the nature of the recognition sequence and cleavage site. This contrasts with the usual situation, where enzyme characterization involves detailed studies of kinetic parameters, catalytic properties, and protein structure. Consequently, great efforts have been made to elucidate the nucleotide sequence which they recognize and the positions of cleavage relative to that sequence, while only limited data are available concerning kinetic parameters, etc. A summary of the most recent data for the well-characterized enzymes is presented in Table I, and the partially characterized enzymes are listed in Table II.

An important first step in the characterization of a new restriction enzyme involves a description of the fragment patterns that are obtained upon digestion of various substrate DNAs. Comparison of these fragment patterns with those obtained using enzymes of known speci-

TABLE I
Type II Restriction Endonucleases and Their Recognition Sequences

Enzyme[a]	Recognition sequence[b]	Ends[c]	Microorganism	References[d,e]
AccI	GT↓(A)(G)AC	2b 5'-ext.	Acinetobacter calcoaceticus	50
AluI	AG↓CT	Flush	Arthrobacter luteus	35
AsuI	G↓GNCC	3b 5'-ext.	Anabaena subcylindrica	21
AvaI	C↓PyCGPuG	4b 5'-ext.	Anabaena variabilis	22, 29
BalI	TGG↓CCA	Flush	Brevibacterium albidum	15
BamHI	G↓GATCC	4b 5'-ext.	Bacillus amyloliquefaciens	37, 46
BbvI	GC(↑)GC	?	Bacillus brevis	16
BclI	T↓GATCA	4b 5'-ext.	Bacillus caldolyticus	5, 39
BglII	A↓GATCT	4b 5'-ext.	Bacillus globigii	33, 47, 52
EcoRI	G↓AATTC	4b 5'-ext.	Escherichia coli RY13	18, 20, 49
EcoRII	↓CC(↑)GG	5b 5'-ext.	Escherichia coli R245	4, 6, 49
FnuDII	CG↓CG	Flush	Fusobacterium nucleatum D	26
HaeI	(↑)GG↓CC(↓)	Flush	Haemophilus aegyptius	31
HaeII	PuGCGC↓Py	4b 3'-ext.	Haemophilus aegyptius	3, 34, 45
HaeIII	GG↓CC	Flush	Haemophilus aegyptius	7, 27
HgaI	5'-GACGC(N)$_5$↓ 3'-CTGCG(N)$_{10}$↑	5b 5'-ext.	Haemophilus gallinarum	9, 43, 44
HhaI	GCG↓C	2b 3'-ext.	Haemophilus haemolyticus	36
HindII	GTPy↓PuAC	Flush	Haemophilus influenzae R$_d$	24, 42
HindIII	A↓AGCTT	4b 5'-ext.	Haemophilus influenzae R$_d$	32
HinfI	G↓ANTC	3b 5'-ext.	Haemophilus influenzae R$_f$	23, 28, 30
HpaI	GTT↓AAC	Flush	Haemophilus parainfluenzae	13, 19, 40
HpaII	C↓CGG	2b 5'-ext.	Haemophilus parainfluenzae	13, 19, 40
HphI	5'-GGTGA(N)$_8$↓ 3'-CCACT(N)$_7$↑	1b 3'-ext.	Haemophilus parahaemolyticus	25, 28
KpnI	GGTAC↓C	4b 3'-ext.	Klebsiella pneumoniae	41, 48
MboI	↓GATC	4b 5'-ext.	Moraxella bovis	14
MboII	5'-GAAGA(N)$_8$↓ 3'-CTTCT(N)$_7$↑	1b 3'-ext.	Moraxella bovis	10, 11, 14
MnlI	5'-CCTC(N) 5–10↓ 3'-GGAG(N)	?	Moraxella nonliquefaciens ATCC 17953	51

Enzyme	Sequence		Organism	Ref.
PstI	CTGCA↓G	4b 3'-ext.	*Providencia stuartii* 164	8, 41
PvuII	CAG↓CTG	Flush	*Proteus vulgaris*	16
SacI	GAGCT↓C	4b 3'-ext.	*Streptomyces achromogenes*	2
SacII	CCGC↓GG	2b 3'-ext.	*Streptomyces achromogenes*	2
SalI	G↓TCGAC	4b 5'-ext.	*Streptomyces albus* G	1
TaqI	T↓CGA	2b 5'-ext.	*Thermus aquaticus* YTI	38
XbaI	T↓CTAGA	4b 5'-ext.	*Xanthomonas badrii*	53
XhoI	C↓TCGAG	4b 5'-ext.	*Xanthomonas holcicola*	17
XmaI	C↓CCGGG	4b 5'-ext.	*Xanthomonas malvacearum*	12

[a] Restriction enzymes are named in accordance with the proposal of Smith and Nathans (1973).

[b] Recognition sequences are written from 5' → 3'; one strand only is presented and the cleavage site is indicated by an arrow. For example, G↓GATCC is the abbreviation for

5'-G↓G-A-T-C-C-3'
3'-C-C-T-A-G↑G-5'

[c] b, Base; ext., extension.

[d] Where more than one reference is given, the one underlined gives the purification procedure.

[e] Key to references: 1. Arrand et al. (1978); 2. J. R. Arrand, P. A. Myers, and R. J. Roberts, unpublished results; 3. B. G. Barrell and P. Slocombe, unpublished results; 4. Bigger et al. (1973); 5. A. H. A. Bingham, R. J. Sharp, and A. Atkinson, unpublished results; 6. Boyer et al. (1973); 7. Bron and Murray (1975); 8. Brown and Smith (1976); 9. Garfin Brown and Smith (1977); 10. Brown et al. (1979); 11. Endow (1977); 12. Endow and Roberts (1977); 13. Garfin and Goodman (1974); 14. Gelinas et al. (1977a); 15. Gelinas et al. (1977b); 16. T. R. Gingeras and R. J. Roberts, unpublished results; 17. Gingeras et al. (1978); 18. Greene et al. (1974); 19. Gromkova and Goodgal (1972); 20. Hedgpeth et al. (1972); 21. S. G. Hughes, T. Bruce, and K. Murray, unpublished results; 22. S. G. Hughes and K. Murray, unpublished results; 23. C. A. Hutchison, III, and B. G. Barrell, unpublished results; 24. Kelly and Smith (1970); 25. Kleid et al. (1976); 26. A. Lui, B. C. McBride, and M. Smith, unpublished results; 27. Middleton et al. (1972); 28. J. H. Middleton, P. V. Stankus, M. H. Edgell, and C. A. Hutchison, III, unpublished results; 29. Murray et al. (1976); 30. K. Murray, A. Morrison, H. W. Cooke, and R. J. Roberts, unpublished results; 31. K. Murray, A. Morrison, and R. J. Roberts, unpublished results; 32. Old et al. (1975); 33. Pirrotta (1976); 34. Roberts et al. (1975); 35. Roberts et al. (1976a); 36. Roberts et al. (1976b); 37. Roberts et al. (1977); 38. Sato et al. (1977); 39. D. Sciaky and R. J. Roberts, unpublished results; 40. Sharp et al. (1973); 41. Smith et al. (1976); 42. Smith and Wilcox (1970); 43. Takanami (1973); 44. Takanami (1974); 45. Tu et al. (1976); 46. Wilson and Young (1975); 47. Wilson and Young (1976); 48. R. Wu and R. J. Roberts, unpublished results; 49. Yoshimori (1971); 50. M. Zabeau and R. J. Roberts, unpublished results; 51. M. Zabeau, R. Greene, P. A. Myers, and R. J. Roberts, unpublished results; 52. B. S. Zain and R. J. Roberts, unpublished results; 53. Zain and Roberts (1977).

TABLE II
PARTIALLY CHARACTERIZED TYPE II RESTRICTION ENDONUCLEASES[a]

Enzyme	Microorganism	Source	λ	Ad2	SV40	Reference[b,c]
AcaI	Anabaena catanula	K. Murray	?	?	?	10
AtuAI	Agrobacterium tumefaciens ATCC 15955	E. Nester	>30	>30	?	18
AvaII	Anabaena variabilis	K. Murray	>25	>30	5	14
AvaIII	Anabaena variabilis	K. Murray	?	?	?	17, 21
BamNx	Bacillus amyloliquefaciens N	T. Ando	?	?	1	19, 20
BglI	Bacillus globigii	G. A. Wilson	22	12	1	24
BmeI	Bacillus megaterium	J. Upcroft	10	20	4	5
BpuI	Bacillus pumilus AHU1387	T. Ando	6	30	2	11
BstEI	Bacillus stearothermophilus ET	N. Welker	?	?	?	13
BstEII	Bacillus stearothermophilus ET	N. Welker	11	8	0	13
BstEIII	Bacillus stearothermophilus ET	N. Welker	?	?	?	13
CauI	Chloroflexus aurantiacus	A. Bingham	>30	>30	15	4
CauII	Chloroflexus aurantiacus	A. Bingham	>30	>30	0	4
EclI	Enterobacter cloacae	H. Hartmann	14	?	?	9
Fnu84I	Fusobacterium nucleatum 84	B. C. McBride	>50	?	?	12
HapI	Haemophilus aphrophilus	ATCC 19415	>30	?	?	16
Hin1056II	Haemophilus influenzae 1056	J. Stuy	>50	>50	0	15
Hind-GLU	Haemophilus influenzae Rd123	V. Tanyashin	?	?	?	22
MnoII	Moraxella nonliquefaciens	ATCC 19975	>10	>6	2	16
MnlII	Moraxella nonliquefaciens	ATCC 17953	?	?	?	16
PfaI	Pseudomonas faecalis	M. Van Montagu	>30	>30	?	23
PovI	Proteus vulgaris	ATCC 13315	3	7	0	7

Enzyme	Microorganism	Source				Ref.
RspI	Rhodopseudomonas spheroides	A. Bingham	0	12	0	3
SacIII	Streptomyces achromogenes	ATCC 12767		>30	?	1
SalII	Streptomyces albus G	J. M. Ghuysen	>2	>3	0	1
SfaNI	Streptococcus faecalis ND547	D. Clewell	>50	>30	4	18
SgrI	Streptomyces griseus	ATCC 23345	0	7	0	1
TaqII	Thermus aquaticus YTI	J. I. Harris	>50	>50	1	2, 16
XhoII	Xanthomonas holcicola	ATCC 13461	20	20	4	6
XniI	Xanthomonas nigromaculans	ATCC 23390	4	10	?	8

[a] Restriction enzymes have also been detected in the following strains, but the amounts of enzyme activity present might be too low to warrant further characterization: *AimI* from *Achromobacter immobilis*, ATCC 15934 (S. A. Endow and R. J. Roberts, unpublished observations); *BstAI* from *Bacillus stearothermophilus* 240 (A. H. A. Bingham, P. A. Sharp, and A. Atkinson, unpublished observations); *CviI* from *Chromobacterium violacum* (S. A. Endow and R. J. Roberts, unpublished observations); *MglI* and *MglII*, respectively, from *Moraxella gluedii* strains LG1 and LG2 (Smith *et al.*, 1976); *MviI* and *MviII* from *Myxococcus virescens* (Morris and Parish, 1976); *SspI* from *Serratia* species SAI (B. Torheim, unpublished observations). In addition restriction enzymes have been identified in a number of *Bacillus* strains, but it is not known which, if any, show new specificities (Shibata *et al.*, 1976).

[b] Where more than one reference is given, the one underlined gives the purification procedure.

[c] Key to references: 1. J. R. Arrand, P. A. Myers, and R. J. Roberts, unpublished results; 2. Bickle *et al.* (1977); 3. A. H. A. Bingham, A. Atkinson, and J. Darbyshire, unpublished results; 4. A. H. A. Bingham and J. Darbyshire, unpublished results; 5. R. E. Gelinas, P. A. Myers, and R. J. Roberts, unpublished results; 6. Gingeras *et al.* (1978); 7. T. R. Gingeras and R. J. Roberts, unpublished results; 8. F. Hanberg, P. A. Myers, and R. J. Roberts, unpublished results; 9. Hartmann and Goebel (1977); 10. S. G. Hughes, T. Bruce, and K. Murray, unpublished results; 11. Ikawa *et al.* (1976); 12. A. Lui, B. C. McBride, and M. Smith, unpublished results; 13. R. B. Meagher, unpublished results; 14. Murray *et al.* (1976); 15. J. A. Olson, P. A. Myers, and R. J. Roberts, unpublished results; 16. R. J. Roberts and P. A. Myers, unpublished results; 17. G. Roizes, unpublished results; 18. D. Sciaky and R. J. Roberts, unpublished results; 19. Shibata and Ando (1975); 20. Shibata and Ando (1976); 21. H. Shimatake and M. Rosenberg, unpublished results; 22. Tanyashin *et al.* (1976); 23. M. Van Montagu, unpublished results; 24. Wilson and Young (1976).

ficity can indicate whether the new enzyme recognizes some new sequence or is likely to recognize a sequence identical with that of an enzyme already characterized. Enzymes from different bacterial sources which recognize the same sequence are termed "isoschizomers" (Roberts, 1976). When two enzymes are suspected to have identical sequence specificity, this is best tested by digesting several substrate DNAs sequentially with both the new and the known enzyme (double digests) and showing that no additional DNA fragments appear. A list of the isoschizomers discovered to date is compiled in Table III.

Although isoschizomers can be quickly identified in this way, one cannot assume that the location of the cleavage sites within the recognition sequence will necessarily be identical. Only the enzymes *Hpa*II, *Hap*II, *Mno*I (Sugisaki and Takanami, 1973; Garfin and Goodman, 1974; B. R. Baumstark, R. J. Roberts, and U. L. RajBhandary, unpublished results), *Hae*III, *Bsu*RI (Bron and Murray, 1975), *Blu*I, *Xho*I, *Xpa*I (Gingeras *et al.*, 1978), *Mbo*I, and *Sau*3A (Sussenbach *et al.*, 1976; Gelinas *et al.*, 1977a) have been rigorously shown to cleave at identical positions within the recognition sequence. In contrast, the enzymes *Sma*I and *Xma*I appear to cleave the sequence CCCGGG at different sites, generating flush ends and four base 5'-extensions, respectively (see Table II; Endow and Roberts, 1977).

The most straightforward way to characterize recognition sequences and cleavage points involves direct sequence analysis of the ends of DNA fragment mixtures produced by restriction enzyme cleavage. The easiest approach, first used for the identification of the *Hin*dII recognition sequence (Kelly and Smith, 1970), consists of labeling the 5'-ends of restriction fragments with polynucleotide kinase (Richardson, 1965), followed by digestion with pancreatic DNase. This procedure generates a heterogeneous array of products, including labeled 5'-terminal di-, tri-, tetranucleotides, etc., and the nucleotide sequence common to the 5'-ends of this mixture of restriction fragments can be deduced (Murray, 1973). Subsequent analysis of individual products with venom phosphodiesterase gives the 5'-terminal mononucleotide, whereas the use of exonuclease I (Lehman, 1960) will give the 5'-terminal dinucleotide. For most restriction enzymes that generate either flush ends or 5'-extensions, these data will allow deduction of the recognition sequence. Repair reactions using DNA polymerases can then be used to confirm this sequence. For enzymes producing 5'-extensions, repair reactions using α-^{32}P-labeled deoxyribonucleoside triphosphates and any DNA polymerase will permit the analysis of the incorporated bases and the identification of their nearest neighbors (Wu and Taylor, 1971). In the case of flush-ended fragments, nearest-neighbor sequence information can be obtained by incubating the DNA fragments with T4 DNA polymerase in the presence of only one α-^{32}P-labeled deoxyribonucleoside triphosphate (Englund, 1971).

An analogous approach can be used for fragments with 3'-extensions. Here the 3'-ends are labeled using terminal deoxynucleotidyl-transferase and [α-^{32}P]ribonucleoside triphosphates (Roychoudhury and Kossel, 1971; Roychoudhury et al., 1976) and the products are digested with either micrococcal nuclease to obtain the 3'-dinucleotide or with spleen phosphodiesterase to obtain the 3'-mononucleotide. Because all known type II restriction enzymes leave termini bearing a 5'-monophosphate and a 3'-hydroxyl group, the 3'-dinucleotide can occasionally be identified by digesting uniformly ^{32}P-labeled restriction fragments with micrococcal nuclease and identifying the dinucleoside monophosphate which is derived from the 3'-end of the fragments (Roberts et al., 1976b). As can be seen later in Table V, a number of restriction enzymes recognize a sequence which is a subset of sequences recognized by other enzymes. For example, the hexanucleotide sequences recognized by the enzymes BamHI, BglII, and BclI are subsets of the tetranucleotide sequence recognized by MboI (see Table V). This provides a means, in some cases, by which one can check a deduced sequence or obtain preliminary evidence of relatedness if no sequence information is available. Double digestion between an enzyme recognizing a hexanucleotide (e.g., BamHI, BglII) and one recognizing only the central tetranucleotide of that sequence (e.g., MboI) will produce a pattern indistinguishable from that produced by MboI alone. Care should be taken when interpreting such data because small differences are easily overlooked, especially as one of the two enzymes will usually produce few fragments (e.g., BamHI) while the other (MboI) will produce many fragments. This test is, therefore, best repeated on several different DNA substrates.

Now that several viral genomes have been completely sequenced (Sanger et al., 1977; Fiers et al., 1978; Reddy et al., 1978; Schaller et al., 1978), it is possible to identify recognition sequences by the approximate mapping of cleavage sites and comparison of sequences from the regions to find the common features. This approach, first used to determine the sequence recognized by the enzyme HphI (Kleid et al., 1976), is actually the only route by which restriction enzymes which cleave at sites remote from their recognition sequence can be characterized. Recently, an elegant approach, based upon the plus–minus method for sequencing DNA (Sanger and Coulson, 1975)

TABLE III
ISOSCHIZOMERS

Prototype	Sequence recognized	Isoschizomers[a]
*Bam*HI	GGATCC	*Bam*FI, *Bam*KI, *Bam*NI, *Bst*I
*Bcl*I	TGATCA	*Atu*CI, *Cpe*I
*Eco*RII	CC($_A^T$)GG	*Atu*BI
*Fnu*DII	CGCG	*Acc*II, *Bce*R, *Hin*1056I, *Tha*I
*Hae*II	PuGCGCPy	*Hin*HI, *Ngo*I
*Hae*III	GGCC	*Bsp*RI, *Bsu*R, *Bsu*1076, *Bsu*1114, *Blu*II, *Fnu*DI, *Hhg*I, *Pal*I, *Sfa*I
*Hha*I	GCGC	*Fnu*DIII, *Mnn*III
*Hind*II	GTPyPuAC	*Chu*II, *Hinc*II, *Mnn*I
*Hind*III	AAGCTT	*Bbr*I, *Chu*I, *Hin*bIII, *Hin*fII, *Hsu*I
*Hin*fI	GANTC	*Fnu*AI, *Hha*II
*Hpa*II	CCGG	*Hap*II, *Mno*I
*Mbo*I	GATC	*Dpn*II, *Fnu*CI, *Fnu*EI, *Mos*I, *Sau*3A
*Pst*I	CTGCAG	*Bce*170, *Bsu*1247, *Sal*PI, *Xma*II
*Sac*I	GAGCTC	*Sst*I
*Sac*II	CCGCGG	*Sst*II, *Tgl*I
*Sac*III	?	*Sst*III
*Sal*I	GTCGAC	*Xam*I
*Xma*I	CCCGGG	*Sma*I
*Xho*I	CTCGAG	*Blu*I, *Xpa*I

[a] *Acc*II is from *Acinetobacter calcoaceticus* (M. Zabeau and R. J. Roberts, unpublished results).

*Atu*BI is from *Agrobacterium tumefaciens* B6 (Roizes *et al.*, 1977; D. Sciaky and R. J. Roberts, unpublished results; M. Van Montagu, unpublished results).

*Atu*CI is from *Agrobacterium tumefaciens* C58 (D. Sciaky and R. J. Roberts, unpublished results).

*Bam*FI, *Bam*KI, and *Bam*NI, respectively, are from *Bacillus amyloliquefaciens*, strains F, K, and N (Shibata and Ando, 1976; Shibata *et al.*, 1976).

*Bbr*I is from *Bordetella bronchiseptica* ATCC 19395 (R. J. Roberts and P. A. Myers, unpublished results).

*Bce*170 and *Bce*R, respectively, are from *Bacillus cereus* strains 170 and Rf, sm, st (Shibata *et al.*, 1976).

*Blu*I and *Blu*II are from *Brevibacterium luteum* ATCC 15830 (Gingeras *et al.*, 1978; M. Van Montagu, unpublished results).

*Bsp*RI is from *Bacillus sphaericus* R (Kiss *et al.*, 1977).

*Bst*I is from *Bacillus stearothermophilus* 1503-4R (Catterall and Welker, 1977).

*Bsu*R is from *Bacillus subtilis* strain X5 (Bron *et al.*, 1975; Bron and Murray, 1975).

*Bsu*1076, *Bsu*1114, and *Bsu*1247, respectively, are from *Bacillus subtilis* strains IAM 1076, IAM 1114, and IAM 1247 (Shibata *et al.*, 1976).

*Chu*I and *Chu*II are from *Corynebacterium humiferum* ATCC 21108 (S. A. Endow and R. J. Roberts, unpublished results).

*Cpe*I is from *Corynebacterium petrophilum* ATCC 19080 (J. Fisherman, T. Gingeras, and R. J. Roberts, unpublished results).

*Dpn*II is from *Diplococcus pneumoniae* (Lacks and Greenberg, 1977).

*Fnu*AI, *Fnu*CI, *Fnu*DI, *Fnu*DIII, and *Fnu*EI, respectively, are from *Fusobacterium*

has been used to characterize the cleavage sites of the enzymes *Pst*I (Brown and Smith, 1976), *Taq*I (Sato *et al.*, 1977), *Hga*I (Brown and Smith, 1977), and *Mbo*II (Brown *et al.*, 1979).

So far, relatively few type II restriction enzymes have been extensively purified to near homogeneity and characterized as proteins. These include *Eco*RI (Greene *et al.*, 1974; Modrich and Zabel, 1976),

---

TABLE III (*Continued*)

nucleatum strains A, C, D, and E (A. Lui, B. C. McBride, and M. Smith, unpublished results).

*Hap*II is from *Haemophilus aphrophilus* ATCC 14385 (Takanami, 1973; Sugisaki and Takanami, 1973).

*Hha*II is from *Haemophilus haemolyticus* ATCC 10014 (Mann *et al.*, 1978).

*Hhg*I is from *Haemophilus haemoglobinophilus* ATCC 19416 (R. J. Roberts and P. A. Myers, unpublished results).

*Hin*10561 is from *Haemophilus influenzae* 1056 (J. A. Olson, P. A. Myers, and R. J. Roberts, unpublished results).

*Hin*bIII is from *Haemophilus influenzae* strains 1076 and R$_b$ (J. H. Middleton, P. V. Stankus, M. H. Edgell, and C. A. Hutchison, III, unpublished results; J. A. Olson, P. A. Myers, and R. J. Roberts, unpublished results).

*Hinc*II is from *Haemophilus influenzae* strains 1160, 1161, and R$_c$ (J. A. Olson, P. A. Myers, and R. J. Roberts, unpublished results; Landy *et al.*, 1974b).

*Hinf*II is from *Haemophilus influenzae* R$_f$ (M. Mann and H. O. Smith, unpublished results).

*Hin*HI is from *Haemophilus influenzae* H-I (Takanami and Kojo, 1973; Takanami, 1974).

*Hsu*I is from *Haemophilus suis* ATCC 19417 (R. J. Roberts and P. A. Myers, unpublished results).

*Mno*I, *Mnn*I, and *Mnn*III, respectively, are from *Moraxella nonliquefaciens* strains ATCC 19975 (*Mno*) and ATCC 17954 (*Mnn*) (R. J. Roberts and P. A. Myers, unpublished results; F. Hanberg, P. A. Myers, and R. J. Roberts, unpublished results).

*Mos*I is from *Moraxella osloensis* ATCC 19976 (Gelinas *et al.*, 1977a).

*Ngo*I is from *Neisseria gonorrhoea* (G. A. Wilson and F. E. Young, unpublished results).

*Pal*I is from *Providencia alcalifaciens* ATCC 9886 (R. E. Gelinas, P. A. Myers, and R. J. Roberts, unpublished results).

*Sau*3A is from *Staphylococcus aureus* strains 3A, 3C, 55, 71, 879, and 1030 (Sussenbach *et al.*, 1976; Stobberingh *et al.*, 1977).

*Sfa*I is from *Streptomyces faecalis* var. *zymogenes* (R. Wu, C. King, and E. Jay, unpublished results).

*Sal*PI is from *Streptomyces albus* CM1 52766 (Chater, 1977).

*Sma*I is from *Serratia marcescens* S$_b$ (C. Mulder and P. Greene, unpublished results).

*Sst*I, *Sst*II, and *Sst*III are from *Streptomyces stanford* (S. Goff and A. Rambach, unpublished results).

*Tha*I is from *Thermoplasma acidophilum* (McConnell *et al.*, 1978).

*Tgl*I is from *Thermopolyspora glauca* ATCC 15345 (T. R. Gingeras and R. J. Roberts, unpublished results).

*Xam*I is from *Xanthomonas amaranthicola* ATCC 11645 (Arrand *et al.*, 1978).

*Xma*II is from *Xanthomonas malvacearum* ATCC 9924 (Endow and Roberts, 1977).

*Xpa*I is from *Xanthomonas papavericola* ATCC 14180 (Gingeras *et al.*, 1978).

*Eco*RII (Roulland-Dussoix *et al.*, 1975), *Bst*I (Catterall and Welker, 1977), and *Bsp*I (P. Venetianer, unpublished results). The active forms of these enzymes are dimers, composed of two identical subunits with molecular weights of 28,500 for *Eco*RI, 40,000 for *Eco*RII, and 46,000 for *Bst*I. All type II enzymes require only magnesium as a cofactor, and for many enzymes rather broad pH and magnesium optima have been observed. The catalytic properties of both the *Eco*RI endonuclease and methylase have been studied in great detail (Modrich and Zabel, 1976; Modrich and Rubin, 1977; Rubin and Modrich, 1977).

Two phosphodiester bonds, one per strand, have to be cleaved in order to produce a double-strand scission, and it is perhaps surprising that vastly different K_m values have been measured for the two events with both *Eco*RI (Modrich and Zabel, 1976) and *Hpa*II (Ruben *et al.*, 1977). In both cases, cleavage of the second strand occurred at a reduced rate relative to that of the first strand, a feature which can be utilized to introduce site-specific nicks into duplex DNAs. Such substrates should be very useful for *in vitro* mutagenesis around restriction enzyme cleavage sites by nick translation using *E. coli* DNA polymerase in the presence of mutagenic base analogs.

Chemically synthesized self-complementary oligonucleotides containing restriction enzyme recognition sites have been used for studying the catalytic and recognition properties of restriction enzymes. Both octa- and decanucleotide duplex fragments have been shown to act as substrates for the enzymes *Eco*RI, *Bam*HI, and *Alu*I (Greene *et al.*, 1975; Bahl *et al.*, 1976; Scheller *et al.*, 1977a), although the affinity of *Eco*RI for the octanucleotide duplex was much lower than that observed for intact SV40 DNA (Greene *et al.*, 1975). In contrast, much longer duplexes were required before cleavage could occur with *Hin*dIII (Scheller *et al.*, 1977a).

The effects of base analogs upon restriction enzyme cleavage have also been examined. Many natural DNAs contain unusual bases, such as uracil, hydroxymethyluracil, or hydroxymethylcytosine (HMC), and their susceptibility to cleavage was found to vary considerably (Ito *et al.*, 1975; Kaplan and Nierlich, 1975; Berkner and Folk, 1977). Only one enzyme has been reported to cleave native T4 DNA, which contains glucosylated HMC residues (Tanyashin *et al.*, 1976). Removal of the glucose residues renders the DNA accessible to some enzymes but not others (Li *et al.*, 1975; Kaplan and Nierlich, 1975). Of particular interest are the effects of base substitution on the *Eco*RI restriction and modification enzymes. *In vivo*, methylation of one adenosine residue (A*) within the recognition sequence (GA*ATTC) (Dugaiczyck *et al.*, 1974) prevents cleavage by the restriction enzyme. However, when substrates were prepared *in vitro* which contained inosine in place of guanosine, it was observed that although endonuclease cleavage occurred at a normal rate, the modification enzyme was exquisitely sensitive to this substitution and methylation proceeded extremely slowly (Modrich and Rubin, 1977). Similar differential effects on cleavage and methylation have been observed with substitutions at the 5′-position of the thymidine ring (Berkner and Folk, 1978). Thus, the basis for recognition of the nucleic acid must differ between the restriction endonuclease and the methylase, a conclusion which, if valid, may have important implications for theories of protein–nucleic acid interactions.

A number of enzymes, including *Hae*III, *Hha*I, *Hpa*II, *Mbo*II, *Hin*fI, and *Hga*I, have been shown to cleave single-stranded DNA, although much more enzyme is required than would be needed to cut a corresponding amount of duplex DNA (Blakesley and Wells, 1975; Horiuchi and Zinder, 1975; Godson and Roberts, 1976). Only for the enzyme *Hae*III has this single-strand cleavage been shown to occur at the same sites as those which are recognized in duplex DNA substrates (Blakesley and Wells, 1975; Horiuchi and Zinder, 1975). Although the detailed mechanism of single-strand cleavage is not understood, the most reasonable explanation is that the secondary structure of single-stranded molecules allows the formation of transient duplexes which are then cleaved in the usual manner. Consistent with this explanation is the effect of temperature on the reaction, which showed that loss of the ability of *Hae*III and *Hha*I to cleave ϕX174 single strands paralleled the thermal denaturation of secondary structure (Blakesley *et al.*, 1977).

It is of some importance to ascertain if RNA : DNA hybrids are an acceptable substrate for any restriction enzyme, for this may give some insight into the nature of the recognition process and will be of value for the manipulation of cDNAs produced by reverse transcription.

3. *Recognition Specificities*

Presently, the recognition and cleavage specificities of enzymes recognizing 36 different nucleotide sequences have been determined (see Table I). They may be classified conveniently into three different groups, as indicated in Table IV. The majority of these enzymes belong to a group recognizing sequences which contain a twofold axis of symmetry, a feature usually referred to as a palindrome. It has been speculated that enzymes of this group might contain subunits ar-

TABLE IV
RECOGNITION SPECIFICITIES

Palindromic sites			Asymmetric sites		
Tetranucleotides	Pentanucleotides	Hexanucleotides	Tetranucleotides	Pentanucleotides	"Relaxed" sites
AGCT (AluI)	GGNCC (AsuI)	TGGCCA (BalI)	CCTC (MnlI)	GACGC (HgaI)	GT($\frac{A}{C}$)($\frac{G}{T}$)AC (AccI)
CGCG (FnuDII)	GC($\frac{A}{T}$)GC (BbvI)	GGATCC (BamHI)		GGTGA (HphI)	CPyCGPuG (AvaI)
GGCC (HaeIII)	GANTC (HinfI)	TGATCA (BclI)		GAAGA (MboII)	($\frac{A}{T}$)GGCC($\frac{T}{A}$) (HaeI)
GCGC (HhaI)	CC($\frac{A}{T}$)GG (EcoRII)	AGATCT (BglII)			PuGCGCPy (HaeII)
CCGG (HpaII)		GAATTC (EcoRI)			GTPyPuAC (HindII)
GATC (MboI)		AAGCTT (HindIII)			
TCGA (TaqI)		GTTAAC (HpaI)			
		GGTACC (KpnI)			
		CTGCAG (PstI)			
		CAGCTG (PvuII)			
		GAGCTC (SacI)			
		CCGCGG (SacII)			
		GTCGAC (SalI)			
		TCTAGA (XbaI)			
		CTCGAG (XhoI)			
		CCCGGG (XmaI)			

ranged symmetrically (Kelly and Smith, 1970). The finding that the active forms of some type II enzymes contain two identical subunits (see above) is consistent with this proposal.

Several enzymes have been found to recognize asymmetric sequences (see Table IV), and they all share the property of cleaving at sites 5–10 base pairs to one side of the recognition site. One clue to the nature of the protein–nucleic acid interaction comes from the cleavage specificity of *Hph*I and *Mbo*II. In both cases the sites of cleavage lie 10 base pairs away from the midpoint of the recognition sequence, and, of course, 10 base pairs are required for one turn of the double helix. In these cases, it therefore seems likely that the enzyme will interact with one face of the DNA by binding at one point and cleaving at a corresponding point on the next turn of the helix.

A last group of enzymes displays a recognition specificity which we have termed "relaxed." These enzymes do not distinguish between the purine or pyrimidine bases at certain positions in their recognition sequences. Because the enzymes exhibiting relaxed specificities apparently recognize all possible purine–pyrimidine combinations, their recognition sequences can be either palindromic or nonpalindromic. For example, two of the three possible *Hin*dII recognition sites (the palindromic sequence GTTAAC and the asymmetric but complementary sequences GTCAAC or GTTGAC) occur in the sequence of bacteriophage φX174 DNA (Sanger *et al.*, 1977), whereas recognition of the third sequence, GTCGAC, has been inferred from the coincidence of *Hin*dII and *Sal*I cleavage sites (Arrand *et al.*, 1978). An unusual type of relaxed specificity has been found for the enzyme *Acc*I, which is unable to differentiate C from A at one position or T from G at another (M. Zabeau and R. J. Roberts, unpublished observations). This case should be of some interest to the connoisseur of protein–nucleic acid interactions.

As mentioned earlier, a number of restriction enzymes have recognition sequences which are related one to another. The most common relationship arises when one enzyme recognizes a tetranucleotide sequence that is the core of a hexanucleotide sequence recognized by one or more other enzymes. The most extensive set known at the moment contains *Mbo*I (GATC), *Bam*HI (GGATCC), *Bgl*II (AGATCT), and *Bcl*I (TGATCA) and probably *Xho*II, because although the precise recognition sequence is unknown, all *Xho*II sites appear to coincide with *Mbo*I sites (R. J. Roberts and P. A. Myers, unpublished observations). Other examples are shown in Table V. Another kind of sequence relationship involves enzymes recognizing relaxed sequences. In this case, it sometimes happens that one of the alternative

TABLE V

RESTRICTION ENZYMES RECOGNIZING
RELATED SEQUENCES

Common central tetranucleotide core

AGCT

    AGCT   (*Alu*I)
    C**AGCT**G (*Pvu*II)
    G**AGCT**C (*Sac*I)
    A**AGCT**T (*Hind*III)

TCGA

    TCGA   (*Taq*I)
    C**TCGA**G (*Xho*I)
    G**TCGA**C (*Sal*I)

GATC

    GATC   (*Mbo*I)
    G**GATC**C (*Bam*HI)
    T**GATC**A (*Bcl*I)
    A**GATC**T (*Bgl*II)

CGCG

    CGCG   (*Fnu*DII)
    C**CGCG**G (*Sac*II)

GGCC

    GGCC   (*Hae*III)
    (^A_T) **GGCC** (^T_A) (*Hae*I)
    T**GGCC**A (*Bal*I)

GCGC

    GCGC   (*Hha*I)
    Pu**GCGC**Py (*Hae*II)

Relaxed cores

GTPyPuAC

    GTPyPuAC (*Hind*II)
    GTT A **AC** (*Hpa*I)
    GTC G **AC** (*Sal*I)
    GT^C_A ^T_G **AC** (*Acc*I)

CPyCGPuG

    CPyCGPuG (*Ava*I)
    CT CGA G (*Xho*I)
    CC CGG G (*Xma*I)

sequences is identical with that recognized by some other enzyme. For instance, *Sal*I (GTCGAC) sites are also recognized by *Hind*II (GTPyPuAC) and *Acc*I (GT(^C_A)(^T_G)AC).

Although the recognition sequences shown in Table I contain all the necessary information required for cleavage to occur, sequences outside the recognition site can markedly influence the rate of cleavage. For instance, the rates of cleavage by *Eco*RI at different sites in λ or adenovirus-2 DNA can vary by an order of magnitude (Thomas and Davis, 1975; Forsblom *et al.*, 1976). Variations in the rate of cleavage can be induced by performing digests in the presence of DNA-binding agents. Actinomycin D and distamycin A both inhibit cleavage by *Hind*III or *Eco*RI at particular sites in a highly specific manner (Braga *et al.*, 1976; Nosikov *et al.*, 1976; Kuroyedov *et al.*, 1977; Nosikov and Sain, 1977). A similar though less pronounced effect has been described for the DNA binding ligand 6,4′-diamidino-2-phenylindole (Kania and Fanning, 1976). These effects may prove useful as a means of generating specific partially digested fragments for molecular cloning (Maizels, 1976) or mapping (Braga *et al.*, 1976).

In contrast, a relaxation of the recognition specificity of *Eco*RI has been obtained by changing the pH and ionic conditions of the reaction (Polisky *et al.*, 1975). At low ionic strength, *Eco*RI will cleave many, but maybe not all, sequences containing AATT which is the central tetranucleotide of its usual recognition sequence, GAATTC. A similar phenomenon has been reported for *Bsu*RI, although the specificity of its isoschizomer, *Hae*III, could not be relaxed under the same conditions (Heininger *et al.*, 1977).

From Table I it is apparent that the site of cleavage by these enzymes, relative to the location of the recognition sequence, can vary enormously. Presumably, this reflects the bifunctionality of these enzymes in that they must possess two active features—a specificity or recognition function and a cleavage function. The position of cleavage is probably determined by the spatial arrangement of the two active sites, which may not necessarily be constrained to one another. This may have interesting implications for genetic studies in that mutant enzymes might be expected which have varied either their recognition specificity or their mode of cleavage.

D. SOME BIOLOGICAL ASPECTS

Although restriction enzymes have been found to occur widely in various species throughout the bacterial kingdom (Roberts, 1976), very little is known about their function. In part this is because many of the 140 enzymes listed in Tables I–III have been isolated from poorly studied strains. Rather few of these enzymes have been shown to participate in a restriction and modification system, since this demands the identification of a companion modification enzyme and a genetic analysis of its involvement in processes of host-controlled restriction. Demonstration of this last property usually requires the correlation of the presence of a restriction enzyme with phage restriction phenomena and thus depends upon the availability of related strains differing in their host-specificity systems. However, the failure to detect phage restriction *in vivo* cannot be considered as evidence that the enzyme does not mediate restriction because many phages have devised strategies to escape restriction. For instance, resistance to cleavage can result from (a) the absence of restriction enzyme recognition sites in the phage genome [T7 DNA only becomes susceptible to *Eco*RI cleavage, either *in vitro* or *in vivo*, when a mutation is present which creates an *Eco*RI site (Tsujimoto and Ogawa, 1977)], (b) non-specific base modifications, such as glucosylations (T-even phages), and (c) the occurrence in nonrestricting bacterial strains of DNA methylases with a sequence specificity identical with or overlapping that shown by a particular restriction enzyme (the *E. coli dam* and *mec* methylases, which modify *Mbo*I and *Eco*RII sites, respectively). An additional mechanism by which phages can overcome host-controlled restriction is shown by phages T3 and T7 which counteract the *Eco*K and *Eco*B, but not the *Eco*RI, restriction systems by encoding an inhibitory protein (Studier, 1975; Studier and Movva, 1976; Tsujimoto and Ogawa, 1977). In phage λ the *ral* gene affects both restriction and modification, although the mechanism is not yet understood (Zabeau *et al.*, 1979). Although bacteria might use similar mechanisms to protect their own chromosomal DNA against the deleterious activities of restriction enzymes, no example has yet been described. In all cases studied, protection is achieved by means of a companion modification methylase.

Rigorous proof for the involvement of site-specific nucleases in *in vivo* restriction processes should include the isolation of mutant strains deficient in their ability to restrict *in vivo* and a demonstration of the concomitant loss of the restriction enzyme activity *in vitro*. Good evidence is available to show that the following type II enzymes participate in *in vivo* restriction and modification systems: *Bam*HI (Wilson and Young, 1976); *Bam*NI, *Bam*Nx (Shibata and Ando, 1974, 1976), *Bst*I (Catterall *et al.*, 1976); *Bsu*RI (Bron *et al.*, 1975), *Eco*RI, *Eco*RII (Roulland-Dussoix *et al.*, 1974), *Mvi*I (Morris and Parish, 1976); *Sal*I (Chater and Wilde, 1976); and *Sau*3A (Sussenbach *et al.*, 1976; Stobberingh *et al.*, 1977). The existence of corresponding modification methylases has been identified for the enzymes *Bpu*I (Ikawa *et al.*, 1976), *Bsp*I (Kiss *et al.*, 1977), *Dpn*II (Lacks and Greenberg, 1977), *Hae*III, *Hpa*II (Mann and Smith, 1977), *Hind*II, and *Hind*III (Roy and Smith, 1973; Roszczyk and Goodgal, 1975).

Despite the fact that some restriction and modification systems, such as the *Eco*K, *Eco*B, *Eco*P1, and *Eco*RI, have been intensively studied both genetically and biochemically, very little is known about their physiological roles. It has been suggested that their primary function is to serve as a barrier to prevent genetic exchange between unrelated organisms (Arber and Linn, 1969; Arber, 1974). This conclusion is based upon the finding that restriction systems can be very efficient in blocking bacteriophage growth (at least if the phage has not developed a mechanism to overcome restriction). However, genetic studies on the transformation of chromosomal markers revealed that *in vivo* restriction has rather little effect on marker exchange between closely related strains (Gromkova and Goodgal, 1974, 1976; Trautner *et al.*, 1974; Stuy, 1976).

It is, of course, always difficult to assess the biological role(s) of any enzyme of this sort from either its *in vitro* properties or even a single phenotype. The finding that a strain possessing a restriction enzyme can prevent bacteriophage growth cannot be taken as *a priori* evidence that this is the main function of that enzyme. For any strain containing both a restriction enzyme and a companion modification enzyme must surely be expected to degrade unmodified DNA. Could it be that the very different properties of the type I and type II enzymes reflect quite different functions? Recent experiments (Chang and Cohen, 1977) have demonstrated that the *Eco*RI enzyme can promote site-specific recombination *in vivo*, presumably by a mechanism that parallels a current technique for preparing recombinant DNAs *in vitro* (see Section V). Although the frequency of these events was rather low, this may reflect the experimental conditions chosen rather than the inherent ability of the organism to catalyze such processes. If this phenomenon is of widespread occurrence among bacteria containing the appropriate type II enzymes, it may provide a facile mechanism for bacterial evolution and perhaps an alternative route for the preparation of recombinant DNAs.

Alternatively, restriction enzymes could be involved in some other aspects of DNA metabolism. For instance, the restriction systems in *Haemophilus* and *Bacillus* strains may play a role in transformation (Gromkova and Goodgal, 1974; Wilson and Young, 1976). It should also be noted that other cellular enzymes, such as the *E. coli* K12 *mec* and *dam* methylases, have been found to recognize sequences similar to those recognized by certain restriction enzymes. The *mec* methylase, which specifically methylates cytosine residues, has recently been shown to exhibit the same sequence specificity as the *Eco*RII methylase (Schlagman *et al.*, 1976; Hattman, 1977). This explains why phage DNA isolated from bacteriophages grown on *E. coli* K12 strains is always partially or completely resistant to *Eco*RII cleavage (Hughes and Hattman, 1975). On the other hand, the *dam* methylase, which is specific for adenosine residues, has a sequence specificity that overlaps the recognition sequence of *Mbo*I and hence renders phage DNA partially resistant to digestion by that enzyme (Vovis and Lacks, 1977). Although the function of these methylases is unknown, it seems likely that *E. coli* strains will possess other proteins which recognize these sequences. The recent observation that the *Eco*RII sites in the *lac*I gene are hot spots for mutagenesis might be relevant in this connection (J. Miller, personal communication).

III. Basic Aspects of Restriction Enzyme Technology

A. fractionation and isolation of DNA fragments

Both polyacrylamide (Danna and Nathans, 1971) and agarose (Aaij and Borst, 1972; Sharp *et al.*, 1973) gel electrophoresis systems have proved invaluable for the fractionation of restriction fragments obtained from small DNA substrates such as viral genomes, plasmids, and organelle DNAs. The choice of an appropriate gel system depends upon a number of factors, including the molecular weight range of the fragments, the complexity of the digest, and the particular experimental purpose. Good resolution of low molecular weight fragments ranging up to 3×10^6 daltons can be obtained on polyacrylamide, mixed agarose–polyacrylamide, or 1.4 to 4% agarose gels. Low percentage (0.3–0.7%) agarose gels, run horizontally (Kaplan *et al.*, 1977; McDonell *et al.*, 1977), allow the separation of high molecular weight fragments ranging up to 5×10^7 daltons. In general, a higher resolution of small DNA fragments is obtained on polyacrylamide or mixed agarose–polyacrylamide gels as compared to agarose gels on which diffusion often causes band broadening. Optimal separation of fragments over a wide molecular weight range can be obtained by using gradient polyacrylamide gels (Jeppesen, 1974), which give extremely sharp bands (Allet, 1973). On most gel systems the mobility of the DNA fragments is found to be a smooth function of their molecular weight (Danna and Nathans, 1971; Edgell *et al.*, 1972; Helling *et al.*, 1974; Maniatis *et al.*, 1975; Thomas and Davis, 1975; McDonell *et al.*, 1977). Occasionally, other factors, such as base composition, can influence fragment mobility (Zeiger *et al.*, 1972). Such anomalies are most frequently observed with polyacrylamide gels and to a much lesser extent with agarose gels (Thomas and Davis, 1975). Alteration of electrophoretic conditions can lead to extensive mobility shifts, and this feature has recently been exploited to develop an analytical two-dimensional gel system, suitable for the fractionation of complex digests of small genomes (Derynck and Fiers, 1977).

The complete fractionation of the extremely complex digests of cellular chromosomes cannot be accomplished in a single step by any existing technique. Gel systems can be used, but only as part of some more extensive separation scheme. In one procedure, the restriction enzyme digest is first fractionated on an agarose gel using a continuous electroelution device (Lee and Sinsheimer, 1974a; Polsky *et al.*, 1978). Fragments from individual fractions are then digested with a second enzyme and fractionated on a multislot slab gel (Potter and Newbold, 1976). This method could prove useful for the analysis of the less complex chromosomes, such as those of bacteria (Potter *et al.*, 1977) and *Drosophila* (Potter and Thomas, 1977). Reverse-phase chromatography (RPC), originally developed for the fractionation of tRNAs (Pearson *et al.*, 1971), can also provide a separation of restriction fragments (Hardies and Wells, 1976; Landy *et al.*, 1976). This separation presumably involves both ion exchange and hydrophobic effects due to interaction of the DNA fragments with the quaternary alkylammonium salt embedded in the inert support. On the whole, fragments separate according to size, but sometimes fragments of similar size show large differences in their elution properties. The high capacity of RPC columns offers an important practical advantage, particularly if it is the first step during fractionation of a complicated digest. A two-step fractionation procedure involving RPC-5 column chromatography and agarose gel electrophoresis should be of general use for the enrichment of single copy genes from mammalian chromosomes (Tilghman *et al.*, 1977).

The preparative isolation of individual restriction fragments is currently achieved by fractionating large quantities (up to a few milligrams) of digested DNA on either polyacrylamide or agarose gels and then recovering the DNA fragment from the gel slices. Most of the procedures used for isolating DNA from gel matrices revolve around the problem of removing impurities, such as nonpolymerized gel material, from the eluted DNA. The variety of techniques used to elute DNA fragments from gel slices fall into three major classes.

1. Electrophoretic elution: In this procedure the fragments are usually collected in a dialysis bag (Allet *et al.*, 1973). Small fragments are sometimes lost through the pores of the bag, whereas large fragments elute slowly and sometimes become irretrievably absorbed to the membrane. Impurities are relatively low with this method, since the gel essentially remains intact.

2. Diffusion: This commonly used procedure involves shredding the gel slice into very small pieces, suspending them in aqueous solution, and allowing the DNA to diffuse out. It is simple and usually gives high recoveries of the DNA fragments. The major disadvantage lies in the high recovery of impurities and the often extensive manipulations required to remove them.

3. Chaotropic agents: This method is specific for agarose gels. In the presence of sodium perchlorate (Fuke and Thomas, 1970), potassium iodide (Blin *et al.*, 1975), or some similar chaotropic agent, agarose gels will dissolve and release the entrapped DNA. Subsequent purification of the DNA can be achieved by isopycnic centrifugation (Blin *et al.*, 1975) or by chromatography on hydroxyapatite (Southern, 1975a). This latter method gives high yields of fragments and is capable of removing most agarose-derived impurities, but sometimes results in fragments which are difficult to redigest, perhaps as a result of other impurities derived from the hydroxyapatite.

Minor variants of these basic techniques have been reported, including procedures such as the "freeze–squeeze" technique (Thuring *et al.*, 1975), which is a form of active diffusion. In all cases, additional purification is required to remove impurities that impede further manipulation. Repeated phenol extractions and ethanol precipitations can help, as can binding to and elution from small columns of DEAE-cellulose. Nevertheless, the isolation of small restriction fragments, which often necessitates several digestion and fractionation steps, is extremely time-consuming, and further technical improvements will be required if some of the newer applications of restriction enzyme technology are to yield their true potential.

In addition to the gel electrophoretic methods, procedures involving velocity or buoyant density centrifugation and column fractionation have also been used to isolate DNA fragments. Although these methods avoid the contamination problems described above, their usefulness is rather limited. Isopycnic centrifugation has been used to separate DNA fragments with different base compositions (Fritsch *et al.*, 1975) and could be useful to reisolate cloned DNA fragments (Chang and Cohen, 1974). Velocity sedimentation using sucrose gradients or differential sedimentation (Blattner *et al.*, 1977) can be used to separate very large from very small fragments. RPC-5 columns (see above) which give poor resolution but have a large loading capacity might be very useful for isolating large quantities of a particular fragment from a simple restriction digest (Tiemeier *et al.*, 1978). Finally, differential precipitation techniques (Lis and Schleif, 1975) might also be exploited as a first step when handling very large quantities of DNA.

The detection of DNAs in gels is easily accomplished using autoradiography in the case of ^{32}P-labeled DNA or ethidium bromide as a fluorescent stain for unlabeled DNA (Aaij and Borst, 1972; Sharp *et al.*, 1973). The sensitivity of ethidium bromide is such that a few nanograms of DNA can easily be detected in a band. Quantitation of the amount of DNA can be accomplished by densitometer tracings of films from both radioactive and ethidium bromide-stained DNA (Shen *et al.*, 1976; Prunell *et al.*, 1977a) or by direct fluorescence scanning of the stained gels (Oliver and McLaughlin, 1977). Another stain, acridine orange, can also be used to stain DNA and has the advantage that double- and single-stranded DNA can be distinguished by the color of their fluorescence (McMaster and Carmichael, 1977). Dyes, such as methylene blue (Allet, 1973), toluidine blue (Streeck *et al.*, 1974), and Stains-All (Landy *et al.*, 1974b), have also been used to detect DNA fragments in polyacrylamide gels.

B. PHYSICAL MAPPING

One of the major uses of restriction enzymes has been the construction of physical maps of chromosomes, in which restriction enzyme cleavage sites serve as defined reference points relative to which genes and other functional sequences can accurately be located. Cleavage maps have been derived for a large number of small genomes, such as viral, organelle, and plasmid DNAs, and the available information is summarized in Table VI. In addition, physical maps of a large number of chromosomal genes from different organisms have been derived. The development of new mapping techniques and the availability of a large number of restriction enzymes (Tables I–III) have greatly facilitated the once arduous task of physical mapping. A primary step in the construction of a cleavage map involves the establishment of one or more reference points, and for linear genomes, the ends of the molecule provide these directly. Similarly, a circular genome can be converted into a linear molecule if an enzyme is available which makes only a single cut. In general, enzymes which make only a limited number of cuts are most useful for the construction of a primary map, because these cleavage sites can be positioned readily and provide valuable internal landmarks.

A large variety of mapping methods have been used to derive the relative order of restriction fragments. Common approaches involve the analysis of the products of partial digestion, reciprocal redigestion of individual restriction fragments, and simultaneous digestions with two or more enzymes (Danna *et al.*, 1973; Griffin *et al.*, 1974; Lee and Sinsheimer, 1974b; Subramanian *et al.*, 1974). Fragments have also been ordered by pulse labeling using fragments of known map positions as primers (Seeburg and Schaller, 1975; Jeppesen *et al.*, 1976) and by limited exonuclease digestion of linear genomes (Thomas *et*

TABLE VI

RESTRICTION ENZYME MAPS OF VIRAL, PLASMID, AND ORGANELLE DNAS

DNA	Maps and references[a]
E. coli phages	
λ	*Ava*I (54), *Bam*HI (47, 83), *Bgl*II (61, 84), *Eco*RI (103), *Hind*II (87), *Hind*III (1, 78, 87), *Hpa*I (1, 87), *Pst*I (22, 94), *Kpn*I, *Sac*I, *Sac*II, *Xba*I (61), *Sal*I (14), *Sma*I (59, 68), *Xho*I (54, 61)
λ dv	*Eco*RI, *Hind*II, *Hind*III, *Hpa*I, *Bsu*R (96)
P22	*Eco*RI (58)
φ80	*Eco*RI (50), *Sma*I (59)
T5	*Bam*HI, *Hpa*I (49), *Eco*RI, (85, 112), *Hind*III (112) *Sal*I, *Sma*I (49, 112)
T7	*Dpn*II, *Hpa*I (67)
P1	*Bam*HI, *Bgl*II, *Eco*RI, *Hind*III, *Pst*I (5)
P2	*Eco*RI (23)
P4	*Eco*RI (39)
Mu	*Eco*RI, *Hind*II, *Hind*III (1), *Bal*I, *Bam*HI, *Bgl*II, *Kpn*I, *Pst*I, *Sal*I (60)
φX174 RF[b]	
G4 RF	*Eco*RI, *Hae*II + III, *Hind*II, *Hpa*II (36), *Pst*I, *Kpn*I (38)
S13 RF	*Alu*I, *Ava*I, *Hae*II, *Hha*I, *Hinf*I, *Hpa*I, *Hpa*II, *Hph*I, *Mbo*I, *Mbo*II, *Pst*I, *Xho*I (37), *Hind*II, *Hae*III (37, 46)
f1 RF	*Alu*I (108), *Hae*II, *Hae*III, *Hind*II (52), *Hap*II (90), *Eco*RII (113)
fd RF[b]	
M13	*Alu*I, *Hae*II (109), *Hae*III (107), *Hap*II (90, 107), *Hind*II (99, 107)
ZJ/2	*Alu*I, *Hae*II, *Hae*III, *Hap*II, *Hind*II (108)
Other bacteriophages	
PM2	*Hind*III (15, 82), *Hpa*II (15)
φ15	*Eco*RI (57)
φ29	*Eco*RI (55, 57), *Hind*III, *Hpa*I, *Hpa*II (56)
Animal viruses	
Adenoviruses	
Ad1, Ad6	*Eco*RI (33)
Ad2	*Eco*RI, *Hpa*I (76), *Bal*I, *Bam*HI, *Bgl*II, *Hind*III, *Kpn*I, *Sal*I, *Sma*I, *Xba*I (4)
Ad5	*Eco*RI (76), *Hpa*I (63, 76), *Hind*III (98), *Bam*HI, *Sma*I (4)
Ad3, Ad7	*Eco*RI, *Hind*III (91, 104), *Bam*HI, *Hpa*I, *Sma*I, *Sal*I, *Kpn*I, *Xba*I, *Xho*I (104)
Ad12	*Bam*HI, *Eco*RI (77), *Sal*I (45)
Ad16	*Bam*HI, *Eco*RI, *Hpa*I, *Sal*I (111)
AdFL	*Bam*HI, *Bgl*II, *Eco*RI, *Hind*III, *Hpa*I, *Sal*I (65)

TABLE VI (*Continued*)

DNA	Maps and references[a]
Parvoviruses	
AAV	*Bam*HI (19), *Eco*RI (20), *Hin*dII, *Hin*dIII (11), *Hae*II, *Sal*I, *Pst*I (27), *Hae*III, *Hpa*II (28, 95), *Alu*I, *Hha*I (28)
H1	*Eco*RI, *Hae*II, *Hae*III, *Hin*dII, *Hin*dIII, *Hpa*II (86)
Papovaviruses	
SV40[b]	
BK	*Eco*RI, *Hin*dII + III (53)
Polyoma	*Bam*HI (116), *Eco*RI, *Hpa*II (43), *Hae*II (42), *Hae*III (97), *Hin*dII (24, 32, 41), *Hin*dIII (24, 43), *Hga*I (92), *Hha*I (41), *Kpn*I, *Pst*I (26)
HPV1	*Bam*HI (31, 34), *Eco*RI, *Hin*dII, *Hin*dIII (30, 34), *Hae*III (35), *Hpa*I (30), *Hpa*II (31, 35)
HPV2, HPV3	*Bam*HI, *Eco*RI, *Hae*III, *Hin*dII, *Hin*dIII, *Hpa*II (35)
Herpesviruses	
HSV1	*Bgl*II (25, 75), *Eco*RI (93), *Hin*dIII, *Xba*I (93, 114), *Hpa*I (114)
HSV2	*Bgl*II, *Eco*RI, *Hin*dIII, *Hpa*I, *Kpn*I, *Xba*I (75)
Poxviruses	
VV	*Hin*dIII, *Sst*I (115)
RPV	*Hin*dIII, *Sst*I (115)
RNA tumor viruses	
MSV	*Hae*II, *Hin*dII, *Hin*dIII (18)
Plant viruses	
CMV	*Bam*HI, *Eco*RI, *Hin*dIII (69)
E. coli plasmids	
Col E1	*Hae*II, *Hae*III (80, 105)
pBR313	*Alu*I, *Bam*HI, *Bgl*I, *Eco*RI, *Eco*RII, *Hae*II, *Hae*III, *Hin*dII, *Hpa*I, *Pst*I, *Sal*I, *Xma*I (12)
pBR322	*Alu*I, *Bam*HI, *Bgl*I, *Eco*RI, *Eco*RII, *Hae*II, *Hae*III, *Hin*dII, *Hin*dIII, *Pst*I, *Sal*I (13)
pCR1	*Eco*RI, *Hin*dII, *Hin*dIII, *Kpn*I, *Pst*I, *Xma*I, *Sal*I (2, 3)
pKB158	*Bam*HI, *Bgl*II, *Eco*RI, *Hin*dII, *Hpa*I, *Pst*I, *Sal*I (6)
pMB9	*Bam*HI, *Eco*RI, *Hin*dIII, *Sal*I (88)
pNT1	*Hae*II, *Hae*III, *Hha*I, *Hin*fI (105)
pSC101	*Eco*RI, *Hin*dIII (21), *Bam*HI, *Sal*I (48)
F	*Bam*HI, *Eco*RI, *Hin*dIII (79)
R100	*Bam*HI, *Eco*RI, *Hin*dIII (79)
NRI	*Eco*RI (101)
R538.1	*Bam*HI, *Eco*RI, *Hin*dIII (110)
RK2	*Bam*HI, *Eco*RI, *Hin*dIII, *Hpa*I, *Sal*I (70), *Bgl*II, *Pst*I, *Sma*I (71)
RP1	*Bam*HI, *Bgl*II, *Eco*RI, *Pst*I (44)
RP4	*Bam*HI, *Eco*RI, *Hin*dIII (7, 29), *Bgl*II, *Hpa*I, *Kpn*I, *Pst*I, *Sac*II, *Sal*I, *Sma*I, *Xho*I (29)

(*Continued*)

TABLE VI (*Continued*)

DNA	Maps and references[a]
Yeast plasmids	
Scp1	*Eco*RI (9, 17, 51, 66), *Hind*II (66), *Hind*III (9, 17, 51), *Hpa*I (9, 17), *Pst*I (17, 66)
Scp2	*Eco*RI, *Hind*III, *Hpa*I, *Pst*I (17)
Mitochondrial DNA	
Saccharomyces cervisiae	*Eco*RI (73, 89), *Hind*II + III (89), *Bam*HI, *Hind*III, *Hpa*I, *Pst*I, *Sal*I (73), *Hha*I, *Xba*I (74)
Neurospora crassa	*Bam*HI (102), *Eco*RI (10, 102)
Drosophila melanogaster	*Hae*III, *Hind*III (62)
Mouse	*Eco*RI, *Hind*III (16, 81), *Bam*HI, *Hind*II, *Hpa*I, *Pst*I (72, 81), *Hae*II, *Hha*I (81)
Rat	*Eco*RI, *Hind*III (64, 81), *Bam*HI, *Hae*II, *Hha*I, *Hind*II, *Hpa*I (81)
Sheep–goat	*Eco*RI, *Hind*III (106)
Monkey	*Eco*RI, *Hind*III (16)
Human	*Eco*RI, *Hind*III (16)
Chloroplast DNAs	
Euglena gracilis	*Bam*HI, *Sal*I (40)
Zea mays	*Bam*HI, *Eco*RI, *Hind*III (8)

[a] Key to references (in parentheses): 1. Allet and Bukhari (1975); 2. Armstrong and Helinski (1977); 3. Armstrong *et al*. (1977); 4. J. R. Arrand, H. Delius, R. E. Gelinas, M. B. Matthews, C. Mulder, R. J. Roberts, J. Sambrook, M. Zabeau, and B. S. Zain (unpublished results); 5. Bachi and Arber (1977); 6. Backman *et al*. (1977); 7. Barth and Grinter (1977); 8. Bedbrook and Bogorad (1976); 9. Beggs *et al*. (1976); 10. Bernard *et al*. (1976); 11. Berns *et al*. (1975); 12. Bolivar *et al*. (1977a); 13. Bolivar *et al*. (1977b); 14. Botchan (1976); 15. Brack *et al*. (1976a); 16. Brown and Vinograd (1974); 17. Cameron *et al*. (1977); 18. Canaani *et al*. (1977); 19. Carter *et al*. (1975); 20. Carter *et al*. (1976); 21. Chang *et al*. (1975); 22. Chater (1977); 23. Chattoraj *et al*. (1977); 24. Chen *et al*. (1975); 25. Clements *et al*. (1977); 26. Crawford and Robbins (1976); 27. de la Maza and Carter (1976); 28. de la Maza and Carter (1977); 29. DePicker *et al*. (1977); 30. Favre *et al*. (1975); 31. Favre *et al*. (1977); 32. Folk *et al*. (1975); 33. Forsblom *et al*. (1976); 34. Gissmann and zur Hausen (1976); 35. Gissmann *et al*. (1977); 36. Godson (1975); 37. Godson (1976); 38. Godson and Roberts (1976); 39. Goldstein *et al*. (1975); 40. Gray and Hallick (1977); 41. Griffin and Fried (1975); 42. Griffin and Fried (1976); 43. Griffin *et al*. (1974); 44. Grinsted *et al*. (1977); 45. Groneberg *et al*. (1977); 46. Grosveld *et al*. (1976); 47. Haggerty and Schleif (1976); 48. Hamer and Thomas (1976); 49. Hamlett *et al*. (1977); 50. Helling *et al*. (1974); 51. Hollenberg *et al*. (1976); 52. Horiuchi *et al*. (1975); 53. Howley *et al*. (1975); 54. Hughes (1977); 55. Inciarte *et al*. (1976); 56. Ito and Kawamura (1976); 57. Ito *et al*. (1976); 58. E. N. Jackson, D. A. Jackson, and R. J. Deans, quoted in Susskind and Botstein (1977); 59. James *et al*. (1976); 60. Kahmann *et al*. (1977); 61. Kamp *et al*. (1977); 62. Klukas and Dawid (1976); 63. Kozlov *et al*.

al., 1976; McDonell *et al.*, 1977). Several electron microscope approaches have been described (Morrow and Berg, 1972; Sharp *et al.*, 1973; Mulder *et al.*, 1974; Thomas and Davis, 1975; Brack *et al.*, 1976a). Hybridization procedures using DNA fragments immobilized on nitrocellulose filters (Gillespie and Spiegelman, 1965) can be used to map fragments (Lebowitz *et al.*, 1974b), and recently an ingenious modification of the Southern (1975b) blotting procedure (see below) has been developed that allows a complete set of restriction fragments to be mapped simultaneously (Sato *et al.*, 1977). Comparative restriction enzyme analysis of deletion or substitution mutants can often be used to advantage (Allet *et al.*, 1973; Kamp *et al.*, 1977; Robinson and Landy, 1977a,b).

End-labeling procedures can be useful to identify terminal fragments (by labeling before cleavage) and to detect small fragments which are easily overlooked (by labeling after cleavage). A label can be introduced at the 5'-end of each strand using polynucleotide kinase (Richardson, 1965) or at the 3'-ends of strands using either a DNA polymerase to repair DNA previously resected with exonuclease III

Table VI (*Continued*)
(1975);   64. Kroon *et al.* (1977);   65. Larsen and Nathans (1977);   66. Livingston and Klein (1977);   67. McDonell *et al.* (1977);   68. McParland *et al.* (1976);   69. Meagher *et al.* (1977);   70. Meyer *et al.* (1977a);   71. Meyer *et al.* (1977b);   72. Moore *et al.* (1977);   73. Morimoto *et al.* (1977);   74. Morimoto *et al.* (1978); 75. Morse *et al.* (1977);   76. Mulder *et al.* (1974);   77. C. Mulder and H. Delius, quoted in Ortin *et al.* (1976);   78. Murray and Murray (1975);   79. Ohtsubo and Ohtsubo (1977);   80. Oka and Takanami (1976);   81. Parker and Watson (1977); 82. Parker *et al.* (1977);   83. Perricaudet and Tiollais (1975);   84. Pirrotta (1976); 85. Rhoades (1975);   86. Rhode (1977);   87. Robinson and Landy (1977a, b); 88. Rodriguez *et al.* (1976);   89. Sanders *et al.* (1975);   90. Seeburg and Schaller (1975);   91. Sekikawa and Fujinaga (1977);   92. Shishido and Berg (1976); 93. Skare and Summers (1977);   94. Smith *et al.* (1976);   95. Spear *et al.* (1977); 96. Streeck and Hobom (1975);   97. Summers (1975);   98. Sussenbach and Kuijk (1977);   99. Tabak *et al.* (1974);   100. Takanami *et al.* (1975);   101. Tanaka *et al.* (1976);   102. Terpstra *et al.* (1977);   103. Thomas and Davis (1975);   104. Tibbetts (1977);   105. Tomizawa *et al.* (1977);   106. Upholt and Dawid (1977); 107. Van den Hondel and Schoenmakers (1975);   108. Van den Hondel and Schoenmakers (1976);   109. Van den Hondel *et al.* (1976);   110. Vapnek (1977);   111. Varsanyi *et al.* (1977);   112. von Gabain *et al.* (1976);   113. Vovis *et al.* (1975); 114. Wilkie (1976);   115. Wittek *et al.* (1977);   116. Yaniv *et al.* (1975).
   [b] Since the complete nucleotide sequences of the genomes φX174 (Sanger *et al.*, 1977), fd (Schaller *et al.*, 1978), and SV40 (Fiers *et al.*, 1978; Reddy *et al.*, 1978) are known, the cleavage maps for enzymes of known recognition sequence can be derived from these. References to previously determined maps can be found in the papers listed.

(Roberts *et al.*, 1974) or by the addition of labeled nucleotides with terminal deoxynucleotidyltransferase (Roychoudhury *et al.*, 1976).

Thus far, the accurate determination of fragment size has been a major stumbling block in constructing self-consistent cleavage maps, and is the usual source of discrepancies between maps derived independently. Several methods have been used to estimate fragment sizes, including the electron microscope measurement of the lengths of large fragments (Mulder *et al.*, 1974; Thomas and Davis, 1975) or digestion of uniformly labeled DNA and quantitation of the radioactivity present in each fragment (Danna and Nathans, 1971). Perhaps the easiest and most widely used method involves a comparison of the electrophoretic mobilities of fragments with those of standards whose sizes have been determined accurately. Since the complete nucleotide sequences of ϕX174 (Sanger *et al.*, 1977), fd (Schaller *et al.*, 1978), and SV40 (Fiers *et al.*, 1978; Reddy *et al.*, 1978) have recently been determined, fragments derived from these genomes provide the most accurate low molecular weight standards. Fragments of higher molecular weight whose size has been determined accurately include the *Eco*RI and *Hin*dIII fragments of phage λ (Wellauer *et al.*, 1974; Thomas and Davis, 1975; Parker *et al.*, 1977), the *Dpn*II (*Mbo*I) and *Hpa*II fragments of phage T7 (McDonell *et al.*, 1977), the *Eco*RI fragments of phage T5 (von Gabain *et al.*, 1976), and the permuted linears of SV40 and phage PM2 (Parker *et al.*, 1977). It should be emphasized that this method of determining fragment size is not absolutely reliable because of the frequent occurrence of anomalous mobilities due to the effects of base composition or sequence arrangement. More accurate molecular weight determinations can be made if denaturing conditions are used (Maniatis *et al.*, 1975; McDonell *et al.*, 1977; McMaster and Carmichael, 1977).

Primary maps for enzymes which generate relatively few cuts are easily obtained, but the methods become extremely laborious when constructing cleavage maps for many different enzymes, especially for enzymes that cleave frequently. Two new mapping techniques (Smith and Birnstiel, 1976; Parker *et al.*, 1977) obviate this difficulty and provide a simple means to construct self-consistent maps. Once a primary map is available, the strategy devised by Smith and Birnstiel (1976) seems ideally suited to the derivation of more detailed maps. Their procedure exploits the principle originally used in the plus–minus method of DNA sequencing (Sanger and Coulson, 1975). DNA fragments labeled at only one of their ends are subjected to mild restriction enzyme digestion so that, on average, only one cut is made per molecule. The mixture of partially digested DNA fragments is then fractionated by gel electrophoresis, and those fragments containing a labeled end are identified by autoradiography. They form an overlapping set of fragments extending from a common labeled end, which serves as a reference point, to positions determined by the restriction enzyme cleavage sites. From the sizes of the labeled fragments, the positions of corresponding cleavage sites can be deduced directly. If several samples, each partially digested with a different enzyme are run in adjacent slots of one slab gel, the map positions of all the cleavage sites can be determined relative to one another, and a totally self-consistent map results. The sizes of the different fragments expected upon complete digestion of the starting DNA can then be deduced from the map, although fragments of similar size can not be assigned readily. Nevertheless, the speed of this method, combined with the ease of producing an *accurate* map, more than compensate for this disadvantage. Clearly, the accuracy of the map is dependent upon the distance of the cleavage sites from the labeled end. By repeating the process using overlapping sets of DNA fragments, this limitation is easily overcome.

The second mapping procedure, developed by Parker *et al.* (1977), again depends upon partial digestion, but is limited to circular DNA substrates. It takes advantage of the observation that ethidium bromide, at the appropriate concentrations, will allow the conversion of covalently closed circular DNA molecules to linear molecules but will prevent further cleavage of the resulting linears. By using an enzyme for which several sites are present in the circular DNA, a permuted set of unit length linear molecules is generated. If this set is now cleaved by a second enzyme, which cleaves the original DNA only once, each linear of this permuted set will give two fragments whose molecular weights will sum to that of the original DNA. Gel fractionation of such a digest will give an overlapping series of fragments such that the combined molecular weights of the largest and the smallest fragments will equal those of the second largest and second smallest, etc. If the molecular weights of the fragments produced by complete digestion with the multicut enzyme are known, then it is possible to deduce the locations of its cleavage sites relative to that of the single-cut enzyme. This same procedure can be used for other single- or multicut enzymes and self-consistent maps can be generated in each case. Although this method is limited to circular DNA substrates, its main advantage over the former method is that no prior map information is required and that the technical manipulations of fragment isolation and end labeling are avoided. This procedure should be particularly useful for mapping DNA fragments cloned in plasmid vectors.

IV. APPLICATIONS OF RESTRICTION ENZYME TECHNOLOGY TO THE STUDY OF CHROMOSOME STRUCTURE AND FUNCTION

The extraordinary value of restriction enzymes results from their specificity, because the sites of cleavage provide useful reference points for mapping genes and their controlling elements. Thus the acquisition of detailed physical maps serves as a prelude to further studies which attempt to correlate sequence content and function. Many of the approaches detailed in the literature involve the application of a few basic techniques to a host of different systems. This section will describe the various techniques that exploit restriction enzymes to study gene structure and function. Other uses of these enzymes, which among other things can greatly simplify the isolation of individual genes, concern recombinant DNA technology and will be treated in Section V. DNA sequence analysis which also depends heavily on restriction enzymes is the subject of a separate chapter (Biro and Weissman, Chapter V, this volume).

A. COMPARISON OF RELATED DNAs

When a small genome (phage, virus, mitochondrion, etc.) is cleaved with a restriction endonuclease, a number of specific fragments are produced which give a characteristic banding pattern when displayed by gel electrophoresis. Comparison of such patterns can be used to assess the degree of relatedness of two DNAs. The analysis of substitution and deletion mutants is straightforward and offers a significant advantage over electron microscopic heteroduplex analysis in that fragments of interest can be identified immediately. Thus small deletions have been detected in SV40 and polyoma strains by this method (Nathans and Danna, 1972a; Fried et al., 1974) and have allowed the accurate mapping of the origin of replication in phage λ (Denniston-Thompson et al., 1977). It is important to note the limitations of this method. When only small changes are involved, the ability to detect them will depend upon the size of the fragment in which they are located. A small deletion may significantly affect the mobility of a small fragment, but may pass unnoticed in a large fragment. Occasionally single base changes have been detected, as was shown for mutations affecting promotors and operators in phage λ (Allet and Solem, 1974; Maurer et al., 1974; Kleid et al., 1976), but more often such mutations will lie between restriction sites and escape attention. Several different restriction enzymes should be employed to reduce the possibility of erroneous interpretation due to gross rearrangements in sequences lying between restriction sites.

Despite the simplicity of this approach, it has often proved valuable. The genome structures of hybrid viruses (Lebowitz *et al.*, 1974a; Allet and Bukhari, 1975; Magazin *et al.*, 1977) and insertion mutants resulting from gene transposition (Barth and Grinter, 1977) were easily determined. Other examples include the studies of the evolutionary variants of SV40-containing insertions of host DNA and those containing multiple origins of replication (Brockman *et al.*, 1973; Khoury *et al.*, 1974; Lee *et al.*, 1975; Rao and Singer, 1977). Similarly, the papovaviruses isolated from patients with progressive multifocal leukoencephalopathy were found to be related to SV40 (Sack *et al.*, 1973; Osborn *et al.*, 1974, 1976).

In a similar vein, restriction sites have been used as physical markers to examine the products of genetic crosses. The maternal inheritance of mitochondria (Hutchison *et al.*, 1974; Levings and Pring, 1976) and the demonstration that most, if not all, yeast ribosomal genes are located on one chromosome (Petes and Botstein, 1977) illustrate the power of this approach for the biochemical characterization of genetic events. Genes on adenovirus types 2 and 5 have been physically mapped by the analysis of interserotypic recombinants (Grodzicker *et al.*, 1974, 1977; Sambrook *et al.*, 1975), while other applications include the study of mismatch repair in polyoma (Miller *et al.*, 1976) and recombination processes such as the specific integration of phage λ (Mizuuchi and Nash, 1976) and the promiscuous insertion of transposons (Rubens *et al.*, 1976; Kretschmer and Cohen, 1977).

In many instances, rather complex situations have been successfully analyzed. The herpes simplex virus genome was shown to consist of four types of permuted molecules (Hayward *et al.*, 1975; Wilkie and Cortini, 1976; Skare and Summers, 1977), and other types of sequence heterogeneities in populations of DNA molecules have proved tractable. These range from the relatively simple yeast plasmids, containing inverted repeats (Guérineau *et al.*, 1976; Hollenberg *et al.*, 1976; Livingston and Klein, 1977), to the more complex cases presented by the plant mitochondrial DNAs (Quétier and Vedel, 1977), the large mitochondrial DNA of yeast which has not yet been isolated intact (Bernard *et al.*, 1975; Morimoto *et al.*, 1975; Sanders *et al.*, 1975; Prunell *et al.*, 1977b), and the kinetoplast DNA of trypanosomes (Kleisen *et al.*, 1976a,b).

One rather specialized use concerns the analysis of linear DNAs which contain unusual structural features at their ends. Using end-labeled DNA, the inverted terminal repetition of adenovirus-2 has been sized by finding restriction enzymes which cleave either within or outside the repetition (Roberts *et al.*, 1974). A similar approach has identified several different forms of the ends of adeno-associated virus (de la Maza and Carter, 1977; Spear *et al.*, 1977). Finally, lengthy cohesive ends in a linear DNA usually lead to the submolar recovery of terminal fragments and the concomitant appearance of a new fragment, which can be dissociated by heating (Allet *et al.*, 1973; Scher *et al.*, 1977).

Although cleavage of eukaryotic DNA usually gives a digest too complicated to be analyzed directly, distinct banding patterns frequently occur resulting from the presence of highly reiterated sequences. Thus, if a cleavage site falls within the basic repeat of a satellite DNA or that of a multicopy gene family, such as those encoding ribosomal RNAs, a series of identical fragments are formed which can be visualized directly above a continuous background of unresolved fragments. In some cases the repeated sequences themselves can be purified by buoyant density centrifugation and analyzed separately. Many examples have been reported, and it is clear that most satellite DNAs contain a hierarchical arrangement of repeats within repeats, a structure not readily analyzed by other means (Botchan, 1974; Southern, 1975a; Roizes, 1976; Carlson and Brutlag, 1977; Endow, 1977; Horz and Zachau, 1977). Often the presence of cryptic satellites can be demonstrated by restriction enzyme analysis (Roizes, 1974), an interesting example being the specific satellite found associated with the human Y chromosome (Cooke, 1976).

The ribosomal RNA genes of yeast (Cramer *et al.*, 1976; Meyerink and Retel, 1976; Nath and Bollon, 1976, 1977) and *Drosophila* (Wellauer and Dawid, 1977; White and Hogness, 1977) have been shown to be tandemly repeated. Although a similar arrangement has been reported for the amplified ribosomal RNA genes of *Xenopus laevis* (Wellauer *et al.*, 1974), those of *Physarum* and *Tetrahymena* consist of giant palindromes each containing two rRNA gene sets (Engberg *et al.*, 1976; Karrer and Gall, 1976; Molgaard *et al.*, 1976; Vogt and Braun, 1976).

B. HYBRIDIZATION TECHNIQUES

When two complementary strands of RNA or DNA are mixed, they will anneal to form a duplex. This property of hybridization has been of immense value in studies of nucleic acids, since the single strands and the duplexes can be separated easily and their relative abundance measured. By studying the rate with which a well-defined single-stranded sequence (probe) forms a duplex structure, the concentration and hence abundance of its complement can be determined. Alternatively, the extent of complementarity can be assessed by measuring the fraction of the probe recoverable in duplex form. Various modifications of these basic procedures, in which restriction fragments are used as probes, have been applied to obtain transcription maps of viral genomes (Khoury et al., 1973; Sambrook et al., 1973; Carter et al., 1976; Hayashi et al., 1976; Kamen and Shure, 1976; Pettersson et al., 1976) and to study the viral and plasmid sequences present in transformed cell Gallimore et al., 1974; Sambrook et al., 1974; Chilton et al., 1977).

One method for detecting hybridization depends upon the observation that single-stranded DNA can be immobilized on nitrocellulose filters and yet retain its capacity to anneal to complementary sequences (Gillespie and Spiegelman, 1965). An imaginative extension of this technique of tremendous practical importance has been devised by Southern (1975b). Restriction fragments are first fractionated by agarose gel electrophoresis and then denatured in situ. The single strands are eluted from the gel onto a sheet of nitrocellulose, where they become bound and form an exact replica of the original gel pattern. When a labeled probe is hybridized to this filter, many different fragments can be assayed simultaneously for their homology with that specific probe. While this method is clearly economical in the case of a viral genome, which may give rise to relatively few fragments, the technique is of immense value in the study of more complicated genomes. For despite the fact that a restriction enzyme digest of chromosomal DNA contains so many fragments that they cannot be resolved from one another on a single gel, this procedure allows the detection of a single fragment by virtue of its hybridization to the probe. Numerous uses of this technique have appeared in the literature, emphasizing its value. For instance, it can provide an alternative approach to map transcripts, as was shown for herpes simplex virus (Clements et al., 1977) and bacteriophage T7 (McAllister and Barrett, 1977). The analysis of Bacillus subtilis mRNAs led to the isolation of a gene that is activated early during sporulation (Segall and Losick, 1977). A more powerful application results when pure RNA species are available. Thus, the structural organization of repeated ribosomal RNA genes has been studied in many organisms, such as Bacillus (Potter et al., 1977), Dictyostelium (Cockburn et al., 1976; Maizels, 1976), Drosophila (Tartof and Dawid, 1976), and mouse (Arnheim and Southern, 1977; Cory and Adams, 1977). Similarly, tRNA genes (Olson et al., 1977) and several small RNAs encoded by adenovirus-2 (Mathews, 1975; Pettersson and Mathews, 1977) have been mapped. With the availability of high specific activity DNA probes, prepared in vitro by nick translation, single copy genes can be detected in a digest of chromosomal DNA. Elegant examples include studies of integrated SV40 genomes (Botchan *et al.*, 1976; Ketner and Kelly, 1976), the rabbit β-globin gene (Jeffreys and Flavell, 1977a), and the single copy ribosomal RNA gene present within the micronucleus of *Tetrahymena* (Yao and Gall, 1977).

Recent studies of the genes for ovalbumin (Breathnach *et al.*, 1977; Doel *et al.*, 1977) and β-globin (Jeffreys and Flavell, 1977b; Tilghman *et al.*, 1978) revealed that restriction enzyme cleavage maps of the chromosomal genes differ from those of the cDNAs prepared from their corresponding mRNAs. In both cases, sequences were found within the structural region of the gene which were not present in the mRNA. A similar situation has been reported for some of the 28 S ribosomal RNA genes of *Drosophila melanogaster* (Glover and Hogness, 1977). These findings are probably related to the phenomenon first observed with adenovirus-2 (Berget *et al.*, 1977; Chow *et al.*, 1977), where it was shown that most mRNAs, coded by the virus, are composed of sequences derived from widely different parts of the genome.

One variation of the Southern hybridization procedure has been described, and is termed sandwich hybridization (Dunn and Hassell, 1977). It was developed to study a particular type of mRNA derived from adeno–SV40 hybrid viruses, which contain both SV40 and adenovirus sequences. If unlabeled mRNA is hybridized to a filter containing only adenovirus sequences, the SV40 sequences within the mRNA will form single-stranded tails which can be used to form tertiary hybrids with labeled SV40 DNA. In this way only hybrid mRNAs will be detected, and the adenovirus sequences contained within them can be mapped. This technique is extremely sensitive because the labeled probe can be much longer than the sequences to which it hybridizes resulting in amplification of the signal. Quite general applications can be envisaged, since the 3'-poly(A) tails on mRNA could be used to form tertiary hybrids with labeled poly(dT), and if other eukaryotic mRNAs resemble the spliced Ad2 mRNAs, various parts of their coding sequences could be mapped using appropriate probes.

C. GENE MAPPING—GENETIC PROCEDURES

Much of the previous discussion has centered around the biochemical aspects of isolating and characterizing restriction fragments. While emphasis has been placed upon their sequence organization, a most important question concerns their biological information content. Any given fragment may contain an entire gene, part of a gene, a controlling element, etc. Several techniques are available to test the biological activity of a restriction fragment. One method which involves transformation takes advantage of the fact that many cells can take up naked DNA and integrate it within their genomes. In this way the segments of the SV40 and adenovirus genomes which contain the genes for malignant transformation have been identified (Graham *et al.*, 1974; Yano *et al.*, 1977). The thymidine kinase gene specified by herpes simplex virus has been mapped using viral DNA fragments to transform a mouse cell line carrying a defect in this enzyme activity (Maitland and McDougall, 1977; Wigler *et al.*, 1977). A transfection assay has been used to demonstrate that the proviral DNA of spleen necrosis virus integrates at a specific site in the host chromosome during chronic infection (Battula and Temin, 1977).

Another approach termed "marker rescue" takes advantage of the fact that cellular repair or recombination processes enable wild-type genetic markers on restriction fragments to replace mutant alleles within genomes, thus providing a means to identify genes on restriction fragments. In the procedure originally developed to map mutations in phase ϕX174 (Hutchison and Edgell, 1971), single-stranded DNA from the mutant phage was annealed to individual restriction fragments obtained from wild-type viral DNA and used to transfect *E. coli*. By scoring the production of wild-type viruses, the physical location of the mutation could be demonstrated. This technique of marker rescue has been used extensively to map amber mutations in single-stranded DNA phages of *E. coli* (Hayashi and Hayashi, 1974; Seeburg and Schaller, 1975; Van den Hondel *et al.*, 1975, 1976; Weisbeek *et al.*, 1976; Ravetch *et al.*, 1977) and temperature-sensitive mutations in SV40 (Lai and Nathans, 1974, 1975) and polyoma (Feunteun *et al.*, 1976; Miller and Fried, 1976). Some bacterial strains, such as *Bacillus* and *Haemophilus*, which can be transformed efficiently, allow marker rescue from wild-type fragments to be measured directly (Harris-Warrick *et al.*, 1975; Inciarte *et al.*, 1976; Potter *et al.*, 1977).

D. GENE MAPPING—BIOCHEMICAL APPROACHES

If a DNA fragment contains the necessary signals for the initiation of transcription and translation, it should, in theory, be possible to characterize it by examining the polypeptides made from it *in vitro*. Although this is often possible for fragments of prokaryotic origin, no cell-free system has yet been described which will correctly accomplish both processes for a fragment of eukaryotic origin. Consequently, restriction fragments have been used more often as a means of purifying, by hybridization, specific mRNAs for subsequent translation *in vitro* or to serve as templates for RNA polymerases.

The use of mRNA, purified by hybridization to restriction fragments, to direct translation *in vitro* (Prives *et al.*, 1974) has proved valuable to locate the DNA sequences encoding both early and late functions in adenovirus-2 (Lewis *et al.*, 1975, 1976, 1977). It should be noted that this technique does not rigorously map structural gene sequences, since the RNA transcript might have been isolated by hybridization to noncoding sequences located in other parts of the genome (Lewis *et al.*, 1977) A simple, but indirect, procedure which takes advantage of the fact that DNA–RNA hybrids are unable to program protein synthesis has recently been described (Paterson *et al.*, 1977). Coupled transcription–translation systems of prokaryotic origin have been used extensively, and the level of sophistication reached is illustrated by the studies of the organization of the gene clusters encoding the transcriptional and translational machinery of *E. coli* (Lindahl *et al.*, 1976, 1977a,b). A eukaryotic coupled system has been described (Rozenblatt *et al.*, 1976) which allowed the mapping of the coding regions for the late SV40 structural proteins. This system has not been shown to initiate transcription correctly.

Many facets of transcription, including its regulation in prokaryotes, can now be studied using highly purified *in vitro* systems. Run-off RNAs synthesized *in vitro* on overlapping fragments have been used to map initiation sites for transcription in phage fd (Okamoto *et al.*, 1975; Seeburg and Schaller, 1975) and within the *E. coli* arabinose operon (Hirsh and Schleif, 1977), among others. Fragments obtained by cleavage within sequences located near to the promotor regions of certain genes have allowed the identification of target sites for regulatory proteins. Examples include the mapping of the arabinose repressor binding site (Hirsh and Schleif, 1977) and the catabolite repressor protein binding site within the galactose operon (Sklar *et al.*, 1977) and studies of the control of phage λ repressor synthesis (Meyer *et al.*, 1975; Walz *et al.*, 1976).

Two eukaryotic systems, which approach *in vivo* conditions, can be used for either translation or coupled transcription–translation. By injection of DNA into the nuclei of *Xenopus* oocytes correct transcription and translation can occur in certain cases (De Robertis and Mertz, 1977), while similar injection of mRNA into the cytoplasm can lead to accurate translation (Mertz and Gurdon, 1977). An alternative approach involves microinjection of either RNA or DNA into tissue culture cells (Graessmann *et al.*, 1976). Both systems, but particularly the latter, are necessarily limited in the quantities of material that can be introduced. Nevertheless they afford an opportunity to conduct many interesting experiments, which are not yet possible by any other technique.

E. FILTER BINDING ASSAYS

One very useful property of nitrocellulose membranes is their ability to retain protein DNA complexes while allowing free DNA to pass through them. Thus fragments that contain specific sequences to which a protein will bind can be separated and characterized. The specific fragments can be identified by gel electrophoresis, for they will either be present in filter-bound material or selectively absent from the nonretained material. This approach has been used to map the *E. coli* RNA polymerase binding sites on different phage genomes, including ϕX174 (Chen *et al.*, 1973), fd (Okamoto *et al.*, 1975; Seeburg and Schaller, 1975), and λ (Allet and Solem, 1974; Maurer *et al.*, 1974; Jones *et al.*, 1977). Binding sites for other regulatory proteins, such as the λ repressor (Maniatis and Ptashne, 1973), the *lac* repressor (Landy *et al.*, 1974a), and the catabolite repressor protein (Majors, 1975), have also been mapped by this method.

The observation that *E. coli* RNA polymerase–DNA complexes can inhibit cleavage by the enzyme *Hin*dII at a small number of unique sites in phages λ and T7 (Allet *et al.*, 1974; Maurer *et al.*, 1974) suggested that in some cases the recognition sequences for both enzymes were mutually overlapping. This conclusion was supported by the finding that mutations affecting two λ promotors concomitantly destroyed *Hin*dII recognition sites (Allet and Solem, 1974; Maurer *et al.*, 1974). Several promotors displaying this property have been identified and mapped by using RNA polymerase to protect against *Hin*dII cleavage or by observing the loss of RNA polymerase binding following *Hin*dII cleavage (Hsieh and Wang, 1976; Jones and Reznikoff, 1977; Jones *et al.*, 1977).

F. PULSE LABELING TECHNIQUES

Pulse labeling techniques have been used widely to study various aspects of macromolecular synthesis. The ingenious approach, originally developed by Dintzis (1961) to determine the direction of chain growth during polypeptide synthesis by pulse labeling nascent chains, has also been applied to the study of nucleic acid synthesis. For instance, the primary transcripts of adenovirus-2 DNA have been characterized (Bachenheimer and Darnell, 1975). Nascent transcripts were pulse labeled under conditions where only sequences close to the 3'-end of the RNA chains were labeled. After fractionation of the transcripts by velocity sedimentation, fractions from across the gradient were hybridized to restriction fragments derived from different regions of the genome. By comparing the sizes of the RNAs which hybridized to different restriction fragments, it was shown that the most abundant transcripts were initiated at a unique site and could extend for most of the length of the genome. The initiation and termination sites of this giant late transcript were subsequently determined (Evans *et al.*, 1977).

A procedure which exploits the same principle has been extensively used to map the origins and termini of DNA replication and to determine the direction of synthesis. In this technique, fully replicated molecules, which have been pulse labeled for a short period of time, are isolated and subjected to restriction enzyme cleavage, and the relative amounts of radioactivity incorporated into various DNA fragments are determined. In this way, depending upon the mode of replication, one (or two) gradients of radioactivity are obtained along the entire genome. Regions containing low and high amounts of radioactivity correspond to those in which replication has been, respectively, initiated or terminated. This approach has proved useful to map DNA replication in SV40 (Danna and Nathans, 1972; Nathans and Danna, 1972b), polyoma (Crawford *et al.* 1974), adenovirus (Schilling *et al.*, 1975; Tolun and Pettersson, 1975), adeno-associated virus (Hauswirth and Berns, 1977), herpes simplex virus (Hirsch *et al.*, 1977), and bacteriophages ϕX174 (Godson, 1974), fI (Horiuchi and Zinder, 1976), and M13 (Suggs and Ray, 1977). Separated strands of restriction fragments were used to show that each of the two ends of the adenovirus genome served as a replication origin for one of the complementary strands (Horwitz, 1976; Weingarter *et al.*, 1976; Sussenbach and Kuijk, 1977).

V. Recombinant DNA Technology

Perhaps the most exciting technical advance of recent years has been the development of methods for isolating and amplifying DNA fragments from any source. This recombinant DNA technology has two main aspects. The first concerns the purification of one specific DNA fragment from a complex mixture of fragments with an ease that is unsurpassed. This aspect has been exploited widely, since a eukaryotic chromosome can now be dissected and individual genes examined at leisure. Although restriction enzymes are a key ingredient of this procedure, their greatest potential lies within the second aspect of the technology. For these methods can also be adapted to the construction of new precisely defined sequence combinations and permit feats of synthetic biology which may soon outstrip those of synthetic chemistry. Since various aspects of *in vitro* genetic engineering have been covered extensively in other reviews (Cohen, 1975; Glover, 1976; Murray, 1976; Collins, 1977; Timmis *et al.*, 1977), this section will concentrate upon the special role of restriction enzymes within this technology.

A. MOLECULAR CLONING

The isolation of particular DNA segments by molecular cloning involves the insertion of DNA fragments at particular sites in suitable receptor DNA molecules, termed "vectors," which are capable of replicating autonomously. A large variety of vectors, derived by genetic or biochemical manipulation of bacteriophage genomes and plasmid DNAs, are now available for cloning DNA segments in *E. coli* (for a recent review, see Timmis *et al.*, 1977), and the potential use of eukaryotic viral genomes such as SV40 (Ganem *et al.*, 1976; Goff and Berg, 1976; Nussbaum *et al.*, 1976; Hamer *et al.*, 1977) and adenovirus (Larsen and Nathans, 1977) is now being explored.

Originally, two different methods were used to join DNA fragments *in vitro*. The first method took advantage of the property of certain restriction enzymes to generate fragments with single-stranded extensions—cohesive ends—that are able to reanneal spontaneously and can be joined by using either the *E. coli* or T4 polynucleotide ligases (Mertz and Davis, 1972; Dugaiczyk *et al.*, 1975). The second procedure involved the use of complementary poly(dA) and poly(dT) homopolymer blocks which were added to the 3′-ends of DNA fragments using the enzyme terminal deoxynucleotidyltransferase (Jackson *et al.*, 1972; Lobban and Kaiser, 1973). Both procedures rely upon the presence of complementary single-stranded extensions at the ends of DNA fragments.

More recently, the need for cohesive termini has been alleviated because T4 ligase has the ability to join flush-ended fragments (Sgaramella *et al.*, 1970; Sgaramella and Khorana, 1972; Mottes *et al.*, 1977). Since flush ends can be generated either directly by cleavage with the appropriate restriction enzyme (see Table VII) or from a 5′-extension by repair synthesis with a DNA polymerase (Backman *et al.*, 1976; Ita-

44   MARC ZABEAU AND RICHARD J. ROBERTS

kura *et al.*, 1977) or from both 5′- and 3′-extensions by trimming with the single-strand-specific endonuclease S_1 (Scheller *et al.*, 1977b; Ullrich *et al.*, 1977), all fragments should be amenable to this procedure. Another recent innovation involves the joining of fragments by means of "linkers" or "adaptors" (Bahl *et al.*, 1976; Scheller *et al.*, 1977a). These are chemically synthesized self-complementary oligonucleotides constructed to contain the recognition sites for one or more restriction enzymes. Thus, the decanucleotide CCGAATTCGG, which contains an *Eco*RI site, may be linked by flush-end ligation, to both ends of any flush-ended fragment. After cleavage with *Eco*RI, the original flush-ended fragment will be converted to one with cohesive termini which can be inserted into an *Eco*RI vector. This technique allows DNA fragments obtained by most methods to be cloned in a readily reisolable form (Heyneker *et al.*, 1976; Marians *et al.*, 1976; Scheller *et al.*, 1977b; Ullrich *et al.*, 1977), thus overcoming the main disadvantage of the homopolymer tailing method, which usually results in the destruction of the restriction enzyme site.

A large number of prokaryotic and eukaryotic genes have now been isolated using recombinant DNA methods, and clones of interest have been identified by either genetic or biochemical means. Selection based upon complementation has been used to identify clones containing specific bacterial genes (Hershfield *et al.*, 1974; Clarke and Carbon, 1975, 1976; Borck *et al.*, 1976; Collins *et al.*, 1976; Struhl and Davis, 1976; Kozlov *et al.*, 1977) and certain yeast genes (Struhl *et al.*, 1976; Ratzkin and Carbon, 1977). Antibiotic resistance genes provide easily selectable markers which have proved useful for constructing new plasmid vectors (Cohen *et al.*, 1973; Tanaka and Weisblum, 1975; Hamer and Thomas, 1976; Rodriguez *et al.*, 1976; Bolivar *et al.*, 1977a,b) and for isolating specific DNA segments such as replicator regions (Timmis *et al.*, 1975; Lovett and Helinski, 1976). Hybridization techniques that allow the detection of particular DNA sequences in bacterial colonies (Grunstein and Hogness, 1975) or in virus plaques (Kramer *et al.*, 1976; Benton and Davis, 1977; Villarreal and Berg, 1977) provide a biochemical method for the identification of clones containing genes whose RNA transcripts are readily available. These include the genes for ribosomal RNA (Artanavis-Tsakonas *et al.*, 1977; Hershey *et al.*, 1977), transfer RNA (Kramer *et al.*, 1976; Beckmann *et al.*, 1977; Yen *et al.*, 1977), rabbit hemoglobin (Maniatis *et al.*, 1976; Tilghman *et al.*, 1977), sea urchin histone (Kedes *et al.*, 1975; Clarkson *et al.*, 1976), and mouse immunoglobulin (Tonegawa *et al.*, 1977).

B. GENE CONSTRUCTION

The complete chemical synthesis of a gene, for a tRNA, was first accomplished by Khorana and his colleagues (Agarwal *et al.*, 1970; Khorana *et al.*, 1972). Although the gene was extremely small, being only 75 base pairs long, the route chosen was very time-consuming. Nevertheless, important principles were established. Small blocks of oligonucleotides were constructed such that each contained two protruding single-strand extensions. When two blocks, with complementary extensions, were treated with DNA ligase, a specific and directed joining occurred, resulting in a new, longer oligonucleotide. The specificity provided by the requirement for exact complementarity allowed several blocks to be joined simultaneously in a directed fashion. Of course, there is no reason this approach should be limited to short oligonucleotides, and indeed, as the synthesis progressed, ever-longer building blocks were used. With the present availability of a host of restriction enzymes, many of which have rather different single-stranded extensions (Table VII), it is apparent that similar synthetic feats could be accomplished using restriction fragments as the building blocks. This principle has often been used for the construction of recombinant DNAs, but its real potential remains to be exploited.

A fundamental difference between the purely chemical synthesis of DNA molecules and the construction of recombinant DNAs lies in the nature of the substrates. In the former case, every base in the building block is known and selected for its specific contribution to the final product. In the latter case, the substrates are less well defined because the restriction fragments containing the desired sequences may or may not contain extraneous DNA. Consequently, the construction of the final product may require the further manipulation of the initial recombinant. When devising a synthetic strategy, it is important to realize that a wealth of restriction enzymes is now available and that they differ both in recognition specificity and in mode of cleavage. Although the choice of enzyme is frequently limited by the location of its cleavage sites, joining procedures can be used which lead to the selective loss or preservation of the recognition site and facilitate further manipulations. In some instances, multiple steps can be performed simultaneously in a directed manner or the desired recombinants can be enriched *in vitro* before cloning.

One enzyme, *Hga*I, is worthy of special attention. It recognizes a pentanucleotide sequence, but cleaves at sites five nucleotides away

TABLE VII
Ends Generated by Restriction Enzymes[a]

5'-Extensions				3'-Extensions			
Five base	Four base	Three base	Two base	Four base	Two base	One base	Flush ends
CC$_{\mathrm{T}}^{\mathrm{A}}$GG *Eco*RII	GATC *Mbo*I	GNC *Asu*I	CG *Hpa*II	GCGC *Hae*II	CG *Hha*I	*Hga*I	*Alu*I
	GATC *Bam*HI	ANT *Hinf*I	CG *Taq*I	TGCA *Pst*I	GC *Sac*II	*Mbo*II	*Bal*I
	GATC *Bgl*II		CG	GTAC *Kpn*I			*Fnu*DII
	GATC *Bcl*I		CT (*Acc*I)	AGCT *Sac*I			*Hae*I
	TCGA *Sal*I		AG				*Hae*III
	TCGA *Xho*I		AT				*Hind*II
	PyCGPu *Ava*II						*Hpa*I
	AATT *Eco*RI						*Pvu*II
	AATT *Eco*RI*						*Sma*I
	AGCT *Hind*III						
	CTAG *Xba*I						
	CCGG *Xma*I						

[a] Since all sequences are written 5' → 3', DNA fragments with the 5'-extension GATC (*Mbo*I) or the 3'-extension TGCA (*Pst*I) will have the following structures:

5'-extension
5' G-A-T-C-N-N--N-N-3'
3'-N-N--N-N-C-T-A-G-5'

3'-extension
5'-G-N--N-C-T-G-C-A-3'
3' A-C-G-T-C-N--N-G-5'

from this sequence on one strand and ten nucleotides away on the other strand (Brown and Smith, 1977). Thus, each fragment possesses cohesive termini, five nucleotides long, but which have different sequences, since they do not form part of the recognition sequence. If the 14 fragments generated by cleavage of φX174 DNA were to be joined *in vitro*, it is unlikely that any but the original neighboring fragments would be capable of reannealing correctly. Among the products, one might hope to find intact φX174 DNA! Clearly, the ability to reconstruct whole genomes from many different fragments, in a single step, could have far-reaching consequences (Brown and Smith, 1977). There is no reason to suppose that *Hga*I will be unique in its mode of cleavage, and, as other similar enzymes are found, they should considerably extend the range of manipulations possible.

The selection *in vitro* of certain combinations of restriction fragments by the directed reannealing of cohesive termini was dramatically illustrated above. A similar but less specific selection could occur if a mixture of *Hin*fI fragments joined because here the cohesive termini will have one of four possible sequences (see Table VII). *Acc*I, *Asu*I, *Eco*RII, *Hph*I, and *Mbo*II fragments also share this property. Another approach to *in vitro* selection is provided by the sets of enzymes which recognize related sequences and produce identical cohesive termini (Table VII). Thus *Bam*HI, *Bgl*II, and *Bcl*I each give fragments with a 5'-tetranucleotide extension GATC, which can be joined to one another. However, the hybrid sites generated by such joining are no longer recognized by any of the initial enzymes, and so hybrid recombinants could be selected *in vitro* by including the restriction enzymes in the joining reaction (Roberts, 1977). These hybrid sites would be susceptible to *Mbo*I, which recognizes only the central tetranucleotide GATC. In a similar manner, *Sal*I and *Xho*I fragments could be joined and the hybrid sites cleaved by *Taq*I, which recognizes only their central tetranucleotide.

The ability to excise a cloned fragment from its vector is often an important consideration when planning a synthetic strategy. In addition to the trivial case of joining two fragments generated by the same restriction endonuclease, several other possibilities deserve consideration. The first concerns the use of enzymes, such as *Hga*I, *Hph*I, *Mbo*II, or *Mnl*I, which cleave at sites remote from their recognition sequences. In each case, the ends of the fragments can be manipulated without affecting the recognition sequence, and it might be imagined that fragments generated by these enzymes would automatically be reisolable after cloning. This is not the case, because for each enzyme the cleavage site is oriented to one specific side of the recognition sequence. Thus, any one fragment may contain zero, one, or two recognition sequences, and there is no simple way to determine which is the case. An additional complication arises if two sites for one of these enzymes occur in close proximity, because they may or may not interfere with one another (Endow, 1977).

Two enzymes, *Mbo*I and *Eco*RII, overcome this problem of orientated cleavage. In both cases, they produce fragments with 5′-extensions which contain the entire recognition sequence. Thus, repair with a DNA polymerase followed by flush-end ligation will always allow exact excision of the cloned fragment. In a similar vein, when an enzyme recognizes a hexanucleotide and cleaves to leave a 5′-tetranucleotide extension, repair will always regenerate the central tetranucleotide. If a second enzyme is available which recognizes only this central tetranucleotide, then it will always be possible to recleave at that site. An example would be *Taq*I, which cleaves all *Sal*I and *Xho*I sites. Of course such an enzyme may also cleave at internal sites of the cloned fragment, thus limiting its use.

A more generally useful method, which depends upon flush-end ligation, involves the careful choice of ends to be joined (Backman *et al.*, 1976). When a fragment generated by *Eco*RI is repaired, five of the six nucleotides of the recognition sequence will be present in duplex form at the ends of that fragment. By joining these new flush-ends to an appropriate flush-ended fragment which carries the sixth base pair of the recognition sequence, a functional *Eco*RI site will be restored. Here, a fragment carrying a 5′-terminal C residue is required and could be obtained by (a) direct cleavage with *Alu*I, *Bal*I, *Fnu*DII, *Hae*III, or *Pvu*II; (b) repairing the 5′-extensions generated by *Hpa*II, *Sac*II, *Taq*I, *Xba*I, or *Xma*I; or (c) using S_1 endonuclease to trim the single-stranded extensions produced by *Asu*I, *Hha*I, *Hinf*I, or *Sal*I. This strategy can be applied to fragments produced by many other enzymes (see Table VII).

It should also be possible to reconstruct *Eco*RI sites by adding poly(C) homopolymer blocks to the repaired *Eco*RI fragments, but this procedure has not yet become routine (Rougeon *et al.*, 1975; Rougeon and Mach, 1977). The difficulties encountered in this case may reflect the preference of terminal deoxynucleotidyltransferase for a substrate containing a 3′-single-strand extension. For this same approach has been used successfully with *Pst*I fragments (Mann *et al.*, 1978), which contain a 3′-tetranucleotide extension. Similar procedures should be possible for other fragments carrying 3′-extensions.

Despite the variety of manipulations now available, several problems remain. One is provided by the desirability of guaranteeing expression, into mRNA and protein, of the cloned sequence. In principle this can be achieved by transferring the cloned segment to a vector constructed such that the site of insertion into this vector lies immediately adjacent to a functional promotor and a good ribosome binding site. Even so, because a functional polypeptide will result only when the correct reading frame is in phase with the ribosome binding site, three such vectors should be available, one for each of the three possible reading frames. Although such a strategy may be satisfactory for prokaryotic genes, it is important to note that it may not suffice for eukaryotic genes. With the finding that several eukaryotic mRNAs are not linear transcripts of the genome, it seems unlikely that cloned segments of eukaryotic chromosomes can ever be expressed as correct polypeptides. The cloning of reverse transcripts of mRNAs may thus become an essential element in the characterization of these genes.

The potential importance of these methods is well illustrated by the sophisticated techniques used to transpose the gene for the phage λ repressor into an environment where it became controlled by the regulatory system of the lactose operon (Backman *et al.*, 1976). Another elegant example comes from studies of a promotor in phage fd (Okamoto *et al.*, 1977). Here an *Hha*I cleavage site which lay within the RNA polymerase binding site was used to construct new sequence arrangements. Subsequent analysis demonstrated that sequences located on one side of the *Hha*I site were not involved in RNA polymerase recognition, but were required for binding. Finally, several spectacular feats of synthesis have been described (Heyneker *et al.*, 1976; Marians *et al.*, 1976; Itakura *et al.*, 1977) which combine many of the approaches outlined above. We now stand on the threshhold of a new endeavor in which the proven degradative capabilities of the restriction endonucleases will soon be matched by their synthetic opportunities.

VI. CONCLUDING REMARKS

Although creative ideas are essential to any field of research, their exploitation is usually limited by the tools and techniques available. Consequently, when a new tool is discovered, a burst of knowledge usually ensues. Such has been the case for the restriction endonucleases which, in the few years since their discovery, have reshaped the practice of molecular biology. Three major advances in our technical capabilities derive from their existence. The first concerns the physical mapping of DNA genomes and their correlation to genetic maps, which has led to the finding that some eukaryotic messenger RNAs are not derived from genomic sequences in a trivial way. The concept of a gene may require yet further revision for eukaryotes. The second has been the development of extremely rapid techniques for the determination of DNA sequences which have culminated in the derivation of complete sequences for three small viruses, ϕX174, fd, and SV40. From the ϕX174 sequence came the first observation that two or more genes may sometimes be encoded within the same stretch of DNA. A similar conclusion emerged from the SV40 sequence, which has also confirmed the nonlinear relationship between DNA and mRNA. The third capability lies in the establishment of a new field, genetic engineering, which has released the eukaryotic chromosome from its impenetrable complexity. At last the molecular biologist truly can hope to study biology at the molecular level.

The chemist and biologist, who have enjoyed an occasional mild flirtation, are now engaged in a serious relationship that may soon be consummated in a fulfilling marriage. The degradative phase of chemistry, which resulted in powerful methods for structure determination, is now paralleled in the biologist's ability to determine gene structure. The great triumphs of chemical synthesis, such as that of vitamin B_{12}, may soon pale by comparison with those of the biologist who might hope to construct an organism entirely dedicated to its synthesis. The impact of these developments upon both the scientist and society will be immense.

REFERENCES

Aaij, C., and Borst, P. (1972). *Biochim. Biophys. Acta* **269**, 192.

Agarwal, K. L., Buchi, H., Caruthers, M. H., Gupta, N., Khorana, H. G., Kleppe, K., Kumar, A., Ohtsuka, E., RajBhandary, U. L., Van de Sande, J. H., Sgaramella, V., Weber, H., and Yamada, T. (1970). *Nature (London)* **227**, 27.

Allet, B. (1973). *Biochemistry* **12**, 3972.

Allet, B., and Bukhari, A. I. (1975). *J. Mol. Biol.* **92**, 529.

Allet, B., and Solem, R. (1974). *J. Mol. Biol.* **85**, 475.

Allet, B., Jeppesen, P. G. N., Katagiri, K. J., and Delius, H. (1973). *Nature (London)* **241**, 120.

Allet, B., Roberts, R. J., Gesteland, R. F., and Solem, R. (1974). *Nature (London)* **249**, 217.

Arber, W. (1965). *Annu. Rev. Microbiol.* **19**, 365.

Arber, W. (1971). *In* "The Bacteriophage Lambda" (A. D. Hershey, ed.), p. 83. Cold Spring Harbor Lab., Cold Spring Harbor, New York.

Arber, W. (1974). *Prog. Nucleic Acid Res. Mol. Biol.* **14**, 1.

Arber, W., and Kuhnlein, U. (1967). *Pathol. Microbiol.* **30**, 946.

Arber, W., and Linn, S. (1969). *Annu. Rev. Biochem.* **38**, 467.

Armstrong, K. A., and Helinski, D. R. (1977). *In* "DNA Insertion Elements, Plasmids, and Episomes" (A. I. Bukhari, J. Shapiro, and S. Adhya, eds.), p. 681. Cold Spring Harbor Lab., Cold Spring Harbor, New York.

Armstrong, K. A., Hershfield, V., and Helinski, D. R. (1977). *Science* **196**, 172.
Arnheim, N., and Southern, E. M. (1977). *Cell* **11**, 363.
Arrand, J. R., Myers, P. A., and Roberts, R. J. (1978). *J. Mol. Biol.* **118**, 127.
Artavanis-Tsakonas, S., Schedl, P., Tschudi, C., Pirrotta, V., Steward, R., and Gehring, W. J. (1977). *Cell* **12**, 1057.
Bachenheimer, S., and Darnell, J. E. (1975). *Proc. Natl. Acad. Sci. U.S.A.* **72**, 4445.
Bachi, B., and Arber, W. (1977). *Mol. Gen. Genet.* **153**, 311.
Backman, K., Ptashne, M., and Gilbert, W. (1976). *Proc. Natl. Acad. Sci. U.S.A.* **73**, 4174.
Backman, K., Hawley, D., and Ross, M. J. (1977). *Science* **196**, 182.
Bahl, C. P., Marians, K. J., Wu, R., Stawinsky, J., and Narang, S. A. (1976). *Gene* **1**, 81.
Barth, P. T., and Grinter, N. J. (1977). *J. Mol. Biol.* **113**, 455.
Battula, N., and Temin, H. M. (1977). *Proc. Natl. Acad. Sci. U.S.A.* **74**, 281.
Beckmann, J. S., Johnson, P. F., and Abelson, J. (1977). *Science* **196**, 205.
Bedbrook, J. R., and Bogorad, L. (1976). *Proc. Natl. Acad. Sci. U.S.A.* **73**, 4309.
Beggs, J. D., Guérineau, M., and Atkins, J. F. (1976). *Mol. Gen. Genet.* **148**, 287.
Benton, W. D., and Davis, R. W. (1977). *Science* **196**, 180.
Berget, S. M., Moore, C., and Sharp, P. A. (1977). *Proc. Natl. Acad. Sci. U.S.A.* **74**, 3171.
Berkner, K. L., and Folk, W. R. (1977). *J. Biol. Chem.* **252**, 3185.
Bernard, U., Bade, E., and Kuntzel, H. (1975). *Biochem. Biophys. Res. Commun.* **64**, 783.
Bernard, U., Goldthwaite, C., and Kuntzel, H. (1976). *Nucleic Acids Res.* **3**, 3101.
Berns, K. I., Kort, J., Fife, K. H., Grogan, E. W., and Spear, I. S. (1975). *J. Virol.* **16**, 712.
Bertani, G., and Weigle, J. J. (1953). *J. Bacteriol.* **65**, 113.
Bickle, T. A., Pirrotta, V., and Imber, R. (1977). *Nucleic Acids Res.* **4**, 2561.
Bigger, C. H., Murray, K., and Murray, N. E. (1973). *Nature (London), New Biol.* **244**, 7.
Bingham, A. H. A., Sharman, A. F., and Atkinson, T. (1977). *FEBS Lett.* **76**, 250.
Blakesley, R. W., and Wells, R. D. (1975). *Nature (London)* **257**, 421.
Blakesley, R. W., Dodgson, J. B., Nes, I. F., and Wells, R. D. (1977). *J. Biol. Chem.* **252**, 7300.
Blattner, F. R., Williams, B. G., Blechl, A. E., Denniston-Thompson, K., Faber, H. E., Furlong, L.-A., Grunwald, D. J., Kiefer, D. O., Moore, D. D., Schumm, J. W., Sheldon, E. L., and Smithies, O. (1977). *Science* **196**, 161.
Blin, N., von Gabain, A., and Bujard, H. (1975). *FEBS Lett.* **53**, 84.
Bolivar, F., Rodriguez, R. L., Betlach, M. C., and Boyer, H. W. (1977a). *Gene* **2**, 75.
Bolivar, F., Rodriguez, R. L., Greene, P. J., Betlach, M. C., Heyneker, H. L., Boyer, H. W., Crosa, J. H., and Falkow, S. (1977b). *Gene* **2**, 95.
Borck, K., Beggs, J. D., Brammar, W. J., Hopkins, A. S., and Murray, N. E. (1976). *Mol. Gen. Genet.* **146**, 199.
Botchan, M. (1974). *Nature (London)* **251**, 288.
Botchan, M., Topp, W., and Sambrook, J. (1976). *Cell* **9**, 269.
Botchan, P. (1976). *J. Mol. Biol.* **105**, 161.
Boyer, H. W. (1971). *Annu. Rev. Microbiol.* **25**, 153.
Boyer, H. W., Chow, L. T., Dugaiczyk, A., Hedgpeth, J., and Goodman, H. M. (1973). *Nature (London), New Biol.* **244**, 40.
Brack, C., Eberle, H., Bickle, T. A., and Yuan, R. (1976a). *J. Mol. Biol.* **104**, 305.
Brack, C., Eberle, H., Bickle, T. A., and Yuan, R. (1976b). *J. Mol. Biol.* **108**, 583.
Braga, E. A., Nosikov, V. V., and Polyanovsky, O. L. (1976). *FEBS Lett.* **70**, 91.
Breathnach, R., Mandel, J. L., and Chambon, P. (1977). *Nature (London)* **270**, 314.
Brockman, W. W., Lee, T. N. H., and Nathans, D. (1973). *Virology* **54**, 384.
Bron, S., and Murray, K. (1975). *Mol. Gen. Genet.* **143**, 25.
Bron, S., Murray, K., and Trautner, T. A. (1975). *Mol. Gen. Genet.* **143**, 13.

Brown, N. L., and Smith, M. (1976). *FEBS Lett.* **65**, 284.

Brown, N. L., and Smith, M. (1977). *Proc. Natl. Acad. Sci. U.S.A.* **74**, 3213.

Brown, N. L., Hutchison, C. A., III, and Smith, M. (1979). *J. Mol. Biol.* (in press).

Brown, W. M., and Vinograd, J. (1974). *Proc. Natl. Acad. Sci. U.S.A.* **71**, 4617.

Cameron, J. R., Philippsen, P., and Davis, R. W. (1977). *Nucleic Acids Res.* **4**, 1429.

Canaani, E., Duesberg, P., and Dina, D. (1977). *Proc. Natl. Acad. Sci. U.S.A.* **74**, 29.

Carlson, M., and Brutlag, D. (1977). *Cell* **11**, 371.

Carter, B. J., Khoury, G., and Denhardt, D. T. (1975). *J. Virol.* **16**, 559.

Carter, B. J., Fife, K. H., de la Maza, L. M., and Berns, K. I. (1976). *J. Virol.* **19**, 1044.

Catterall, J. F., and Welker, N. E. (1977). *J. Bacteriol.* **129**, 1110.

Catterall, J. F., Lees, N. D., and Welker, N. E. (1976). In "Microbiology—1976" (D. Schessinger, ed.), p. 358. Am. Soc. Microbiol., Washington, D.C.

Chang, A. C. Y., and Cohen, S. N. (1974). *Proc. Natl. Acad. Sci. U.S.A.* **71**, 1030.

Chang, A. C. Y., Lansman, R. A., Clayton, D. A., and Cohen, S. N. (1975). *Cell* **6**, 231.

Chang, S., and Cohen, S. (1977). *Proc. Natl. Acad. Sci. U.S.A.* **74**, 4811.

Chater, K. F. (1977). *Nucleic Acids Res.* **4**, 1989.

Chater, K. F., and Wilde, L. C. (1976). *J. Bacteriol.* **128**, 644.

Chattoraj, D. K., Oberoi, Y. K., and Bertaini, G. (1977). *Virology* **81**, 460.

Chen, C.-Y., Hutchison, C. A., III, and Edgell, M. H. (1973). *Nature (London), New Biol.* **243**, 233.

Chen, M. C. Y., Chang, K. S. S., and Salzman, N. P. (1975). *J. Virol.* **15**, 191.

Chilton, M.-D., Drummond, M. H., Merlo, D. J., Sciaky, D., Montoya, A. L., Gordon, M. P., and Nester, E. W. (1977). *Cell* **11**, 263.

Chow, L. T., Gelinas, R. E., Broker, T. R., and Roberts, R. J. (1977). *Cell* **12**, 1.

Clarke, L., and Carbon, J. (1975). *Proc. Natl. Acad. Sci. U.S.A.* **72**, 4361.

Clarke, L., and Carbon, J. (1976). *Cell* **9**, 91.

Clarkson, S. G., Smith, H. O., Schaffner, W., Gross, K. W., and Birnstiel, M. L. (1976). *Nucleic Acids Res.* **3**, 2617.

Clements, J. B., Watson, R. J., and Wilkie, N. M. (1977). *Cell* **12**, 275.

Cockburn, A. F., Newkirk, M. J., and Firtel, R. A. (1976). *Cell* **9**, 605.

Cohen, S. N. (1975). *Sci. Am.* **233**, 24.

Cohen, S. N., Chang, A. C. Y., Boyer, H. W., and Helling, R. B. (1973). *Proc. Natl. Acad. Sci. U.S.A.* **70**, 3240.

Collins, C. J. (1977). *Curr. Top. Microbiol. Immunol.* **78**, 121.

Collins, C. J., Jackson, D. A., and deVries, F. A. J. (1976). *Proc. Natl. Acad. Sci. U.S.A.* **73**, 3838.

Cooke, H. (1976). *Nature (London)* **262**, 182.

Cory, S., and Adams, J. M. (1977). *Cell* **11**, 795.

Cramer, J. H., Farrelly, F. W., and Rownd, R. H. (1976). *Mol. Gen. Genet.* **148**, 233.

Crawford, L. V., and Robbins, A. K. (1976). *J. Gen. Virol.* **31**, 315.

Crawford, L. V., Robbins, A. K., Nicklin, P. M., and Osborn, K. (1974). *Cold Spring Harbor Symp. Quant. Biol.* **39**, 219.

Danna, K. J., and Nathans, D. (1971). *Proc. Natl. Acad. Sci. U.S.A.* **68**, 2913.

Danna, K. J., and Nathans, D. (1972). *Proc. Natl. Acad. Sci. U.S.A.* **69**, 3097.

Danna, K. J., Sack, G. H., Jr., and Nathans, D. (1973). *J. Mol. Biol.* **78**, 363.

de la Maza, L. M., and Carter, B. J. (1976). *Nucleic Acids Res.* **3**, 2605.

de la Maza, L. M., and Carter, B. J. (1977). *Virology* **82**, 409.

Denniston-Thompson, K., Moore, D. D., Kruger, K. E., Furth, M. E., and Blattner, F. R. (1977). *Science* **198**, 1051.

DePicker, A., Van Montagu, M., and Schell, J. (1977). In "DNA Insertion Elements,

Plasmids and Episomes" (A. I. Bukhari, J. Shapiro, and S. Adhya, eds.), p. 678. Cold Spring Harbor Lab., Cold Spring Harbor, New York.
De Robertis, E. M., and Mertz, J. E. (1977). *Cell* **12**, 175.
Derynck, R., and Fiers, W. (1977). *J. Mol. Biol.* **110**, 387.
Dintzis, H. M. (1961). *Proc. Natl. Acad. Sci. U.S.A.* **47**, 247.
Doel, M. T., Houghton, M., Cook, E. A., and Carey, N. H. (1977). *Nucleic Acids Res.* **4**, 3701.
Dugaiczyk, A., Hedgpeth, J., Boyer, H. W., and Goodman, H. M. (1974). *Biochemistry* **13**, 503.
Dugaiczyk, A., Boyer, H. W., and Goodman, H. M. (1975). *J. Mol. Biol.* **96**, 171.
Dunn, A. R., and Hassell, J. A. (1977). *Cell* **12**, 23.
Edgell, M. H., Hutchison, C. A., III, and Sclair, M. (1972). *J. Virol.* **9**, 574.
Endow, S. A. (1977). *J. Mol. Biol.* **114**, 441.
Endow, S. A., and Roberts, R. J. (1977). *J. Mol. Biol.* **112**, 521.
Engberg, J., Andersson, P., Leick, V., and Collins, J. (1976). *J. Mol. Biol.* **104**, 455.
Englund, P. T. (1971). *J. Biol. Chem.* **246**, 3269.
Eskin, B., and Linn, S. (1972a). *J. Biol. Chem.* **247**, 6183.
Eskin, B., and Linn, S. (1972b). *J. Biol. Chem.* **247**, 6192.
Evans, R. M., Fraser, N., Ziff, E., Weber, J., Wilson, M., and Darnell, J. E. (1977). *Cell* **12**, 733.
Favre, M., Orth, G., Croissant, O., and Yaniv, M. (1975). *Proc. Natl. Acad. Sci. U.S.A.* **72**, 4810.
Favre, M., Orth, G., Croissant, O., and Yaniv, M. (1977). *J. Virol.* **21**, 1210.
Feunteun, J., Sompayrac, L., Fluck, M., and Benjamin, T. (1976). *Proc. Natl. Acad. Sci. U.S.A.* **73**, 4169.
Fiers, W., Contreras, R., Haegeman, G., Rogiers, R., Van de Voorde, A., Van Heuverswyn, H., Van Herreweghe, J., Volckaert, G., and Ysebaert, M. (1978). *Nature (London)* **273**, 113.
Folk, W. R., Fishel, B. R., and Anderson, D. M. (1975). *Virology* **64**, 277.
Forsblom, S., Rigler, R., Ehrenberg, M., Pettersson, U., and Philipson, L. (1976). *Nucleic Acids Res.* **3**, 3255.
Fried, M., Griffin, B. E., Lund, E., and Robberson, D. L. (1974). *Cold Spring Harbor Symp. Quant. Biol.* **39**, 45.
Fritsch, A., Tiollais, P., and Buc, H. (1975). *FEBS Lett.* **52**, 121.
Fuke, M., and Thomas, C. A., Jr. (1970). *J. Mol. Biol.* **52**, 395.
Gallimore, P. H., Sharp, P. A., and Sambrook, J. (1974). *J. Mol. Biol.* **89**, 49.
Ganem, D., Nussbaum, A. L., Davoli, D., and Fareed, G. C. (1976). *Cell* **7**, 349.
Garfin, D. E., and Goodman, H. M. (1974). *Biochem. Biophys. Res. Commun.* **59**, 108.
Gelinas, R. E., Myers, P. A., and Roberts, R. J. (1977a). *J. Mol. Biol.* **114**, 169.
Gelinas, R. E., Myers, P. A., Weiss, G. H., Murray, K., and Roberts, R. J. (1977b). *J. Mol. Biol.* **114**, 433.
Gillespie, D., and Spiegelman, S. (1965). *J. Mol. Biol.* **12**, 829.
Gingeras, T. R., Myers, P. A., Olson, J. A., Hanberg, F. A., and Roberts, R. J. (1978). *J. Mol. Biol.* **117**, 609.
Gissmann, L., and zur Hausen, H. (1976). *Proc. Natl. Acad. Sci. U.S.A.* **73**, 1310.
Gissmann, L., Pfister, H., and zur Hausen, H. (1977). *Virology* **76**, 569.
Glover, D. M. (1976). *New Tech. Biophys. Cell Biol.* **3**, 125.
Glover, D. M., and Hogness, D. S. (1977). *Cell* **10**, 167.
Godson, G. N. (1974). *J. Mol. Biol.* **90**, 127.
Godson, G. N. (1975). *Virology* **63**, 320.

Godson, G. N. (1976). *Virology* **75**, 263.

Godson, G. N., and Roberts, R. J. (1976). *Virology* **73**, 561.

Goff, S. P., and Berg, P. (1976). *Cell* **9**, 695.

Goldstein, L., Thomas, M., and Davis, R. W. (1975). *Virology* **66**, 420.

Graessmann, A., Graessmann, M., Bobrik, R., Hoffmann, E., Lauppe, F., and Mueller, C. (1976). *FEBS Lett.* **61**, 81.

Graham, F. L., Abrahams, P. J., Mulder, A. C., Heijneker, H. L., Warnaar, S. O., deVries, F. A. J., Fiers, W., and Van der Eb, A. J. (1974). *Cold Spring Harbor Symp. Quant. Biol.* **39**, 637.

Gray, P. W., and Hallick, R. B. (1977). *Biochemistry* **16**, 1665.

Greene, P. J., Betlach, M. C., Goodman, H. M., and Boyer, H. W. (1974). *Methods Mol. Biol.* **7**, 87.

Greene, P. J., Poonian, M. S., Nussbaum, A. L., Tobias, L., Garfin, D. E., Boyer, H. W., and Goodman, H. M. (1975). *J. Mol. Biol.* **99**, 237.

Griffin, B. E., and Fried, M. (1975). *Nature (London)* **256**, 175.

Griffin, B. E., and Fried, M. (1976). *Methods Cancer Res.* **12**, 49.

Griffin, B. E., Fried, M., and Cowie, A. (1974). *Proc. Natl. Acad. Sci. U.S.A.* **71**, 2077.

Grinsted, J., Bennet, P. M., and Richmond, M. H. (1977). *Plasmid* **1**, 34.

Grodzicker, T., Williams, J., Sharp, P. A., and Sambrook, J. (1974). *Cold Spring Harbor Symp. Quant. Biol.* **39**, 439.

Grodzicker, T., Anderson, C., Sambrook, J., and Mathews, M. B. (1977). *Virology* **80**, 111.

Gromkova, R., and Goodgal, S. H. (1972). *J. Bacteriol.* **109**, 987.

Gromkova, R., and Goodgal, S. H. (1974). "Mechanisms of Recombination," p. 209. Plenum, New York.

Gromkova, R., and Goodgal, S. H. (1976). *J. Bacteriol.* **127**, 848.

Gromkova, R., Bendler, J., and Goodgal, S. (1973). *J. Bacteriol.* **114**, 1151.

Groneberg, J., Chardonnet, Y., and Doerfler, W. (1977). *Cell* **10**, 101.

Grosveld, F. G., Ojamaa, K. M., and Spencer, J. H. (1976). *Virology* **71**, 312.

Grunstein, M., and Hogness, D. S. (1975). *Proc. Natl. Acad. Sci. U.S.A.* **72**, 3961.

Guérineau, M., Grandchamp, C., and Slonimski, P. P. (1976). *Proc. Natl. Acad. Sci. U.S.A.* **73**, 3030.

Haberman, A. (1974). *J. Mol. Biol.* **89**, 545.

Haberman, A., Heywood, J., and Meselson, M. (1972). *Proc. Natl. Acad. Sci. U.S.A.* **69**, 3138.

Haggerty, D. M., and Schleif, R. F. (1976). *J. Virol.* **18**, 659.

Hamer, D. H., and Thomas, C. A., Jr. (1976). *Proc. Natl. Acad. Sci. U.S.A.* **73**, 1537.

Hamer, D. H., Davoli, D., Thomas, C. A., Jr., and Fareed, G. (1977). *J. Mol. Biol.* **112**, 155.

Hamlett, N. V., Lange-Gufstafson, B., and Rhoades, M. (1977). *J. Virol.* **24**, 249.

Hardies, S. C., and Wells, R. D. (1976). *Proc. Natl. Acad. Sci. U.S.A.* **73**, 3117.

Harris-Warrick, R. M., Elkana, Y., Ehrlich, S. D., and Lederberg, J. (1975). *Proc. Natl. Acad. Sci. U.S.A.* **72**, 2207.

Hartmann, H., and Goebel, W. (1977). *FEBS Lett.* **80**, 285.

Hattman, S. (1977). *J. Bacteriol.* **129**, 1330.

Hauswirth, W. W., and Berns, K. I. (1977). *Virology* **78**, 488.

Hayashi, M. N., and Hayashi, M. (1974). *J. Virol.* **14**, 1142.

Hayashi, M. N., Fujimura, F. K., and Hayashi, M. (1976). *Proc. Natl. Acad. Sci. U.S.A.* **73**, 3519.

Hayward, G. S., Jacob, R. J., Wadsworth, S. C., and Roizman, B. (1975). *Proc. Natl. Acad. Sci. U.S.A.* **72**, 4243.

Hedgpeth, T., Goodman, H. M., and Boyer, H. W. (1972). *Proc. Natl. Acad. Sci. U.S.A.* **69**, 3448.

Heininger, K., Hörz, W., and Zachau, H. G. (1977). *Gene* **1**, 291.

Helling, R. B., Goodman, H. M., and Boyer, H. W. (1974). *J. Virol.* **14**, 1235.

Hershey, N. D., Conrad, S. E., Sodja, A., Yen, P. H., Cohen, M., Jr., Davidson, N., Ilgen, C., and Carbon, J. (1977). *Cell* **11**, 585.

Hershfield, V., Boyer, H. W., Yanofsky, C., Lovett, M. A., and Helinski, D. R. (1974). *Proc. Natl. Acad. Sci. U.S.A.* **71**, 3455–3459.

Heyneker, H. L., Shine, J., Goodman, H. M., Boyer, H. W., Rosenberg, J., Dickerson, R. E., Narang, S. A., Itakura, K., Lin, S.-Y., and Riggs, A. D. (1976). *Nature (London)* **263**, 748.

Hirsch, I., Cabral, G., Patterson, M., and Biswal, N. (1977). *Virology* **81**, 48.

Hirsh, J., and Schleif, R. (1977). *Cell* **11**, 545.

Hollenberg, C. P., Degelmann, A., Kustermann-Kuhn, B., and Royer, H. D. (1976). *Proc. Natl. Acad. Sci. U.S.A.* **73**, 2072.

Horiuchi, K., and Zinder, N. D. (1972). *Proc. Natl. Acad. Sci. U.S.A.* **69**, 3220.

Horiuchi, K., and Zinder, N. D. (1975). *Proc. Natl. Acad. Sci. U.S.A.* **72**, 2555.

Horiuchi, K., and Zinder, N. D. (1976). *Proc. Natl. Acad. Sci. U.S.A.* **73**, 2341.

Horiuchi, K., Vovis, G. F., Enea, V., and Zinder, N. D. (1975). *J. Mol. Biol.* **95**, 147.

Horwitz, M. S. (1976). *J. Virol.* **18**, 307.

Horz, W., and Zachau, H. G. (1977). *Eur. J. Biochem.* **73**, 383.

Howley, P. M., Khoury, G., Byrne, J. C., Takemoto, K. K., and Martin, M. A. (1975). *J. Virol.* **16**, 959.

Hsieh, T.-S., and Wang, J. C. (1976). *Biochemistry* **15**, 5776.

Hughes, S. G. (1977). *Biochem. J.* **163**, 503.

Hughes, S. G., and Hattman, S. (1975). *J. Mol. Biol.* **98**, 645.

Hutchison, C. A., III, and Edgel, M. H. (1971). *J. Virol.* **8**, 181.

Hutchison, C. A., III, Newbold, J. E., Potter, S. S., and Edgell, M. H. (1974). *Nature (London)* **251**, 536.

Ikawa, S., Shibata, T., and Ando, T. (1976). *J. Biochem. (Tokyo)* **80**, 1457.

Inciarte, M. R., Lazaro, J. M., Salas, M., and Vinuela, E. (1976). *Virology* **74**, 314.

Itakura, K., Hirose, T., Crea, R., Riggs, A. D., Heyneker, H. L., Bolivar, F., and Boyer, H. W. (1977). *Science* **198**, 1056.

Ito, J., and Kawamura, F. (1976). In "Microbiology—1976" (D. Schlessinger, ed.), p. 367. Am. Soc. Microbiol., Washington, D.C.

Ito, J., Kawamura, F., and Duffy, J. J. (1975). *FEBS Lett.* **55**, 278.

Ito, J., Kawamura, F., and Yanofsky, S. (1976). *Virology* **70**, 37.

Jackson, D. A., Symonds, R. H., and Berg, P. (1972). *Proc. Natl. Acad. Sci. U.S.A.* **69**, 2904.

James, P. M., Sens, D., Natter, W., Moore, S. K., and James, E. (1976). *J. Bacteriol.* **126**, 487.

Jeffreys, A. J., and Flavell, R. A. (1977a). *Cell* **12**, 429.

Jeffreys, A. J., and Flavell, R. A. (1977b). *Cell* **12**, 1097.

Jeppesen, P. G. N. (1974). *Anal. Biochem.* **58**, 195.

Jeppesen, P. G. N., Sanders, L., and Slocombe, P. M. (1976). *Nucleic Acids Res.* **3**, 1323.

Jones, B. B., and Reznikoff, W. S. (1977). *J. Bacteriol.* **132**, 270.

Jones, B. B., Chan, H., Rothstein, S., Wells, R. D., and Reznikoff, W. S. (1977). *Proc. Natl. Acad. Sci. U.S.A.* **74**, 4914.

Kahmann, R., Kamp, D., and Zipser, D. (1977). *In* "DNA Insertion Elements, Plasmids, and Episomes " (A. I. Bukhari, J. Shapiro, and S. Adhya, eds.), p. 335. Cold Spring Harbor Lab., Cold Spring Harbor, New York.

Kamen, R., and Shure, H. (1976). *Cell* **7**, 361.

Kamp, D., Kahmann, R., Zipser, D., and Roberts, R. J. (1977). *Mol. Gen. Genet.* **154**, 231.

Kania, J., and Fanning, T. G. (1976). *Eur. J. Biochem.* **67**, 367.

Kaplan, D. A., and Nierlich, D. P. (1975). *J. Biol. Chem.* **250**, 2395.

Kaplan, D. A., Russo, R., and Wilcox, G. (1977). *Anal. Biochem.* **78**, 235.

Karrer, K. M., and Gall, J. G. (1976). *J. Mol. Biol.* **104**, 421–453.

Kedes, L. H., Chang, A. C. Y., Houseman, D., and Cohen, S. N. (1975). *Nature (London)* **255**, 533.

Kelly, T. J., Jr., and Smith, H. O. (1970). *J. Mol. Biol.* **51**, 393.

Ketner, G., and Kelly, T. J., Jr. (1976). *Proc. Natl. Acad. Sci. U.S.A.* **73**, 1102.

Khorana, H. G., Agarwal, K. L., Buchi, H., Caruthers, M. H., Gupta, N. K., Kleppe, K., Kumar, A., Ohtsuka, E., RajBhandary, U. L., van de Sande, J. H., Sgaramella, V., Terao, T., Weber, H., and Yamada, T. (1972). *J. Mol. Biol.* **72**, 209.

Khoury, G., Martin, M. A., Lee, T. N. H., Danna, K. J., and Nathans, D. (1973). *J. Mol. Biol.* **78**, 377.

Khoury, G., Fareed, G. C., Berry, K., Martin, M. A., Lee, T. N. H., and Nathans, D. (1974). *J. Mol. Biol.* **87**, 289.

Kiss, A., Sain, B., Csordás-Tòth, E., and Venetianer, P. (1977). *Gene* **1**, 323.

Kleid, D., Humayun, Z., Jeffrey, A., and Ptashne, M. (1976). *Proc. Natl. Acad. Sci. U.S.A.* **73**, 293.

Kleisen, C. M., Borst, P., and Weijers, P. J. (1976a). *Eur. J. Biochem.* **64**, 141.

Kleisen, C. M., Weislogel, P. O., Fonck, K., and Borst, P. (1976b). *Eur. J. Biochem.* **64**, 153.

Klukas, C. K., and Dawid, I. B. (1976). *Cell* **9**, 615.

Kozlov, J. I., Kalinina, N. A., Gening, L. V., Rebentisch, B. A., Strongin, A. Y., Bogush, V. G., and Debabov, V. G. (1977). *Mol. Gen. Genet.* **150**, 211.

Kozlov, J. V., Olsnes, S., Brøgger, A., and Pihl, A. (1975). *FEBS Lett.* **57**, 153.

Kramer, R. A., Cameron, J. R., and Davis, R. W. (1976). *Cell* **8**, 227.

Kretschmer, P. J., and Cohen, S. N. (1977). *J. Bacteriol.* **130**, 888.

Kroon, A. M., Bakker, H., Holtrop, M., and Terpstra, P. (1977). *Biochim. Biophys. Acta* **474**, 61.

Kuroyedov, A. A., Grokhovsky, S. L., Zhuze, A. L., Nosikov, V. V., and Polyanovski, O. L. (1977). *Gene* **1**, 389.

Lacks, S., and Greenberg, B. (1977). *J. Mol. Biol.* **114**, 153.

Lai, C.-J., and Nathans, D. (1974). *Virology* **60**, 466.

Lai, C.-J., and Nathans, D. (1975). *Virology* **66**, 70.

Landy, A., Olchowski, E., Ross, W., and Reiness, G. (1974a). *Mol. Gen. Genet.* **133**, 273.

Landy, A., Ruedisueli, E., Robinson, L., Foeller, C., and Ross, W. (1974b). *Biochemistry* **13**, 2134.

Landy, A., Foeller, C., Reszelbach, R., and Dudock, B. (1976). *Nucleic Acids Res.* **3**, 2575.

Larsen, S. H., and Nathans, D. (1977). *Virology* **82**, 182.

Lautenberger, J. A., and Linn, S. (1972). *J. Biol. Chem.* **247**, 6176.

Lebowitz, P., Kelly, T. J., Jr., Nathans, D., Lee, T. N. H., and Lewis, A. M., Jr. (1974a). *Proc. Natl. Acad. Sci. U.S.A.* **71**, 441.

Lebowitz, P., Siegel, W., and Sklar, J. (1974b). *J. Mol. Biol.* **88**, 105.

Lee, A. S., and Sinsheimer, R. L. (1974a). *Anal. Biochem.* **60**, 640.

Lee, A. S., and Sinsheimer, R. L. (1974b). *Proc. Natl. Acad. Sci. U.S.A.* **71**, 2882.

Lee, T. N. H., Brockman, W. W., and Nathans, D. (1975). *Virology* **66**, 53.

Lehman, I. R. (1960). *J. Biol. Chem.* **235**, 1479.

Levings, C. S., III, and Pring, D. R. (1976). *Science* **193**, 158.

Lewis, J. B., Atkins, J. F., Anderson, C. W., Baum, P. R., and Gesteland, R. F. (1975). *Proc. Natl. Acad. Sci. U.S.A.* **72**, 1344.

Lewis, J. B., Atkins, J. F., Baum, P. R., Solem, R., Gesteland, R. F., and Anderson, C. W. (1976). *Cell* **7**, 141.

Lewis, J. B., Anderson, C. W., and Atkins, J. F. (1977). *Cell* **12**, 37.

Li, L. I., Tanyashin, V. I., Matvienko, N. I., and Bayev, A. A. (1975). *Dokl. Akad. Nauk SSSR* **223**, 1262.

Lindahl, L., Zengel, J., and Nomura, M. (1976). *J. Mol. Biol.* **106**, 837.

Lindahl, L., Post, L., Zengel, J., Gilbert, S. F., Strycharz, W. A., and Nomura, M. (1977a). *J. Biol. Chem.* **252**, 7365.

Lindahl, L., Yamamoto, M., Nomura, M., Kirschbaum, J. B., Allet, B., and Rochaix, J. D. (1977b). *J. Mol. Biol.* **109**, 23.

Linn, S., and Arber, W. (1968). *Proc. Natl. Acad. Sci. U.S.A.* **59**, 1300.

Lis, J. T., and Schleif, R. (1975). *Nucleic Acids Res.* **2**, 383.

Livingston, D. M., and Klein, H. L. (1977). *J. Bacteriol.* **129**, 472.

Lobban, P. E., and Kaiser, A. D. (1973). *J. Mol. Biol.* **78**, 453.

Lovett, M. A., and Helinski, D. R. (1976). *J. Bacteriol.* **127**, 982.

Luria, S. E. (1953). *Cold Spring Harbor Symp. Quant. Biol.* **18**, 237.

Luria, S. E., and Human, M. L. (1952). *J. Bacteriol.* **64**, 557.

McAllister, W. T., and Barrett, C. L. (1977). *Virology* **82**, 275.

McConnell, D., Searcy, D., and Sutcliffe, G. (1978). *Nucleic Acids Res.* **5**, 1729.

McDonell, M. W., Simon, M. N., and Studier, F. W. (1977). *J. Mol. Biol.* **110**, 119.

McMaster, G. K., and Carmichael, G. G. (1977). *Proc. Natl. Acad. Sci. U.S.A.* **74**, 4835.

McParland, R. H., Brown, L. R., and Pearson, G. D. (1976). *J. Virol.* **19**, 1006.

Magazin, M., Howe, M., and Allet, B. (1977). *Virology* **77**, 677.

Maitland, N. J., and McDougall, J. K. (1977). *Cell* **11**, 233.

Maizels, N. (1976). *Cell* **9**, 431.

Majors, J. (1975). *Nature (London)* **256**, 672.

Maniatis, T., and Ptashne, M. (1973). *Nature (London)* **246**, 133.

Maniatis, T., Jeffrey, A., and van de Sande, H. (1975). *Biochemistry* **14**, 3787.

Maniatis, T., Sim, G. K., Efstratiadis, A., and Kafatos, F. C. (1976). *Cell* **8**, 163.

Mann, M. B., and Smith, H. O. (1977). *Nucleic Acids Res.* **4**, 4211.

Mann, M. B., Rao, R. N., and Smith, H. O. (1978). *Gene* **3**, 97.

Marians, K. J., Wu, R., Stawinski, J., Hozumi, T., and Narang, S. A. (1976). *Nature (London)* **263**, 744.

Mathews, M. B. (1975). *Cell* **6**, 223.

Maurer, R., Maniatis, T., and Ptashne, M. (1974). *Nature (London)* **249**, 221.

Meagher, R. B., Shepherd, R. J., and Boyer, H. W. (1977). *Virology* **80**, 362.

Mertz, J. E., and Davis, R. W. (1972). *Proc. Natl. Acad. Sci. U.S.A.* **69**, 3370.

Mertz, J. E., and Gurdon, J. B. (1977). *Proc. Natl. Acad. Sci. U.S.A.* **74**, 1502.

Meselson, M., and Yaun, R. (1968). *Nature (London)* **217**, 1110.

Meselson, M., Yuan, R., and Heywood, J. (1972). *Annu. Rev. Biochem.* **41**, 447.

Meyer, B. J., Kleid, D. G., and Ptashne, M. (1975). *Proc. Natl. Acad. Sci. U.S.A.* **72**, 4785.

Meyer, R., Figurski, D., and Helinski, D. R. (1977a). *Mol. Gen. Genet.* **152**, 129.

Meyer, R., Figurski, D., and Helinski, D. R. (1977b). *In* "DNA Insertion Elements, Plasmids, and Episomes" (A. I. Bukhari, J. Shapiro, and S. Adhya, eds.), p. 680. Cold Spring Harbor Lab., Cold Spring Harbor, New York.

Meyerink, J. H., and Retel, J. (1976). *Nucleic Acids Res.* 3, 2697.

Middleton, J. H., Edgell, M. H., and Hutchison, C. A., III (1972). *J. Virol.* 10, 42.

Miller, L. K., and Fried, M. (1976). *J. Virol.* 18, 824.

Miller, L. K., Cooke, B. E., and Fried, M. (1976). *Proc. Natl. Acad. Sci. U.S.A.* 73, 3073.

Mizuuchi, K., and Nash, H. A. (1976). *Proc. Natl. Acad. Sci. U.S.A.* 73, 3524.

Modrich, P., and Rubin, R. A. (1977). *J. Biol. Chem.* 252, 7273.

Modrich, P., and Zabel, D. (1976). *J. Biol. Chem.* 251, 5866.

Molgaard, H. V., Matthews, H. R., and Bradbury, E. M. (1976). *Eur. J. Biochem.* 68, 541.

Moore, K. H., Johnson, P. H., Chandler, S. E. W., and Grossman, L. I. (1977). *Nucleic Acids Res.* 4, 1273.

Morimoto, R., Lewin, A., Hsu, H.-J., Rabinowitz, M., and Fukuhara, H. (1975). *Proc. Natl. Acad. Sci. U.S.A.* 72, 3868.

Morimoto, R., Lewin, A., and Rabinowitz, M. (1977). *Nucleic Acids Res.* 4, 2331.

Morimoto, R., Merten, S., Lewin, A., Martin, N. C., and Rabinowitz, M. (1978). *Mol. Gen. Genet.* 163, 241.

Morris, D. W., and Parish, J. H. (1976). *Arch. Microbiol.* 108, 227.

Morrow, J. F., and Berg, P. (1972). *Proc. Natl. Acad. Sci. U.S.A.* 69, 3365.

Morse, L. S., Buchman, T. G., Roizman, B., and Schaffer, P. A. (1977). *J. Virol.* 24, 231.

Mottes, M., Morandi, C., Cremaschi, S., and Sgaramella, V. (1977). *Nucleic Acids Res.* 4, 2467.

Mulder, C., Arrand, J. R., Delius, H., Keller, W., Pettersson, U., Roberts, R. J., and Sharp, P. A. (1974). *Cold Spring Harbor Symp. Quant. Biol.* 39, 397.

Murray, K. (1973). *Biochem. J.* 131, 569.

Murray, K. (1976). *Endeavour* 126, 129.

Murray, K., and Murray, N. E. (1975). *J. Mol. Biol.* 98, 551.

Murray, K., Hughes, S. G., Brown, J. S., and Bruce, S. A. (1976). *Biochem. J.* 159, 317.

Murray, N. E., Batten, P. L., and Murray, K. (1973a). *J. Mol. Biol.* 81, 395.

Murray, N. E., Manduca de Ritis, P., and Foster, L. A. (1973b). *Mol. Gen. Genet.* 120, 261.

Nath, K., and Bollon, A. P. (1976). *Mol. Gen. Genet.* 147, 153–168.

Nath, K., and Bollon, A. P. (1977). *J. Biol. Chem.* 252, 6562.

Nathans, D., and Danna, K. J. (1972a). *J. Mol. Biol.* 64, 515.

Nathans, D., and Danna, K. J. (1972b). *Nature (London), New Biol.* 236, 200.

Nathans, D., and Smith, H. O. (1975). *Annu. Rev. Biochem.* 44, 273.

Nosikov, V. V., and Sain, B. (1977). *Nucleic Acids Res.* 4, 2263.

Nosikov, V. V., Braga, E. A., Karlishev, A. V., Zhuze, A. L., and Polyanovsky, O. L. (1976). *Nucleic Acids Res.* 3, 2293.

Nussbaum, A. L., Davoli, D., Ganem, D., and Fareed, G. C. (1976). *Proc. Natl. Acad. Sci. U.S.A.* 73, 1068.

Ohtsubo, H., and Ohtsubo, E. (1977). *In* "DNA Insertion Elements, Plasmids, and Episomes" (A. I. Bukhari, J. Shapiro, and S. Adhya, eds.), p. 49. Cold Spring Harbor Lab., Cold Spring Harbor, New York.

Oka, A., and Takanami, M. (1976). *Nature (London)* 264, 193.

Okamoto, T., Sugimoto, K., Sugisaki, H., and Takanami, M. (1975). *J. Mol. Biol.* 95, 33.

Okamoto, T., Sugimoto, K., Sugisaki, H., and Takanami, M. (1977). *Nucleic Acids Res.* 4, 2213.

Old, R., Murray, K., and Roizes, G. (1975). *J. Mol. Biol.* **92**, 331.
Oliver, S. G., and McLaughlin, C. S. (1977). *Anal. Biochem.* **82**, 271.
Olson, M. V., Montgomery, D. L., Hopper, A. K., Page, G. S., Horodyski, F., and Hall, B. D. (1977). *Nature (London)* **267**, 639.
Ortin, J., Scheidtmann, K.-H., Greenberg, R., Westphal, M., and Doerfler, W. (1976). *J. Virol.* **20**, 355.
Osborn, J. E., Robertson, S. M., Padgett, B. L., Zu Rhein, G. M., Walker, D. L., and Weisblum, B. (1974). *J. Virol.* **13**, 614.
Osborn, J. E., Robertson, S. M., Pagett, B. L., Walker, D. L., and Weissblum, B. (1976). *J. Virol.* **19**, 675.
Parker, R. C., and Watson, R. M. (1977). *Nucleic Acids Res.* **4**, 1291.
Parker, R. C., Watson, R. M., and Vinograd, J. (1977). *Proc. Natl. Acad. Sci. U.S.A.* **74**, 851.
Paterson, B. M., Roberts, B. E., and Kuff, E. L. (1977). *Proc. Natl. Acad. Sci. U.S.A.* **74**, 4370–4374.
Pearson, R. L., Weiss, J. F., and Kelmers, A. D. (1971). *Biochim. Biophys. Acta* **228**, 770.
Perricaudet, M., and Tiollais, P. (1975). *FEBS Lett.* **56**, 7.
Petes, T. D., and Botstein, D. (1977). *Proc. Natl. Acad. Sci. U.S.A.* **74**, 5091.
Pettersson, U., and Mathews, M. B. (1977). *Cell* **12**, 741.
Pettersson, U., Tibbetts, C., and Philipson, L. (1976). *J. Mol. Biol.* **101**, 479.
Piekarowicz, A., Brzezinski, R., and Kauc, L. (1976). *Acta Microbiol. Pol.* **25**, 307.
Pirrotta, V. (1976). *Nucleic Acids Res.* **3**, 1747.
Polisky, B., Greene, P., Garfin, D. E., McCarthy, B. J., Goodman, H. M., and Boyer, H. W. (1975). *Proc. Natl. Acad. Sci. U.S.A.* **72**, 3310.
Polsky, F., Edgell, M. H., Seidman, J. G., and Leder, P. (1978). *Anal. Biochem.* **87**, 397.
Potter, S. S., and Newbold, J. E. (1976). *Anal. Biochem.* **71**, 452.
Potter, S. S., and Thomas, C. A., Jr. (1977). *Cold Spring Harbor Symp. Quant. Biol.* **42**, 1023.
Potter, S. S., Bott, K. F., and Newbold, J. E. (1977). *J. Bacteriol.* **129**, 492.
Prives, C. L., Aviv, H., Paterson, B. M., Roberts, B. E., Rozenblatt, S., Revel, M., and Winocour, E. (1974). *Proc. Natl. Acad. Sci. U.S.A.* **71**, 302.
Prunell, A., Strauss, F., and Leblanc, B. (1977a). *Anal. Biochem.* **78**, 57.
Prunell, A., Kopecka, H., Strauss, F., and Bernardi, G. (1977b). *J. Mol. Biol.* **110**, 17.
Quétier, F., and Vedel, F. (1977). *Nature (London)* **268**, 365.
Rao, G. R. K., and Singer, M. F. (1977). *J. Biol. Chem.* **252**, 5124.
Ratzkin, B., and Carbon, J. (1977). *Proc. Natl. Acad. Sci. U.S.A.* **74**, 487.
Ravetch, J. V., Horiuchi, K., and Model, P. (1977). *Virology* **81**, 341.
Reddy, V. B., Thimmappaya, B., Dhar, R., Subramanian, K. N., Zain, B. S., Pan, J., Celma, M. L., and Weissman, S. M. (1978). *Science*, **200**, 494.
Reiser, J., and Yuan, R. (1977). *J. Biol. Chem.* **252**, 451.
Rhoades, M. (1975). *Virology* **64**, 170.
Rhode, S. L., III (1977). *J. Virol.* **22**, 446.
Richardson, C. C. (1965). *Proc. Natl. Acad. Sci. U.S.A.* **54**, 158.
Risser, R., Hopkins, N., Davis, R. W., Delius, H., and Mulder, C. (1974). *J. Mol. Biol.* **89**, 517.
Roberts, R. J. (1976). *Crit. Rev. Biochem.* **4**, 123.
Roberts, R. J. (1977). *In* "Recombinant Molecules: Impact on Science and Society" (R. F. Beers and E. G. Bassett, eds.), p. 21. Raven, New York.

Roberts, R. J., Arrand, J. R., and Keller, W. (1974). *Proc. Natl. Acad. Sci. U.S.A.* **71**, 3829.
Roberts, R. J., Breitmeyer, J. B., Tabachnik, N. F., and Myers, P. A. (1975). *J. Mol. Biol.* **91**, 121.
Roberts, R. J., Myers, P. A., Morrison, A., and Murray, K. (1976a). *J. Mol. Biol.* **102**, 157.
Roberts, R. J., Myers, P. A., Morrisson, A., and Murray, K. (1976b). *J. Mol. Biol.* **103**, 199.
Roberts, R. J., Wilson, G. A., and Young, F. E. (1977). *Nature (London)* **265**, 82.
Robinson, L. H., and Landy, A. (1977a). *Gene* **2**, 1.
Robinson, L. H., and Landy, A. (1977b). *Gene* **2**, 33.
Rodriguez, R. L., Bolivar, F., Goodman, H. M., Boyer, H. W., and Betlach, M. C. (1976). *ICN-UCLA Symp. Mol. & Cell. Biol.* **5**, 471.
Roizes, G. (1974). *Nucleic Acids Res.* **1**, 1099.
Roizes, G. (1976). *Nucleic Acids Res.* **3**, 2677.
Roizes, G., Patillon, M., and Kovoor, A. (1977). *FEBS Lett.* **82**, 69.
Roszczyk, E., and Goodgal, S. (1975). *J. Bacteriol.* **123**, 287.
Rougeon, F., and Mach, B. (1977). *Gene* **1**, 229.
Rougeon, F., Kourilsky, P., and Mach, B. (1975). *Nucleic Acids Res.* **2**, 2365.
Roulland-Dussoix, D., and Boyer, H. W. (1969). *Biochim. Biophys. Acta* **195**, 219.
Roulland-Dussoix, D., Yoshimori, R., Greene, P., Betlach, M., Goodman, H. M., and Boyer, H. W. (1974). In "Microbiology—1974" (D. Schlessinger, ed.), p. 187. Am. Soc. Microbiol., Washington, D.C.
Roy, P. H., and Smith, H. O. (1973). *J. Mol. Biol.* **81**, 472.
Roychoudhury, R., and Kossel, H. (1971). *Eur. J. Biochem.* **22**, 310.
Roychoudhury, R., Jay, E., and Wu, R. (1976). *Nucleic Acids Res.* **3**, 863.
Rozenblatt, S., Mulligan, R. C., Gorecki, M., Roberts, B. E., and Rich, A. (1976). *Proc. Natl. Acad. Sci. U.S.A.* **73**, 2747.
Ruben, G., Spielman, P., Tu, C. D., Jay, E., Siegel, B., and Wu, R. (1977). *Nucleic Acids Res.* **4**, 1803.
Rubens, C., Heffron, F., and Falkow, S. (1976). *J. Bacteriol.* **128**, 425.
Rubin, R. A., and Modrich, P. (1977). *J. Biol. Chem.* **252**, 7265.
Sack, G. H., Jr. (1974). Ph.D. Thesis, Johns Hopkins University, Baltimore, Maryland.
Sack, G. H., Jr., Narayan, O., Danna, K. J., Weiner, L. P., and Nathans, D. (1973). *Virology* **51**, 345.
Sambrook, J., Sugden, B., Keller, W., and Sharp, P. A. (1973). *Proc. Natl. Acad. Sci. U.S.A.* **70**, 3711.
Sambrook, J., Botchan, M., Gallimore, P. H., Ozanne, B., Pettersson, U., Williams, J., and Sharp, P. A. (1974). *Cold Spring Harbor Symp. Quant. Biol.* **39**, 615.
Sambrook, J., Williams, J., Sharp, P. A., and Grodzicker, T. (1975). *J. Mol. Biol.* **97**, 369.
Sanders, J. P. M., Borst, P., and Weijers, P. J. (1975). *Mol. Gen. Genet.* **143**, 53.
Sanger, F., and Coulson, A. R. (1975). *J. Mol. Biol.* **94**, 441.
Sanger, F., Air, G. M., Barrell, B. G., Brown, N. L., Coulson, A. R., Fiddes, J. C., Hutchison, C. A., III, Slocombe, P. M., and Smith, M. (1977). *Nature (London)* **265**, 687.
Sato, S., Hutchison, C. A., III, and Harris, J. I. (1977). *Proc. Natl. Acad. Sci. U.S.A.* **74**, 542.
Schaller, H., Nusslein, C., Bonhoeffer, F. J., Kurz, C., and Nietzschmann, I. (1972). *Eur. J. Biochem.* **26**, 474.
Schaller, H., Beck, E., and Takanami, M. (1978). In "Single Stranded DNA Phages" (D. H. Dressler, D. T. Denhart, and D. S. Ray, eds.). Cold Spring Harbor Lab., Cold Spring Harbor, New York.
Scheller, R. H., Dickerson, R. E., Boyer, H. W., Riggs, A. D., and Itakura, K. (1977a). *Science* **196**, 177.

Scheller, R. H., Thomas, T. L., Lee, A. S., Klein, W. H., Niles, W. D., Britten, R. J., and Davidson, E. H. (1977b). *Science* **196**, 197.
Scher, B. M., Dean, D. H., and Garro, A. J. (1977). *J. Virol.* **23**, 377.
Schilling, R., Weingartner, B., and Winnacker, E. L. (1975). *J. Virol.* **16**, 767.
Schlagman, S., Hattman, S., May, M. S., and Berger, L. (1976). *J. Bacteriol.* **126**, 990.
Seeburg, P. H., and Schaller, H. (1975). *J. Mol. Biol.* **92**, 261.
Segall, J., and Losick, R. (1977). *Cell* **11**, 751.
Sekikawa, K., and Fujinaga, K. (1977). *Virology* **82**, 509.
Sgaramella, V., and Khorana, H. G. (1972). *J. Mol. Biol.* **72**, 493.
Sgaramella, V., van de Sande, J. H., and Khorana, H. G. (1970). *Proc. Natl. Acad. Sci. U.S.A.* **67**, 1468.
Shaltiel, S., and Er-El, Z. (1973). *Proc. Natl. Acad. Sci. U.S.A.* **70**, 778.
Sharp, P. A., Sugden, B., and Sambrook, J. (1973). *Biochemistry* **12**, 3055.
Shen, C.-K. J., Wiesehahn, G., and Hearst, J. E. (1976). *Nucleic Acids Res.* **9**, 931.
Shibata, T., and Ando, T. (1975). *Mol. Gen. Genet.* **138**, 269.
Shibata, T., and Ando, T. (1976). *Biochim. Biophys. Acta* **442**, 184.
Shibata, T., Ikawa, S., Kim, C., and Ando, T. (1976). *J. Bacteriol.* **128**, 473.
Shishido, K., and Berg, P. (1976). *J. Virol.* **18**, 793.
Skare, J., and Summers, W. C. (1977). *Virology* **76**, 581.
Sklar, J., Weissman, S., Musso, R. E., DiLauro, R., and de Crombrugghe, B. (1977). *J. Biol. Chem.* **252**, 3538.
Smith, D. I., Blattner, F. R., and Davies, J. (1976). *Nucleic Acids Res.* **3**, 343.
Smith, H. O., and Birnstiel, M. L. (1976). *Nucleic Acids Res.* **3**, 2387.
Smith, H. O., and Nathans, D. (1973). *J. Mol. Biol.* **81**, 419.
Smith, H. O., and Wilcox, K. W. (1970). *J. Mol. Biol.* **51**, 379.
Southern, E. M. (1975a). *J. Mol. Biol.* **94**, 51.
Southern, E. M. (1975b). *J. Mol. Biol.* **98**, 503.
Spear, I. S., Fife, K. H., Hauswirth, W. W., Jones, C. J., and Berns, K. I. (1977). *J. Virol.* **24**, 627.
Stobberingh, E. E., Schiphof, R., and Sussenbach, J. S. (1977). *J. Bacteriol.* **131**, 645.
Streeck, R. E., and Hobom, G. (1975). *Eur. J. Biochem.* **57**, 595.
Streeck, R. E., Philippsen, P., and Zachau, H. G. (1974). *Eur. J. Biochem.* **45**, 489.
Struhl, K., and Davis, R. W. (1976). *ICN-UCLA Symp. Mol. & Cell. Biol.* **5**, 495.
Struhl, K., Cameron, J. R., and Davis, R. W. (1976). *Proc. Natl. Acad. Sci. U.S.A.* **73**, 1471.
Studier, F. W. (1975). *J. Mol. Biol.* **94**, 283.
Studier, F. W., and Movva, N. R. (1976). *J. Virol.* **19**, 136.
Stuy, J. H. (1976). *J. Bacteriol.* **128**, 212.
Subramanian, K. N., Pan, J., Zain, B. S., and Weissman, S. M. (1974). *Nucleic Acids Res.* **1**, 727.
Sugden, B., DeTroy, B., Roberts, R. J., and Sambrook, J. (1975). *Anal. Biochem.* **68**, 36.
Suggs, S. V., and Ray, D. S. (1977). *J. Mol. Biol.* **110**, 147.
Sugisaki, H., and Takanami, K. (1973). *Nature (London), New Biol.* **246**, 138.
Sumegi, J., Breedveld, D., Hossenlopp, P., and Chambon, P. (1977). *Biochem. Biophys. Res. Commun.* **76**, 78.
Summers, J. (1975). *J. Virol.* **15**, 946.
Sussenbach, J. S., and Kuijk, M. G. (1977). *Virology* **77**, 149.
Sussenbach, J. S., Monfoort, C. M., Schiphof, R., and Stobberingh, E. E. (1976). *Nucleic Acids Res.* **3**, 3193.
Susskind, M., and Botstein, D. (1977). *In* "DNA Insertion Elements, Plasmids, and Epi-

somes" (A. I. Bukhari, J. Shapiro, and S. Adhya, eds.), p. 737. Cold Spring Harbor
Lab., Cold Spring Harbor, New York.

Tabak, H. F., Griffith, J., Geider, K., Schaller, H., and Kornberg, A. (1974). *J. Biol. Chem.* **249**, 3049.

Takanami, M. (1973). *FEBS Lett.* **34**, 318.

Takanami, M. (1974). *Methods Mol. Biol.* **7**, 113.

Takanami, M., and Kojo, H. (1973). *FEBS Lett.* **29**, 267.

Takanami, M., Okamoto, T., Sugimoto, K., and Sugisaki, H. (1975). *J. Mol. Biol.* **95**, 21.

Takano, T., Watanabe, T., and Fukasawa, T. (1966). *Biochem. Biophys. Res. Commun.* **25**, 192.

Tanaka, N., Cramer, J. H., and Rownd, R. H. (1976). *J. Bacteriol.* **127**, 619.

Tanaka, T., and Weisblum, B. (1975). *J. Bacteriol.* **121**, 354.

Tanyashin, V. I., Li, L. I., Muidzniex, I. O., and Baev, A. A. (1976). *Dokl. Akad. Nauk SSSR* **231**, 226.

Tartof, K. D., and Dawid, I. B. (1976). *Nature (London)* **263**, 27.

Terpstra, P., Holtrop, M., and Kroon, A. M. (1977). *Biochim. Biophys. Acta* **475**, 571.

Thomas, E. K., Hailparn, E. M., Wilson, G. A., and Young, F. E. (1976). *ICN-UCLA Symp. Mol. & Cell. Biol.* **5**, 605.

Thomas, M., and Davis, R. W. (1975). *J. Mol. Biol.* **91**, 315.

Thuring, R. W. J., Sanders, J. P. M., and Borst, P. (1975). *Anal. Biochem.* **66**, 213.

Tibbetts, C. (1977). *J. Virol.* **24**, 564.

Tiemeier, D. C., Tilghman, S. M., and Leder, P. (1978). *Gene* **2**, 173.

Tilghman, S. M., Tiemeier, D. C., Polsky, F., Edgell, M. H., Seidman, J. G., Leder, A., Enquist, L. W., Norman, B., and Leder, P. (1977). *Proc. Natl. Acad. Sci. U.S.A.* **74**, 4406.

Tilghman, S. M., Tiemeier, D. C., Seidman, J. G., Peterlin, B. M., Sullivan, M., Maizel, J. V., and Leder, P. (1978). *Proc. Natl. Acad. Sci. U.S.A.* **75**, 725.

Timmis, K. N., Cabello, F., and Cohen, S. N. (1975). *Proc. Natl. Acad. Sci. U.S.A.* **72**, 2242.

Timmis, K. N., Cohen, S. N., and Cabello, F. (1977). *Prog. Mol. Subcell. Biol.* **6**, 1.

Tolun, A., and Pettersson, U. (1975). *J. Virol.* **16**, 759.

Tomizawa, J., Ohmori, H., and Bird, R. E. (1977). *Proc. Natl. Acad. Sci. U.S.A.* **74**, 1865.

Tonegawa, S., Brack, C., Hozumi, N., and Schuller, R. (1977). *Proc. Natl. Acad. Sci. U.S.A.* **74**, 3518.

Trautner, T. A., Pawlek, B., Bron, S., and Anagnostopoulos, C. (1974). *Mol. Gen. Genet.* **131**, 181.

Tsujimoto, Y., and Ogawa, H. (1977). *Mol. Gen. Genet.* **150**, 221.

Tu, C.-P. D., Roychoudhury, R., and Wu, R. (1976). *Biochem. Biophys. Res. Commun.* **72**, 355.

Ullrich, A., Shine, J., Chirgwin, J., Pictet, R., Tischer, E., Rutter, W. J., and Goodman, H. M. (1977). *Science* **196**, 1313.

Upholt, W. B., and Dawid, I. B. (1977). *Cell* **11**, 571.

Van den Hondel, C. A., and Schoenmakers, J. G. G. (1975). *Eur. J. Biochem.* **53**, 547.

Van den Hondel, C. A., and Schoenmakers, J. G. G. (1976). *J. Virol.* **18**, 1024.

Van den Hondel, C. A., Weijers, A., Konings, R. N. H., and Schoenmakers, J. G. G. (1975). *Eur. J. Biochem.* **53**, 559.

Van den Hondel, C. A., Pennings, L., and Schoenmakers, J. G. G. (1976). *Eur. J. Biochem.* **68**, 55.

Vapnek, D. (1977). *In* "DNA Insertion Elements, Plasmids, and Episomes" (A. I. Bukhari, J. Shapiro, and S. Adhya, eds.) p. 674. Cold Spring Harbor Lab., Cold Spring Harbor, New York.

Varsanyi, T. M., Winberg, G., and Wadell, G. (1977). *FEBS Lett.* **76**, 151.
Villarreal, L. P., and Berg, P. (1977). *Science* **196**, 183.
Vogt, V. M., and Braun, R. (1976). *J. Mol. Biol.* **106**, 567.
von Gabain, A., Hayward, G. S., and Bujard, H. (1976). *Mol. Gen. Genet.* **143**, 279.
Vovis, G. F., and Lacks, S. (1977). *J. Mol. Biol.* **115**, 525–538.
Vovis, G. F., and Zinder, N. D. (1975). *J. Mol. Biol.* **95**, 557.
Vovis, G. F., Horiuchi, K., and Zinder, N. D. (1974). *Proc. Natl. Acad. Sci. U.S.A.* **71**, 3810.
Vovis, G. F., Horiuchi, K., and Zinder, N. (1975). *J. Virol.* **16**, 674.
Walz, A., Pirrotta, V., and Ineichen, K. (1976). *Nature (London)* **262**, 665.
Weingartner, B., Winnacker, E., Tolun, A., and Pettersson, U. (1976). *Cell* **9**, 259.
Weisbeek, P. J. Vereijken, J. M., Baas, P. D., Jansz, H. S., and Van Arkel, G. A. (1976). *Virology* **72**, 61.
Wellauer, P. K., and Dawid, I. B. (1977). *Cell* **10**, 193.
Wellauer, P. K., Reeder, R. H., Carroll, D., Brown, D. D., Deutch, A., Higashinakagawa, T., and Dawid, I. B. (1974). *Proc. Natl. Acad. Sci. U.S.A.* **71**, 2823.
White, R. L., and Hogness, D. S. (1977). *Cell* **10**, 177.
Wigler, M., Silverstein, S., Lee, L.-S., Pellicer, A., Cheng, Y.-C., and Axel, R. (1977). *Cell* **11**, 223.
Wilkie, N. M. (1976). *J. Virol.* **20**, 222.
Wilkie, N. M., and Cortini, R. (1976). *J. Virol.* **20**, 211.
Wilson, G. A., and Young, F. E. (1975). *J. Mol. Biol.* **97**, 123.
Wilson, G. A., and Young, F. E. (1976). *In* "Microbiology—1976" (D. Schlessinger, ed.), p. 350. Am. Soc. Microbiol., Washington, D.C.
Wittek, R., Menna, A., Schumperli, D., Stoffel, S., Muller, H. K., and Wyler, R. (1977). *J. Virol.* **23**, 669.
Wu, R., and Taylor, E. (1971). *J. Mol. Biol.* **57**, 491.
Yaniv, M., Chestier, A., Dauget, C., and Croissant, O. (1975). *FEBS Lett.* **57**, 126.
Yano, S., Ojima, S., Fujinaga, K., Shiroki, K., and Shimojo, H. (1977). *Virology* **82**, 214.
Yao, M.-C., and Gall, J. G. (1977). *Cell* **12**, 121.
Yen, P. H., Sodja, A., Cohen, M., Jr., Conrad, S. E., Wu, M., Davidson, N., and Ilgen, C. (1977). *Cell* **11**, 763.
Yoshimori, R. N. (1971). Ph.D. Thesis, University of California, Berkeley.
Yuan, R., Heywood, J., and Meselson, M. (1972). *Nature (London), New Biol.* **240**, 42.
Yuan, R., Bickle, T. A., Ebbers, W., and Brack, C. (1975). *Nature (London)* **256**, 556.
Zabeau, M., Friedman, S., Van Montagu, M., and Schell, J. (1979). *Mol. Gen. Genet.* (in press).
Zain, B. S., and Roberts, R. J. (1977). *J. Mol. Biol.* **115**, 249.
Zeiger, R. S., Salomon, R., Dingman, C. W., and Peacock, A. C. (1972). *Nature (London), New Biol.* **238**, 65.

Chapter II

DNA Topoisomerases: Enzymes That Catalyze the Concerted Breaking and Rejoining of DNA Backbone Bonds

JAMES C. WANG AND LEROY F. LIU

I. INTRODUCTION

Enzymes that catalyze the concerted breaking and rejoining of DNA backbone bonds were discovered relatively recently. The first of such enzymes, the *E. coli* ω protein, was initially detected in cell lysates as an activity capable of converting a highly negatively supercoiled DNA to a less supercoiled covalently closed form (Wang, 1969). The conversion between these two forms of a DNA requires at least one *transient* cleavage of the backbone bonds of each DNA molecule (Vinograd and Lebowitz, 1966). At first it was thought that the activity in *E. coli* lysates responsible for this conversion was an endonuclease which caused single-chain scissions in the double-stranded DNA. In the presence of excess DNA ligase in the lysates, the observable effect of an endonuclease would be the removal of the superhelical turns of the DNA substrate. Subsequent purification of the protein, however, led to the conclusion that ω protein itself causes both the breaking and rejoining of the DNA backbone bonds (Wang, 1971, 1973). Furthermore, the protein catalyzes these actions in a *concerted* way in the sense that the breakage of a DNA backbone bond is followed efficiently by the rejoining of the bond; the putative intermediate in which a DNA backbone bond is broken exists only transiently.

65

In 1972, Champoux and Dulbecco reported the finding of an activity from mouse nuclei which untwists superhelical DNA. This activity differs from the *E. coli* ω protein in several aspects, which we will describe in later sections.

It is now clear that activities capable of concerted breaking and rejoining of DNA backbone bonds are ubiquitous in nature; a large number have been detected in or isolated from prokaryotic as well as eukaryotic organisms (Baase and Wang, 1974; Pulleyblank and Morgan, 1975; Champoux and McConaughy, 1976; Rosenberg *et al.*, 1977; Yoshida *et al.*, 1977; Bauer *et al.*, 1977; Kung and Wang, 1977; Poccia *et al.*, 1978). In general, the differences between the prokaryotic activities and the eukaryotic activities are exemplified by the differences between the *E. coli* and the mouse proteins.

The nomenclature of these activities has been a problem. Aside from the noncommital designation ω, various authors have referred to these activities as untwisting enzyme (Champoux and Dulbecco, 1972), swivelase (Wang, 1973), relaxing activity (Keller and Wendel, 1974), relaxation protein (Vosberg *et al.*, 1975), and nicking–closing activity (Vosberg and Vinograd, 1975). Since these activities are invariably detected by their ability to promote the interconversion between different *topological isomers* or *topoisomers* of DNA, we propose that they be called DNA *topoisomerases*. If more than one enzyme is found in an organism, then the different activities are to be designated by Roman numerals.

In this chapter, we will first discuss the *E. coli* ω protein (*Eco* DNA topoisomerase I). This will be followed by discussions on other prokaryotic enzymes and then on eukaryotic enzymes. The possible biological functions of these activities will then be discussed. In a final section, we will summarize studies on other proteins which are in some respects related to the DNA topoisomerases.

II. *ECO* DNA TOPOISOMERASE I

The *E. coli* enzyme has been purified to near homogeneity (Depew *et al.*, 1978). The molecular weight of the enzyme is about 110,000 from the electrophoretic mobility of the sodium dodecyl sulfate-treated protein in polyacrylamide gel (Carlson, 1974). In a 5–20% sucrose gradient the sedimentation coefficient of the enzyme is about the same as that of *E. coli* DNA polymerase I, which has a molecular weight of 109,000 (Wang, 1973; Jovin *et al.*, 1969). Thus *E. coli* ω appears to be a single subunit protein and exists as a monomer in solution, at least at low concentrations. Burrington and Morgan (1976) reported that *E. coli* ω had two subunits of molecular weights 56,000 and

31,000, respectively. We believe that the subunits they observed were probably formed by proteolytic cleavage during purification.

Three types of reactions are known to be catalyzed by *Eco* DNA topoisomerase I: the removal of negative superhelical turns (Wang, 1971, 1973), the interconversion between single-stranded DNA rings with and without topological knots (Liu *et al.*, 1976), and the formation of double-stranded covalently closed DNA rings from single-stranded rings of complementary base sequences (Kirkegaard and Wang, 1978). These reactions are illustrated diagrammatically in Fig. 1.

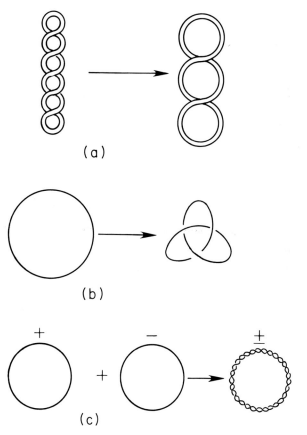

(a)

(b)

(c)

FIG. 1. Types of reactions catalyzed by *Eco* DNA topoisomerase I. (a) Removal of superhelical turns. (b) Interconversion between single-stranded DNA rings with and without topological knots. (c) Combination of two single-stranded DNA rings of complementary base sequences to give a covalently closed double-stranded DNA ring. Reactions (a) and (c) are very similar in that both involve changes in the linking numbers of DNA rings.

The removal of negative superhelical turns has been studied extensively. Early experiments have clearly established that *Eco* DNA topoisomerase I can change the linking number (the topological winding number) of an underwound covalently closed DNA duplex (Wang, 1971, 1973). Since the linking number cannot be changed without interrupting the continuity of one of the DNA strands transiently, it follows that the enzyme must break and rejoin DNA backbone bonds (Wang, 1971).

A quantitative study on the kinetics of *Eco* DNA topoisomerase I-promoted removal of superhelical turns has yet to be carried out. A number of qualitative features of the reaction kinetics have been observed. First, the enzyme is capable of repeated breaking and rejoining of the DNA backbone bonds. Under proper conditions, even when there are fewer enzyme molecules than DNA molecules in the reaction mixture, the linking numbers of all DNA molecules are altered (Carlson, 1974; R. E. Depew and J. C. Wang, unpublished results; cf. also Kung and Wang, 1977, for a similar conclusion with *Micrococcus luteus* DNA topoisomerase I). Second, the processiveness of the enzyme is strongly affected by the medium and the superhelicity of the DNA substrate. For a low-salt medium and a highly negatively supercoiled DNA substrate, the mode of action is processive. Under these conditions if there are fewer enzyme molecules than DNA molecules, a fraction of the DNA molecules lose their superhelical turns rapidly, whereas others retain most of theirs (Wang, 1971). For a medium containing a couple hundred millimoles K^+ and a moderately negatively supercoiled DNA substrate, the mode of action is distributive; when there are fewer enzyme molecules than DNA molecules, all DNA molecules show a gradual reduction in superhelicity (R. E. Depew, L. F. Liu, and J. C. Wang, unpublished results). Finally, the rate of the reaction is strongly dependent on the degree and sense of superhelicity. The removal of the last few negative superhelical turns requires a high level of the enzyme and prolonged incubation; positively supercoiled DNA are hardly affected by the *E. coli* enzyme (Wang, 1971; Kung and Wang, 1977).

The *Eco* DNA topoisomerase I-catalyzed interconversion between single-stranded DNA rings with and without topological knots has been studied recently (Liu *et al.*, 1976). Electron micrographs of single-stranded fd DNA rings with and without topological knots are shown in Fig. 2. That the two forms are indeed topoisomers can be shown by their conversion to a single linear species if each of the molecules is cleaved once by an endonuclease. Aside from electron microscopic appearances, the two forms can also be distinguished by

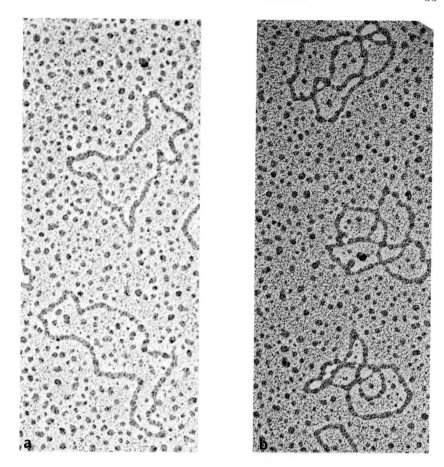

FIG. 2. Electron micrographs showing single-stranded fd DNA rings in two topological forms. (a) Simple rings without knots. (b) Knotted rings. Identical specimen preparation procedures were employed for the samples. (b) is from Liu *et al.* (1976), with permission of the publisher.

their sedimentation coefficients in an alkaline medium and by their electrophoretic mobility in gel.

It should be pointed out that the knotted form contains more than one single species. Knots of various degrees of complexity are present. Two forms, a simple trefoil and a more complex form with a helical region locked into the molecule, are illustrated in Fig. 3. Many different species can be resolved by gel electrophoresis, as depicted in Fig. 4.

The formation of the knotted form is favored at low temperature

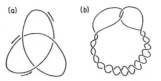

FIG. 3. Two examples of knotted rings. (a) Trefoil. The arrows indicate the polarity of the DNA strand. (b) Knotted ring with a helical region topologically locked into the molecule. (From Liu *et al.*, 1976, with permission of the publisher.)

and high ionic strength; the formation of the simple (unknotted) ring form is favored at high temperature and low ionic strength. It therefore appears that conditions favoring DNA helix formation favors knotted ring formation. This has led Liu *et al.* (1976) to propose the mechanism depicted in Fig. 5.

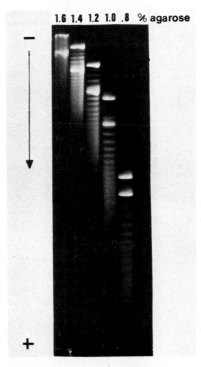

FIG. 4. Electrophoretic pattern of an *Eco* DNA topoisomerase-treated fd DNA sample. The agarose concentration is given in the upper margin of the figure for each cylindrical gel. The electrophoresis medium was 0.03 *M* NaOH. The slowest species in each gel is fd DNA ring without knots. The faster intense band is linear fd DNA. The other bands are species containing topological knots of different degrees of complexity.

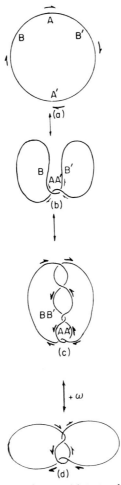

FIG. 5. A model for the knotting of a ring. (a) A simple single-stranded DNA ring. (b) The formation of one or a few turns at AA′ by either base pairing or random fluctuation. (c) The formation of additional helical turns at BB′ by base pairing. Left-handed turns shown above BB′ are introduced because of the topological constraint. (d) The removal of the topological constraint by a DNA topoisomerase allows the formation of a net number of right-handed helical turns at BB′. The species illustrated in (d) is equivalent to the trefoil shown in Fig. 3a. (From Liu *et al.*, 1976, with permission of the publisher.)

The important feature of this mechanism is that knot formation is closely related to helix formation and that the topoisomerase serves the role of removing the topological constraints involved. It immediately follows, then, that if single-stranded rings of complementary base sequences are annealed in the presence of the enzyme, double-

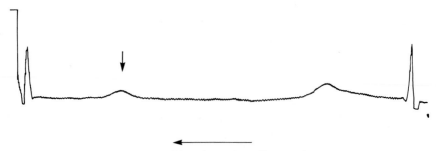

Direction of sedimentation

FIG. 6. The sedimentation pattern in $3\,M$ CsCl–$0.05\,M$ KOH–$0.01\,M$ Na$_3$EDTA of a mixture of single-stranded PM2 DNA rings of complementary base sequences after treatment with *Eco* DNA topoisomerase I. The high sedimentation coefficient of the species marked by an arrow in the figure is characteristic of that of a double-stranded covalently closed DNA ring in this alkaline medium. Single-stranded PM2 DNA rings were obtained by denaturation of double-stranded PM2 DNA molecules containing one random single-chain scission per molecule.

stranded covalently closed rings would form under conditions in which the duplex is more stable than single-stranded rings. This is found to be the case (Kirkegaard and Wang, 1978). Figure 6 depicts such an experiment. When complementary single-stranded DNA rings are annealed in the presence of *Eco* DNA topoisomerase I, sedimentation of the product in an alkaline medium reveals the formation of a species with a sedimentation coefficient characteristic of double-stranded covalently closed DNA. The formation of the double-stranded species has also been confirmed by electron microscopy.

In all three types of reactions, the enzyme catalyzes the conversion of one topoisomer to another. For the last reaction, we can consider two single-stranded rings of complementary base sequences as a special topoisomer of a duplex ring: the single-stranded rings are equivalent to a duplex ring with a linking number of zero.

In catalyzing these reactions, *Eco* DNA topoisomerase I requires no cofactor other than Mg(II). This makes it highly unlikely that the generation of a transient break in the DNA backbone involves the hydrolysis of a phosphodiester bond, since otherwise the reformation of the bond would require an energy cofactor, as exemplified by the ATP or NAD requirements of DNA ligases (Zimmerman *et al.*, 1967; Weiss and Richardson, 1967; Becker *et al.*, 1967; Lindahl and Edelman, 1968). It has, therefore, been proposed that the breakage of a DNA backbone bond is accompanied by the simultaneous formation of a protein–DNA bond, and that the rejoining of the DNA backbone bond

FIG. 7. A diagram illustrating the postulated mechanism that the breaking of a DNA backbone bond is accompanied by the formation of a protein–DNA bond. In the reverse direction, the dissociation of the protein from the DNA is accompanied by the reformation of the DNA backbone bond.

is accomplished by reversing the reaction. These steps are depicted in Fig. 7.

Direct evidence supporting this mechanism is still lacking. We have been able to show, however, that under certain conditions it is possible to demonstrate the cleavage of a DNA chain by the enzyme. Furthermore, when such a cleavage occurs, the protein is found linked, presumably covalently, to the 5′-side of the generated break (Liu, 1977; Depew et al., 1978). Some of the experimental results leading to these conclusions are presented below.

It is found that if a single-stranded DNA is incubated with the E. coli enzyme in the absence of added Mg(II), a complex stable in concentrated salt solutions is formed. Figure 8 illustrates the results of a typical experiment. In this experiment a mixture of single-stranded fd DNA and Eco DNA topoisomerase I was incubated and then analyzed by CsCl and Cs$_2$SO$_4$ density gradient centrifugation. In Fig. 8a, the banding pattern in CsCl is depicted. A group of discrete buoyant species is observed. The buoyant densities of these species, in the order of decreasing densities, agree with the values predicted for fd DNA and fd DNA enzyme complex with one, two, three, and four enzyme molecules per DNA. When banded in Cs$_2$SO$_4$ to allow the detection of species over a wider density range, species with many more protein molecules per fd DNA can be seen in the same reaction mixture (Fig. 8b).

The salt-stable complex is rather unique in that the addition of Mg(II) to the buoyant media, even in the presence of excess EDTA over Mg(II), results in the dissociation of the complex. The dissociated enzyme retains its catalytic activity, and the DNA recovered from the dissociated complex is circular and is indistinguishable from the original DNA used in complex formation (Depew et al., 1978).

If the complex is exposed to alkali, fragmentation of the DNA chain occurs. When the same mixture which gave the banding patterns shown in Fig. 8 was sedimented in an alkaline medium, most of the DNA sedimented as fragments shorter than fd DNA. The fragmentation of DNA was also confirmed by electron microscopy. Banding of

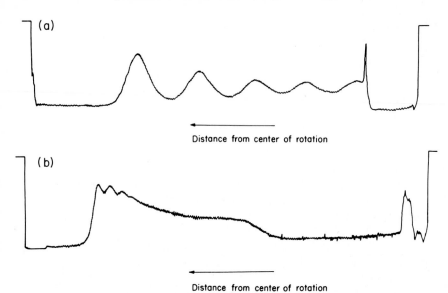

FIG. 8. Density gradient centrifugation of the *Eco* DNA topoisomerase I–fd DNA complex. A reaction mixture (160 μl) containing 10 m*M* Tris–HCl, pH 8, 4 m*M* potassium phosphate, 0.1 m*M* Na$_3$EDTA, 39 μg/ml fd DNA, and 52 μg/ml enzyme was incubated at 37°C for 15 minutes. Ten microliters of 0.2 *M* Na$_3$EDTA was then added and the mixture was incubated at 37°C another 5 minutes. (a) To 90 μl of the mixture, 40 μl of 0.2 *M* Na$_3$EDTA and 370 μl of saturated CsCl solution were added, and the final solution was banded at 42,000 rpm and 20°C. (b) To 70 μl of the remaining mixture, 430 μl of a Cs$_2$SO$_4$ stock solution containing 39.7% (w/w) Cs$_2$SO$_4$ and 88 m*M* Na$_3$EDTA was added, and the final mixture was banded at 45,000 rpm and 20°C for 35 hours. (From Depew *et al.*, 1978, with permission of the publisher.)

the alkali-treated sample in Cs$_2$SO$_4$ containing 0.1 *M* NaOH reveals that fragmentation of the DNA is accompanied by the linking of the protein molecules to the fragments generated, since most of the DNA banded with buoyant densities lower than that of fd DNA (Fig. 9).

Digestion of the fragments generated by alkali treatment of the protein DNA complex with *E. coli* exonuclease I and T4 DNA polymerase exonuclease indicates that the fragments are susceptible to these exonucleases. Therefore, it is most likely that they possess free 3'-hydroxyl termini. This, in turn, indicates that the protein molecules are linked to the 5'-termini. To test whether this is the case, ^{3}H-labeled fd DNA fragments several hundred nucleotides long with ^{32}P-labeled 5'-phosphoryl ends were prepared. This doubly labeled DNA was incubated with the enzyme to form the complex, which was subsequently treated with alkali. It is expected that if a topoisomerase molecule binds to such a fragment, alkali treatment would cleave the

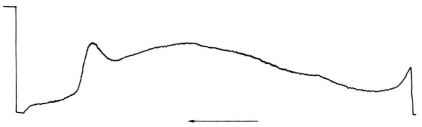

Distance from center of rotation

FIG. 9. Banding pattern of alkali-treated *Eco* DNA topoisomerase I–fd DNA complex. A reaction mixture containing 71 μg/ml of fd DNA and 125 μg/ml of the enzyme, after incubation at 37°C to form the salt-stable complex, was treated with KOH at a concentration of 40 mM. The sample was then banded in an alkaline Cs_2SO_4 density gradient containing 0.1 M NaOH. The banding pattern was recorded after 30 hours at 42,000 rpm. (From Depew *et al.*, 1978, with permission of the publisher.)

fragment into two. If the protein is linked to the 5'-side of the newly generated break, it would be found on the part without the 5'-^{32}P label. On the other hand, if the protein is linked to the 3'-side of the newly generated break, it would be found on the part with the 5'-^{32}P label. When such an experiment was carried out, density gradient centrifugation of the product showed that the ^{32}P radioactivity was only found at the position of free DNA, indicating that the enzyme was linked to the 5'-side of the newly generated break.

The simplest interpretation of these experiments is as follows: when the enzyme–DNA complex is treated with alkali, a phosphodiester bond of the DNA is cleaved. The protein is linked to the 5'-phosphoryl side of the disrupted bond, and a free 3'-hydroxyl end group is generated. Since the protein–DNA linkage is stable in buoyant Cs_2SO_4 containing 0.1 M NaOH, it is most likely a covalent bond. Alternatively, the formation of the salt-stable enzyme–DNA complex might be accompanied by the breakage of a DNA phosphodiester bond, but this scission is revealed only upon denaturing the protein by alkali.

The presence of Mg(II) in the single-stranded DNA and *Eco* DNA topoisomerase I mixture does not prevent the formation of a complex which can be cleaved by alkali. We normally avoid the addition of Mg(II), since otherwise it is not possible to demonstrate the formation of the salt-stable complex prior to alkali treatment, owing to the dissociation of the complex by Mg(II) in a concentrated salt solution. It should be added that, although no *added* Mg(II) is necessary for complex formations, EDTA is inhibitory to complex formation (Liu, 1977). Thus, there might be a minute amount of Mg(II) or other metal ions present in the enzyme stock solution, without which complex formation might not occur.

If a negative superhelical DNA instead of a single-stranded DNA is used, formation of a salt-stable complex cleavable by alkali can also be demonstrated (Liu, 1977). The formation of this double-stranded DNA–enzyme complex is dependent on the degree of superhelicity of the DNA, as shown by the results shown in Fig. 10. PM2 DNA samples of various degrees of superhelicity were incubated with the enzyme in the absence of added Mg(II). The DNA as well as the enzyme concentration was the same in all reaction mixtures. Alkali was then added to the reaction mixtures, and the fraction of DNA which was no longer covalently closed was measured for each mixture. As shown in the figure, the formation of the alkali-cleavable complex is strongly dependent on the superhelicity of the DNA.

This strong dependence is reminiscent of a similar phenomenon in the removal of negative superhelical turns by the *E. coli* enzyme mentioned earlier. This similarity suggests that the alkali-cleavable com-

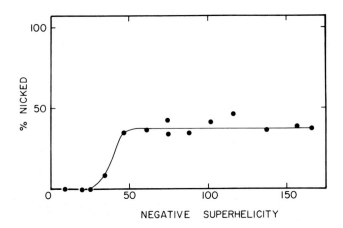

FIG. 10. The formation of the alkali-cleavable complex between *Eco* DNA topoisomerase I and covalently closed PM2 DNA of various degrees of superhelicity. Negative superhelicity is expressed here in the number of bound ethidium per 10^3 DNA nucleotides required for the complete removal of the negative superhelical turns of the DNA in neutral 3 M CsCl at 20°C. PM2 DNA extracted from the phage has a value of 75 on this scale. Reaction mixtures (200 μl each) containing 10 mM Tris–HCl, pH 8, 1.5 mM potassium phosphate, 0.05 mM Na$_3$EDTA, 50 μg/ml bovine plasma albumin, 1 μg of PM2 DNA of various degrees of superhelicity, and 0.15 μg of *Eco* DNA topoisomerase I were incubated at 37°C. After 10 minutes, 8 μl of 1 M KOH was added to each. The percentage of DNA containing one or more single-chain scissions was measured by an ethidium fluorescence assay as described by Morgan and Pulleyblank (1974).

plex is an intermediate in the removal of negative superhelical turns. The lifetime of such an intermediate is presumably very short during the normal course of the reaction.

The strong superhelicity dependence of the rate of removal of negative superhelical turns by the *E. coli* enzyme has been previously attributed to the destabilizing effect exerted by negative superhelical turns on the DNA double helix (Wang, 1971, 1973). If the formation of the alkali-cleavable complex requires an unwinding of the double helix, then its formation would be strongly dependent on the superhelicity of the DNA substrate [for discussions on this point, cf. Davidson (1972), Wang (1974, 1977), and Hsieh and Wang (1975)].

In many ways the alkali-cleavable complex between the enzyme and double-stranded DNA is very similar to the complex between the enzyme and single-stranded DNA: Both are stable in concentrated salt solutions but are dissociated by the addition of Mg(II) to the salt solutions. In either case, upon dissociation of the complex in concentrated salt solutions containing Mg(II), the DNA recovered is identical to the original DNA used in the formation of the complex. Extensive digestion of these complexes with Pronase results in the fragmentation of the DNA. We have also tested the effects of detergents on the enzyme–double-stranded DNA complex. Both sodium dodecyl sulfate and Sarkosyl induce the cleavage of the DNA chain, whereas the nonionic detergents Brigs 58 and Triton X-100 do not induce DNA chain scission. Although we have not tested the detergents on the enzyme–single-stranded DNA complex, we believe that the effects would be similar to those observed for the double-stranded DNA complex.

As we have shown in Fig. 10, it appears that double-stranded DNA without negative superhelical turns does not form the alkali-cleavable complex. Even in the presence of a high concentration of the enzyme there is little formation of the alkali-cleavable complex. This result is shown in Fig. 11. A complex stable in concentrated solutions is formed between the enzyme and nonsuperhelical DNA, however, as illustrated in Fig. 12. Aside from the stability in concentrated salt solutions, this complex bears little resemblance to the other complexes we have described. It is not dissociated upon adding Mg(II) to the solution containing molar quantities of salt. No DNA chain scission occurs upon the addition of alkali; alkali merely causes the dissociation of the complex. Examination of the banding pattern shown in Fig. 12 reveals the curious absence of the band corresponding to one protein molecule per DNA. The band corresponding to three protein molecules per DNA is also noticeably lower in intensity than the one with four protein molecules per DNA. It appears that the formation of

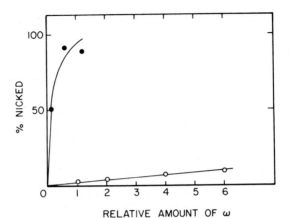

FIG. 11. Effect of superhelicity on the alkali-induced DNA backbone cleavage of the *Eco* DNA topoisomerase I–PM 2 DNA complex. Open symbols, the DNA contained very few negative superhelical turns. Its negative superhelicity is 9 on the horizontal scale shown in Fig. 10. Closed symbols, the DNA was moderately negatively supercoiled. Its negative superhelicity is 52 on the horizontal scale shown in Fig. 10.

(a)

(b)

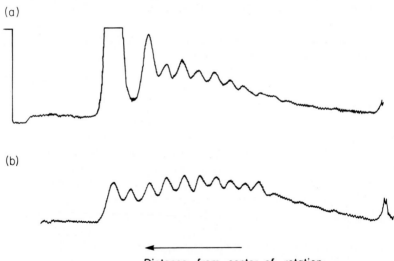

Distance from center of rotation

FIG. 12. (a) Formation of a complex stable in concentrated salt solutions between *Eco* DNA topoisomerase I and nonsuperhelical PM2 DNA. A reaction mixture (62 μl) containing 10 mM Tris–HCl, pH 8, 3 mM potassium phosphate, 0.1 mM Na$_3$EDTA, 32 μg/ml covalently closed PM2 DNA with no superhelical turns, and 45 μg/ml of the enzyme was incubated at 37°C. After 10 minutes 4 μl of 0.2 M Na$_3$EDTA was added and incubation was continued for 5 minutes. A CsCl stock solution was then added and the mixture was centrifuged to equilibrium at 42,000 rpm. (b) The sample is identical to the one shown in (a) except a negatively supercoiled PM2 DNA replaced the PM2 DNA without superhelical turns used in (a).

this complex involves the binding of the enzyme as a dimer or involves a cooperative binding. We have not yet tested whether the protein molecules in this type of complex still retain their catalytic properties.

III. OTHER PROKARYOTIC DNA TOPOISOMERASES

Micrococcus luteus DNA topoisomerase I has been purified to homogeneity (Kung and Wang, 1977). This enzyme is a single subunit protein with a molecular weight of 120,000. Its chromatographic properties on DEAE-cellulose, phosphocellulose, and DNA Sepharose are quite different from those of the *E. coli* enzyme. Rabbit antibodies against the *M. luteus* enzyme has little effect on the *E. coli* enzyme, and *vice versa*. The general catalytic properties of the two enzymes are rather similar. Both show strong dependence on the degree and sense of superhelicity of the DNA substrate in catalyzing the removal of superhelical turns. Both require Mg(II) to convert one DNA topoisomer to another. Similar to the *E. coli* enzyme, the *M. luteus* enzyme can also form a complex with single-stranded DNA which is stable in concentrated salt solutions. Exposure of the complex to alkali also causes breakage of the DNA chain.

A DNA topoisomerase has been purified to near homogeneity from *Bacillus magaterium* (Burrington and Morgan, 1976). It appears to be a single polypeptide of molecular weight 120,000, and is similar to *Eco* and *Mlu* DNA topoisomerase I in its Mg(II) requirement and its preferential action on negatively supercoiled DNA.

A DNA topoisomerase from *Salmonella typhimurium* has been partially purified and appears to be very similar to the *E. coli* enzyme (R. E. Depew and J. C. Wang, unpublished results). The *Salmonella* protein has the same peptide weight as *Eco* DNA topoisomerase I, as indicated by their identical mobilities in sodium dodecyl sulfate–polyacrylamide gels. Both are inhibited by rabbit antibodies against the *E. coli* enzyme. The two enzymes can be resolved by isoelectric focusing, however.

An *E. coli* enzyme termed gyrase has been discovered recently (Gellert *et al.*, 1976a). This enzyme catalyzes the conversion of a covalently closed double-stranded DNA ring without superhelical turns to the negatively supercoiled form. The reaction requires ATP, the hydrolysis of which presumably provides the driving force for carrying out the energetically unfavorable supercoiling of the DNA. A number of drugs have been found to inhibit gyrase. The related antibiotics novobiocin and coumermycin, and the related antibacterial agents nalidixic acid and oxolinic acid, all inhibit gyrase isolated from cells sensitive to these drugs, but not gyrase isolated from resistant cells (Gellert *et al.*, 1976b, 1977; Sugino *et al.*, 1977). Since these drugs all inhibit DNA replication *in vivo*, it is most likely that gyrase is involved in DNA replication. Furthermore, covalently closed λ DNA isolated from cells treated with these drugs is not superhelical; the same DNA isolated from untreated cells is. Thus gyrase also appears to control the supercoiling of DNA in *E. coli* (Gellert *et al.*, 1976b, 1977).

Since gyrase catalyzes the supercoiling of a DNA, it must possess a DNA topoisomerase activity. This activity will be referred to as *Eco* DNA topoisomerase II. *Eco* DNA topoisomerase II can remove both positive and negative superhelical turns (Gellert *et al.*, 1976a, 1977; Sugino *et al.*, 1977).

An apparent DNA gyrase activity has been purified from *Micrococcus luteus* (Liu and Wang, 1978a). The activity separates into two complementing groups upon passing through a hydroxylapatite column. These complementing activities were further purified separately on phosphocellulose columns, yielding two fractions with little cross-contamination. One complementing activity (designated as α) was purified to near homogeneity, as judged by electrophoretic analyses on both sodium dodecyl sulfate–polyacrylamide gel and nondenaturing polyacrylamide gel. In the latter case, it has been found that the activity coincides with the position of the protein band. This α subunit has a peptide molecular weight of 115,000, which is slightly lower than *Mlu* DNA topoisomerase I (molecular weight 120,000). The other complementing activity does not bind to phosphocellulose. Electrophoresis of this partially purified fraction on a nondenaturing polyacrylamide gel, followed by assaying for gyrase activity across the gel in the presence of added purified α subunit, identifies a band as the active component. Electrophoresis of this component on sodium dodecyl sulfate–polyacrylamide gel shows a single band with a molecular weight of 97,000. We designate this polypeptide as subunit β of *M. luteus* gyrase. The partially purified fraction which complements α contains approximately 30% β by mass. The resolution of the nondenaturing polyacrylamide gel is insufficient to exclude the possibility that an additional subunit might be required for gyrase activity. Separately, α or the partially purified β shows little DNA supercoiling activity. Upon mixing the two, an ATP-dependent DNA supercoiling activity is observed. This DNA supercoiling activity, similar to *E. coli* gyrase, is inhibited by novobiocin and coumermycin. The enzyme shows remarkable potassium ion dependence. In the absence of added K^+, very little gyrase activity was observed. The potassium ion requirement cannot be substituted by sodium ion, and the optimal K^+

concentration is around 20 mM. The significance of this requirement is not known. It is interesting to note that K$^+$ has an effect on the degradation of native and single-stranded DNA as well on the hydrolysis of ATP by purified $E. coli$ recBC enzyme (Hermanns and Wackernagel, 1977). The DNA supercoiling activity of $M. luteus$ gyrase is inhibited by rabbit antibodies against Mlu DNA topoisomerase I. The significance of this inhibition is not clear. Neither of the subunits α and β forms a precipitin line with the antibodies by Ouchterlony double-immunodiffusion assays. Under the same conditions a precipitin line is formed with Mlu topoisomerase I.

With either $M. luteus$ or $E. coli$ gyrase, recent experiments indicate that in the enzyme–DNA complex the DNA is wrapped around the enzyme; based on this observation, a model has been proposed for how gyrase catalyzes the supercoiling of DNA (Liu and Wang, 1978a,b).

IV. Eukaryotic DNA Topoisomerases

Champoux and Dulbecco (1972) reported the discovery of an activity in sonicated mouse embryo nuclei which can remove both negative and positive superhelical turns. This activity does not require Mg(II). Since then, similar activities have been found in $Drosophila$ $melanogaster$ eggs (Baase and Wang, 1974), mouse and human tissue culture cells (Vosberg $et\ al.$, 1975; Keller and Wendel, 1974; Keller, 1975; Rosenberg $et\ al.$, 1977; Yoshida $et\ al.$, 1977), rat liver (Champoux and Durnford, 1975; Champoux and McConaughy, 1976), calf thymus (Pulleyblank and Morgan, 1975), vaccinia virus cores (Bauer $et\ al.$, 1977), and sea urchin eggs (Poccia $et\ al.$, 1978).

With the exception of the vaccinia virus activity, the other activities are rather similar in their salt requirements. The mouse activity, for example, has a salt optimum around 0.2 M monovalent cations (Vosberg $et\ al.$, 1975; Champoux and McConaughy, 1976). The salt optimum for the vaccinia virus activity showed a decrease as purification of the activity proceeds. The most purified fraction has a salt optimum of 0.05 M NaCl and is inactive in 0.2 M NaCl (Bauer $et\ al.$, 1977).

The mouse activity has been purified to greater than 90% purity (Champoux and McConaughy, 1976). It is a single subunit protein with a molecular weight of 66,000 (Champoux and McConaughy, 1976). Keller (1975) has presented evidence that the human activity has a peptide molecular weight of 60,000. The most highly purified fraction of the vaccinia activity contains two polypeptides of molecular weights 24,000 and 35,000, and the activity is associated with those fractions enriched in the 35,000 molecular weight polypeptide.

It is now fairly certain that the eukaryotic activities as well as the vaccinia activity act catalytically (Keller, 1975; Vosberg *et al.*, 1975; Champoux and McConaughy, 1976; Bauer *et al.*, 1977). For the highly purified mouse DNA topoisomerase, several hundred DNA molecules can be isomerized per enzyme (Champoux and McConaughy, 1976).

The cellular location of eukaryotic DNA topoisomerases has been studied by several groups. Vosberg *et al.* (1975) found that the bulk of the activity in mouse LA9 cells resided in the nuclei, and that only a small percentage existed in the cytoplasm. Washed mitochondria from LA9 cells also contain a topoisomerase activity, but it is not certain whether this activity is intramitochondrial. Young and Champoux [quoted in Champoux and McConaughy (1976)] also found that there was little topoisomerase activity associated with the postmitochondrial fraction of the cytoplasm of rat liver. By using low ionic strength to avoid the elution of DNA topoisomerase activity from DNA, Yoshida *et al.* (1977) found that the activity in cultured human lymphoid cells was located entirely in two particulate fractions: chromatin and a nuclear membrane fraction containing DNA and possessing endogenous DNA synthesizing activity. DNA topoisomerase activity in isolated sea urchin nuclei appears to be associated with chromatin (Poccia *et al.*, 1978).

V. POSSIBLE BIOLOGICAL ROLES OF DNA TOPOISOMERASES

The ability of this class of enzyme to break and rejoin DNA backbone bonds naturally leads to suggestions that they might be involved in processes which require or might require such events, including replication (Wang, 1971; Champoux and Dulbecco, 1972), transcription (Wang, 1973), and condensation and decondensation of DNA (Baase and Wang, 1974; Bauer *et al.*, 1977). Certain genetic recombination events might also involve such an activity. The λ *int*-promoted recombination, for example, must involve breaking and rejoining of DNA bonds, yet there is no indication that DNA ligase is involved (Syvanen, 1974; Mizuuchi and Nash, 1976).

Direct experimental evidence is lacking for the physiological functions of most of the DNA topoisomerases. For *E. coli* gyrase, however, there is strong evidence that this enzyme (and hence *Eco* DNA topoisomerase II) is involved in DNA replication and the supercoiling of DNA *in vivo*, since gyrase appears to be the target of a number of drugs which inhibit replication and supercoiling (Gellert *et al.*, 1976b, 1977).

Using synchronized cells of a human lymphocytic cell line, Rosenberg *et al.* (1977) examined the level of DNA topoisomerase activity in different growth phases. It is found that the level is high in the S phase and is much lower in the G_0 and G_1 phases. Since the activity measured in G_0 and G_1 cultures could be due to a small fraction of cells out of synchrony, it is plausible that the enzyme might be present only during the S phase. Rosenberg *et al.* (1977) therefore suggested that the human DNA topoisomerase might be required for replication but not for transcription, because the level of transcription is high in G_1. They further suggested that the enzyme might be a control factor whose appearance signaled the beginning of the S phase and whose continued presence was necessary for DNA replication.

During sea urchin embryonic development, the DNA topoisomerase activity in the nuclei shows at least a tenfold variation. It increases during S phase, decreases during G_2, and reaches a minimum at the end of mitosis (Poccia *et al.*, 1978). No abrupt change in activity during mitosis was observed. The topoisomerase activity declines steadily throughout mitosis, while the chromosomes are at first condensing, then decondensing.

It has been suggested that the vaccinia virus DNA topoisomerase, aside from a possible role in virus assembly, might also be involved in the process of transcription from the viral core (Bauer *et al.*, 1977). Such a role in transcription is consistent with the presence in the virion of a number of enzymes required for transcription and modification of the transcript (Bauer *et al.*, 1977). Higashinakagawa *et al.* (1977) have found that there is a DNA topoisomerase activity in amplified nucleoli purified from immature oocytes of *Xenopus laevus*. Amplified nucleoli do not replicate again *in vivo*, and no DNA replication activity is detectable in the purified nucleoli. Transcription of the ribosomal DNA, however, is active in amplified nucleoli, and RNA polymerase I activity is found in purified nucleoli. Higashinakagawa *et al.*, therefore, favor the possibility that the DNA topoisomerase activity in nucleoli is involved in transcription. Further studies are needed to establish the physiological roles of the various DNA topoisomerases.

VI. OTHER PROTEINS WITH PROPERTIES SIMILAR TO DNA TOPOISOMERASES

Aside from catalyzing interconversions between topological isomers of DNA rings, the enzyme *Eco* DNA topoisomerase I is also characterized by its ability to form complexes with single-stranded and negatively supercoiled DNAs. As we have reviewed in an earlier section, these complexes are stable in concentrated salt solution, and treatment of the complexes with alkali, Pronase, or certain ionic detergents leads to breakage of the DNA chain and the linking of the protein to the 5'-side of the break generated. Although studied in less detail, *Mlu* DNA topoisomerase I appears to form similar complexes. *Escherichia coli* gyrase, in the presence of nalidixic or oxolinic acid, which inhibits the *Eco* DNA topoisomerase II activity of gyrase, can also form a complex with double-stranded DNA. Treatment of the complex with sodium dodecyl sulfate (Sugino *et al.*, 1977) or a combination of sodium dodecyl sulfate and proteinase K (Gellert *et al.*, 1977) leads to cleavage of both strands of the DNA.

The cleavage of DNA by chemical or proteolytic enzyme treatment of a DNA–protein complex is reminiscent of the relaxation complexes of plasmid DNAs first studied by Helinski *et al.* (Clewell and Helinski, 1969, 1970; Kline and Helinski, 1971; Blair and Helinski, 1975; Lovett and Helinski, 1975; Guiney and Helinski, 1975; Helinski *et al.*, 1975; Morris *et al.*, 1973). For example, in the relaxation complex of the colicinogenic factor ColE1, the DNA is in the negatively supercoiled form. Treatment of the complex with alkali, Pronase, or sodium dodecyl sulfate converts the DNA to a nicked circular form, with the protein linked to the 5'-side of the nick. The ColE1 relaxation protein contains three major polypeptides of molecular weights 60,000, 16,000, and 11,000. The largest polypeptide is found linked to the DNA upon these treatments. The relaxation complexes examined to date are position and strand specific. Recent DNA sequencing results of Tomizawa *et al.* (1977) indicate that the site at which the ColE1 relaxation protein can cause a scission is separated from the origin of replication by approximately 300 nucleotides. It has been suggested that the relaxation proteins might be involved in replication or conjugal transfer (Helinski *et al.*, 1975).

The coliphage ϕX174 gene *A* product is known to be a protein required for the replication of the double-stranded replicative form of ϕX DNA. The *cis*-A protein introduces a position- and strand-specific single-strand discontinuity into the originally covalently closed replicative form (Francke and Ray, 1971). The coliphage P2 gene *A* product appears to have a similar function (Geisselsoder, 1976). When purified ϕX *cis*-A protein is incubated with negatively supercoiled double-stranded ϕX DNA, at low concentrations of the protein it appears that a complex is formed in which the DNA is no longer superhelical. Treatment of the complex with phenol or a combination of phenol and proteinase K results in the generation of a discontinuity in the viral (or +) strand of the DNA duplex. The discontinuity in the DNA contains a free 3'-hydroxyl terminus and a blocked 5'-end (Ikeda *et al.*, 1976). Presumably the protein itself forms the 5'-end block. Treatment of the complex with 0.2% Sarkosyl or 1% sodium dodecyl sulfate has no effect on its structure (Ikeda *et al.*, 1976).

In vitro studies on the synthesis of the single-stranded circular viral (+) strand of ϕX DNA with purified proteins show that the super-coiled ϕX DNA template is first nicked by *cis*-A protein in the viral strand. Since copies of the strand produced are circular with no inter-ruption in the DNA backbone, and since the other purified proteins required in the *in vitro* system have no known ligase function, the possibility that *cis*-A may function in the manner of a DNA topoiso-merase has been raised (Scott *et al.*, 1977). This possibility has been confirmed recently (Eisenberg *et al.*, 1977). *Cis*-A differs from the other DNA topoisomerases in that the nicked intermediate in which the protein is linked to the 5'-terminus of the DNA appears stable; there is some evidence that the 5'-end can join to a 3'-end of a *differ-ent* molecule (Eisenberg *et al.*, 1977).

A number of other protein–DNA covalent complexes have been reported. Transducing fragments extracted from phage P1 particles by a combination of sodium dodecyl sulfate and phenol extractions con-tain a protein component covalently linked to DNA (Ikeda and Tomi-zawa, 1965). The DNA extracted from the *B. subtilis* phage ϕ29, by treatment with Sarkosyl, Sarkosyl and phenol, or 2 M NaClO$_4$, also contains a tightly associated protein moiety (Ortin *et al.*, 1971; Hiro-kawa, 1972). Sarkosyl-extracted ϕ29 DNA contains about equal amounts of linear and circular forms. The circular form is converted to the linear form by digestion with trypsin, indicating that the protein is likely to be linked to the ends of the DNA molecules and is involved in cyclization of the DNA. Alternatively, the DNA inside the virus is covalently closed and is converted to the other forms by the extraction procedure and proteolytic digestion (Ortin *et al.*, 1971). Proteolytic enzymes also destroy the transfecting activity of ϕ29 DNA but have no effect on that of a number of other *B. subtilis* phage DNAs (Hirokawa, 1972). Studies on the restriction fragments of ϕ29 DNA digested with *Eco*R1 endonuclease indicate that the protein is probably covalently associated with one end of the DNA. This association is not disrupted by heating at 70°C in the presence of 2% sodium dodecyl sulfate or by treatment with other protein denaturants. The protein is also asso-ciated with the other end of the DNA, but this association is disrupted by the heating and sodium dodecyl sulfate treatment (Harding and Ito, 1976). Bellet *et al.* have found that human adenovirus DNA obtained by disrupting the virus with 4 M guanidinium chloride is in a circular or oligomeric form, but can be converted to monomeric linear molecules by digestion with proteolytic enzymes, indicating the association of protein molecules with the ends of the DNA (Robinson *et al.*, 1973; Robinson and Bellet, 1974). Recent results indicate that there is a protein of 55,000 molecular weight directly attached to each 5'-end of the DNA, probably via a covalent linkage (Rekosh *et al.*, 1977). Protein moieties linked to DNA have also been observed in detergent-treated SV40 DNA (Kasamatsu and Wu, 1976a,b), and in HeLa cell mitochondrial DNA (Albring *et al.*, 1977). In both cases, the sites of bound protein are close to the origins of replication of the DNAs. Protein molecules covalently linked to the ends of poliovirus RNA have also been reported recently (Lee *et al.*, 1977; Flanegan *et al.*, 1977). The covalent linkage between the protein and the RNA has been identified as a phosphodiester bond between a tyrosine residue in the protein and the 5'-terminal uridine of the nucleic acid (Ambros and Baltimore, 1978; Rothberg *et al.*, 1978).

For these covalent complexes, since their isolation usually involves treatment with a protein denaturant, it is at present unknown whether the undenatured proteins which can form these complexes might have enzymatic activities. In view of the recent studies on the formation of covalent complexes between DNA and *Eco* topoisomerase I and *E. coli* gyrase, it would not be surprising if the native proteins could catalyze the breaking and rejoining of nucleic acid backbone bonds. The ubiquity of proteins capable of forming covalent complexes with DNA is only beginning to be recognized, and many questions regarding their modes of action and their physiological roles remain to be answered.

Acknowledgments

We would like to acknowledge the support of United States Public Health Service through Grant No. GM24544 (previously GM14621). L.F.L. is an American Cancer Society postdoctoral fellow.

References

Albring, M., Griffith, J., and Attardi, G. (1977). *Proc. Natl. Acad. Sci. U.S.A.* **74,** 1348–1352.

Ambros, V., and Baltimore, D. (1978). *J. Biol. Chem.* **253,** 5263–5266.

Baase, W. A., and Wang, J. C. (1974). *Biochemistry* **13,** 4299–4303.

Bauer, R. W., Ressner, E. C., Kates, J., and Patzke, J. V. (1977). *Proc. Natl. Acad. Sci. U.S.A.* **74,** 1841–1845.

Becker, A., Lyn, G., Gefter, M., and Hurwitz, J. (1967). *Proc. Natl. Acad. Sci. U.S.A.* **58,** 1996–2003.

Blair, D. G., and Helinski, D. R. (1975). *J. Biol. Chem.* **250,** 8790–8795.

Burrington, M. G., and Morgan, A. R. (1976). *Can. J. Biochem.* **54**, 301–306.
Carlson, J. O. (1974). Ph.D. Thesis, University of California at Berkeley.
Champoux, J. J., and Dulbecco, R. (1972). *Proc. Natl. Acad. Sci. U.S.A.* **69**, 143–146.
Champoux, J. J., and Durnford, J. M. (1975). *ICN-UCLA Symp. Mol. & Cell. Biol.* **3**, 83–93.
Champoux, J. J., and McConaughy, B. L. (1976). *Biochemistry* **15**, 4638–4643.
Clewell, D. B., and Helinski, D. R. (1969). *Proc. Natl. Acad. Sci. U.S.A.* **62**, 1159–1166.
Clewell, D. B., and Helinski, D. R. (1970). *Biochem. Biophys. Res. Commun.* **40**, 608–613.
Davidson, N. (1972). *J. Mol. Biol.* **66**, 307–309.
Depew, R. E., and Wang, J. C. (1975). *Proc. Natl. Acad. Sci. U.S.A.* **72**, 4275–4279.
Depew, R. E., Liu, L. F., and Wang, J. C. (1978). *J. Biol. Chem.* **253**, 511–518.
Eisenberg, S., Griffith, J., and Kornberg, A. (1977). *Proc. Natl. Acad. Sci. U.S.A.* **74**, 3198–3202.
Flanegan, J. B., Petterson, R. F., Ambros, V., Hewlett, M. J., and Baltimore, D. (1977). *Proc. Natl. Acad. Sci. U.S.A.* **74**, 961–965.
Francke, B., and Ray D. S. (1971). *J. Mol. Biol.* **61**, 565–586.
Geisselsoder, J. (1976). *J. Mol. Biol.* **100**, 13–22.
Gellert, M., Mizuuchi, K., O'Dea, M. H., and Nash, H. A. (1976a). *Proc. Natl. Acad. Sci. U.S.A.* **73**, 3872–3876.
Gellert, M., O'Dea, M. H., Itoh, T., and Tomizawa, J. (1976b). *Proc. Natl. Acad. Sci. U.S.A.* **73**, 4474–4478.
Gellert, M., Mizuuchi, K., O'Dea, M. H., Itoh, T., and Tomizawa, J. (1977). *Proc. Natl. Acad. Sci. U.S.A.* **74**, 4772–4776.
Guiney, D. G., and Helinski, D. R. (1975). *J. Biol. Chem.* **250**, 8796–8803.
Harding, N. E., and Ito, J. (1976). *Virology* **73**, 389–401.
Helinski, D. R., Lovett, M. A., Williams, P. H., Katz, L., Collins, J., Kupersztoch-Portnoy, Y., Sato, S., Leavitt, P. W., Sparks, R., Hershfield, V., Guiney, D. G., and Blair, D. G. (1975). *In* "DNA Synthesis and Its Regulation" (M. Goulian and P. Hanawalt, eds.), pp. 514–536. Benjamin, New York.
Hermanns, U., and Wackernagel, W. (1977). *Eur. J. Biochem.* **76**, 425–432.
Higashinakagawa, T., Wahn, H., and Reeder, R. H. (1977). *Dev. Biol.* **55**, 375–386.
Hirokawa, H. (1972). *Proc. Natl. Acad. Sci. U.S.A.* **69**, 1555–1559.
Hsieh, T.-S., and Wang, J. C. (1975). *Biochemistry* **14**, 527–535.
Ikeda, H., and Tomizawa, J. (1965). *J. Mol. Biol.* **14**, 110–119.
Ikeda, J. E., Yudelvich, A., and Hurwitz, J. (1976). *Proc. Natl. Acad. Sci. U.S.A.* **73**, 2669–2673.
Jovin, T. M., Englund, P. T., and Bertsch, L. L. (1969). *J. Biol. Chem.* **244**, 2996–3008.
Kasamatsu, H., and Wu, M. (1976a). *Biochem. Biophys. Res. Commun.* **68**, 927–936.
Kasamatsu, H., and Wu, M. (1976b). *Proc. Natl. Acad. Sci. U.S.A.* **73**, 1945–1949.
Keller, W. (1975). *Proc. Natl. Acad. Sci. U.S.A.* **72**, 2550–2554.
Keller, W., and Wendel, I. (1974). *Cold Spring Harbor Symp. Quant. Biol.* **39**, 199–208.
Kirkegaard, K., and Wang, J. C. (1978). *Nucleic Acid Res.* **5**, 3811–3820.
Kline, B. C., and Helinski, D. R. (1971). *Biochemistry* **10**, 4975–4980.
Kung, V., and Wang, J. C. (1977). *J. Biol. Chem.* **252**, 5398–5402.
Lee, Y. F., Nomoto, A., Detjin, B. M., and Wimmer, E. (1977). *Proc. Natl. Acad. Sci. U.S.A.* **72**, 4157–4161.
Lindahl, T., and Edelman, G. M. (1968). *Proc. Natl. Acad. Sci. U.S.A.* **61**, 680–687.
Liu, L. F. (1977). Ph.D. Thesis, University of California, Berkeley.
Liu, L. F., and Wang, J. C. (1978a). Submitted for publication.

Liu, L. F., and Wang, J. C. (1978b). *Cell* **15**, 979–984.
Liu, L. F., Depew, R. E., and Wang, J. C. (1976). *J. Mol. Biol.* **106**, 439–452.
Lovett, M. A., and Helinski, D. R. (1975). *J. Biol. Chem.* **250**, 8790–8795.
Mizuuchi, K., and Nash, H. A. (1976). *Proc. Natl. Acad. Sci. U.S.A.* **73**, 3524–3528.
Morgan, A. R., and Pulleyblank, D. E. (1974). *Biochem. Biophys. Res. Commun.* **61**, 396–403.
Morris, C. F., Hershberger, C. L., and Rownd, R. (1973). *J. Bacteriol.* **114**, 300–308.
Ortin, J., Vinuela, E., and Salas, M. (1971). *Nature (London), New Biol.* **234**, 275–277.
Poccia, D. L., LeVine, D., and Wang, J. C. (1978). *Dev. Biol.* **64**, 273–283.
Pulleyblank, D. E., and Morgan, A. R. (1975). *Biochemistry* **14**, 5205–5209.
Rekosh, D. M. K., Russell, W. C., and Bellet, A. J. D. (1977). *Cell* **11**, 283–299.
Robinson, A. J., and Bellet, A. J. D. (1974). *Cold Spring Harbor Symp. Quant. Biol.* **39**, 523–531.
Robinson, A. J., Younghusband, H. B., and Bellett, A. J. D. (1973). *Virology* **56**, 54–69.
Rosenberg, B. H., Uncers, G., and Deutch, J. F. (1977). *Nucleic Acids Res.* **3**, 3305–3311.
Rothberg, P. G., Harris, T. J. R., Nomoto, A., and Wimmer, E. (1978). *Proc. Natl. Acad. Sci. U.S.A.* **75**, 4868–4872.
Scott, J. F., Eisenberg, S., Bertsch, L. L., and Kornberg, A. (1977). *Proc. Natl. Acad. Sci. U.S.A.* **74**, 193–197.
Sugino, A., Peebles, C. L., Kreuzer, K. N., and Cozzarelli, N. R. (1977). *Proc. Natl. Acad. Sci. U.S.A.* **74**, 4767–4771.
Syvanen, M. (1974). *Proc. Natl. Acad. Sci. U.S.A.* **71**, 2496–2499.
Tomizawa, J. I., Ohmori, H., and Bird, R. E. (1977). *Proc. Natl. Acad. Sci. U.S.A.* **74**, 1865–1869.
Vinograd, J., and Lebowitz, J. (1966). *J. Gen. Physiol.* **49**, Suppl., 103–125.
Vosberg, H. P., and Vinograd, J. (1975). *In* "DNA Synthesis and Its Regulation" (M. Goulian and P. Hanawalt, eds.), pp. 94–120. Benjamin, New York.
Vosberg, H. P., Grossman, L. I., and Vinograd, J. (1975). *Eur. J. Biochem.* **55**, 79–93.
Wang, J. C. (1969). *J. Mol. Biol.* **43**, 263–272.
Wang, J. C. (1971). *J. Mol. Biol.* **55**, 523–533.
Wang, J. C. (1973). *In* "DNA Synthesis in Vitro" (R. D. Wells and R. B. Inman, eds.), pp. 163–174. Univ. Park Press, Baltimore, Maryland.
Wang, J. C. (1974). *J. Mol. Biol.* **87**, 797–816.
Wang, J. C. (1977). *In* "DNA Synthesis" (I. Molineux and M. Kohiyama, eds.), pp. 347–366. Plenum, New York.
Weiss, B., and Richardson, C. C. (1967). *Proc. Natl. Acad. Sci. U.S.A.* **57**, 1021–1028.
Yoshida, S., Ungers, G., and Rosenberg, B. H. (1977). *Nucleic Acids Res.* **4**, 223–228.
Zimmerman, S. B., Little, J. W., Oshinsky, C. K., and Gellert, M. (1967). *Proc. Natl. Acad. Sci. U.S.A.* **57**, 1841–1848.

Chapter III

Enzymatic Methylation of DNA: Patterns and Possible Regulatory Roles

J. HERBERT TAYLOR

I. INTRODUCTION

Enzymatic methylation of DNA is a modification which usually occurs immediately following replication or repair. The enzymes involved are not yet well characterized, but the process appears to be very efficient. The donor molecule in the reaction is S-adenosylmethionine. The methyl group from methionine is transferred either to the 6-amino group in adenine or to the fifth ring carbon in cytosine. In the native DNA both methyl groups lie in the large groove, and therefore the exterior surface of the double helix is modified by the change. Demethylation probably does not occur at the polymer level, but there seems to be constant renewal by addition of methyl groups on new chains whether produced in replication or repair. However, if DNA should replicate two rounds without replacement of methyl groups on the new chains, one-half of the molecules would lack methyl groups on both chains and therefore be effectively demethylated. In bacterial

89

MOLECULAR GENETICS, PART III

and phage DNAs the modified sites frequently contain N^6-methyla-denine, although there are species with an excess of methylated cyto-sines. On the other hand, the modifications in eukaryotes involve cy-tosine almost exclusively. One notable exception has been reported in the protozoan, *Tetrahymena pyriformis*, where the DNA in the mac-ronucleus has a relatively high level of N^6-methyl adenine while that in the metabolically inactive micronucleus has tenfold less (Gorovsky *et al.*, 1973).

The role of methylation is clearly understood in only a few in-stances where it is part of host-controlled modification systems in bac-teria that protect DNA from site-specific endonucleases as described below. In eukaryotes the role of methylation has been a matter of much interest and speculation, but definitive evidence is lacking. Nevertheless, the increased information available from bacterial sys-tems is providing a renewal of interest and some valuable new tools in the search for possible regulatory functions in higher cells.

There are many features of methylation which make it an ideal mechanism for regulation by a modification of DNA which can be maintained through numerous replications. If methylation was asso-ciated with palindromic sequences, as it frequently is in bacterial DNA, maintenance at a high fidelity by specific enzymes can be visu-alized. The sequence below is susceptible to *Hpa*II (a site-specific en-donuclease from *Haemophilus parainfluenzae*), unless the 5-methyl-cytosine is present on one or both chains. In that case the DNA is completely protected. The arrows show the enzymatic cleavages sites (see the footnote to Table I for an explanation of symbols).

$$5' \cdots p\ C \downarrow p\ ^mC\ p\ G\quad p\ G \cdots 3'$$
$$3' \cdots G\quad p\quad G\ p^mC \uparrow p\ C\ p \cdots 5'$$

After replication each duplex will be half-methylated, but still fully protected until an appropriate enzyme can methylate the sym-metrically located cytosines, usually within a minute or two following replication. In *Escherichia coli*, where a modification system has been studied, an unmethylated DNA which enters the cell can be slowly methylated, apparently by the same enzyme which maintains the cell's methylated sites. However, the initiation event at an unmeth-ylated sequence is much slower. It can take hours (Meselson *et al.*, 1972), whereas the methylation of half-methylated sites following rep-lication can occur in minutes.

In eukaryotes enzymatic methylation of DNA following replication appears to be a universal phenomenon, at least for chromosomal DNA. So far no role has been demonstrated for the modification, but a number of proposals have been made for a role in differentiation. In the operation of such a system the initiation events, i.e., the original modification, would be expected to be the crucial event. The maintenance could be very efficient if methylation were confined to palindromic sequences. After each replication, an enzyme would recognize the half-methylated sites and add a methyl group to the cytosine in the symmetrical position on the new chain. Whether there are two different enzymes involved or not, we will tentatively use the terms initiation methylase and maintenance methylase. One might expect the initiation type to recognize a certain palindromic sequence and methylate one or both chains in the absence of any previously existing methylated base at the site. In addition, these special methylases might be rare and limited to certain stages of development, whereas the maintenance type would be present in cells at most if not all stages of development. If there are two distinct species, one might predict that the maintenance type would be less discriminating in sequence requirements. For example, all sequences containing the CCGG sequence with one of the cytosines methylated might be maintained in the methylated state. On the other hand, the original modification of the sites might depend on the flanking sequences, so that the initial modifications could have a higher degree of specificity, while the maintenance could be accomplished by a few very efficient activities operating soon after replication or repair synthesis and be guided by the methylated base on the parental chain. For such a system to operate efficiently, the cell should not have any demethylase activity operating at the polymer level, and incorporation of 5-methylcytosine containing nucleotides into DNA should be prevented by a fail-safe mechanism. Whether these two criteria are met has not been rigorously proved.

There is, however, recent evidence that a pattern of methylated and unmethylated sites within a common sequence in a cell can be inherited by descendents of that cell, i.e., a maintenance system similar to that described above may operate. Bird and Southern (1978) have shown that somatic cell DNA of *Xenopus laevis*, which codes for ribosomal RNA (rDNA), is cleaved by the restriction enzymes *Hpa*II, *Ava*I, and *Hha*I (see Chapter I for nomenclatures of restriction enzymes) to only a limited extent. After digestion with *Hpa*II the segments showed a complex distribution of sizes in electrophoresis with most of the DNA concentrated in the region where 9 to 11 kb (kilobase) fragments would be found on a 1% agarose gel. On the other hand, if the amplified rDNA from oocytes was digested in a similar manner the fragments were small (less than 500 base pairs). This was interpreted to mean that amplified DNA is not methylated whereas somatic cell DNA is methylated at nearly all *Hpa*II sites. However, about 30 to 60% of the repeat units in rDNA in somatic cells had one unmethylated site susceptible to this enzyme.

By using the two DNAs and a cloned fragment of rDNA, Bird (1978) obtained evidence that the base-paired sequence 5'CG/GC3', which is in each of the restriction sites for the enzymes mentioned above, is methylated either on both cytosines or on neither one. In other words, in nonreplicating DNA half-methylated sites are not present. He produced ribosomal DNA restriction fragments from somatic cells by the use of the enzyme, *Eco*RI. When these cells were denatured and reannealed with a large excess of unmethylated fragments in order to expose any half-methylated *Hpa*II sites, he found that all of the fragments remained resistant, a property expected if the DNA which had been denatured was methylated on both chains.

In a second experiment, somatic cell rDNA was denatured and allowed to self-reassociate (anneal). This procedure made the 4.65 kb *Eco*RI fragments resistant to digestion with *Hpa*II at all sites except the original sites susceptible to the enzyme. The result was interpreted to indicate that most sites with paired CGs were symmetrically methylated whereas the *Hpa*II sensitive sites represent a specific unmethylated *Hpa*II sequence. Since only the new chains in cultured cells were detectably labeled with [methyl-^{3}H]methionine, a methyl donor, he concluded that patterns of methylated and unmethylated sites with paired CGs will be inherited by descendents of such cells. The evidence is certainly suggestive, but one caveat seems necessary. Since the sequence specificities of the methylating enzymes in *Xenopus laevis* cells are not known, it may be necessary to reserve the possibility that the differences in pattern observed depends not on the previous methylated condition of the sites, but on the flanking nucleotides that could be different at the unmethylated sites.

II. RESTRICTION AND MODIFICATION IN BACTERIAL CELLS

Restriction and modification have been reviewed by Meselson *et al.* (1972) and Arber (1974). The restriction enzymes (site-specific endonucleases) have been reviewed by Roberts (1976) and in this volume by Zabeau and Roberts in Chapter I. It will only be necessary to summarize some of the features of the system as a background for a discussion of methylation in eukaryotes.

Modification and restriction in bacteria have been distinguished as two types, referred to as class I and class II. The class I systems are exemplified by B and K enzymes of the corresponding strains of *Escherichia coli*, and by the PI modification–restriction system, which is also found in certain strains of *E. coli*. For the operation of these systems, the bacteria have the genetic code for producing two related enzymes. One is a site-specific endonuclease, and the other is a methylase which binds to the same site if it is half-methylated, or even if unmethylated, and adds a methyl group to one or both chains. This effectively protects the site from the corresponding nuclease. In these class I systems, cleavage is remote from the original recognition or binding site for the methylase and the endonuclease; therefore, the precise cleavage sites are not predictable. The nucleases of the B and K systems, as well as the methylases, require ATP, Mg^{2+} and S-adenosylmethionine as cofactors. The PI system is similar to the B and K systems in that the cleavage products are heterogeneous, but the PI nuclease requires only ATP and Mg^{2+} as cofactors. The B and K endonucleases hydrolyze ATP, but the endonuclease of the PI system does not. The sequence recognized by the PI endonuclease is AGATCT, in which the second A in the sequence is the methylated base if the site has been modified. These enzymes (B, K, and PI) apparently function in protecting the cell from foreign DNA, particularly of viral origin. The cell's own DNA is modified and protected. If the invading DNA somehow escapes the cellular endonucleases, it may become modified and then be able to survive in other host cells of the strain in which it was modified.

The class II enzymes are the ones that have become well known as restriction endonucleases even though there is still some doubt concerning their role in the cell (see Chapter I). The specific endonucleases recognize and cleave specific sites in both chains of the DNA which frequently bracket the modified bases. However, some of the nucleases cut both chains between adjacent complementary base pairs and produce ends without unpaired bases. Those most useful in splicing DNA in recombinant DNA work leave four or five bases unpaired at each end of every fragment. Since all these ends are alike or complementary, the fragments can be joined together in any order. The specific sequences required by the nucleases are short enough so that there is a good chance that one will occur at intervals in most DNAs of a few thousand base pairs. For example, the sequence GGCC cleaved by *Hae*III occurs in 18 places in SV40 DNA which has about 5000 base pairs. On the other hand, the sequence CCGG cleaved by *Hpa*II is found only at a single site. One would expect the two sequences to be present at the same frequency unless the DNA has been under selective pressure in evolution to exclude one sequence. The sequence recognized by the *Hha*I endonuclease, GCGC, also is present in fewer copies than should occur by chance. The SV40 genome has only two of these sites, but it has the same chance of occurrence as GGCC or CCGG.

One enzyme of the class II group has been discovered that cleaves DNA at a specific site, but this site is located eight nucleotides away from the specific recognition site (Klied *et al.*, 1976). Two other nucleases discovered in *Diplococcus pneumoniae* are interesting in that both cleave the same site, —GATC—, but *Dpn*I cleaves the site only if it is methylated, while *Dpn*II from the same organism operates like most other class II restriction endonucleases; it cleaves only if the site is unmethylated (Lacks and Greenberg, 1975).

These class II enzymes are important tools in the analysis of eukaryotic DNAs. With the purified enzymes one can identify specific sequences in any DNA if these are not methylated. As described below, these enzymes can also be used for constructing specific sites for assaying methylase activities.

III. SITES OF METHYLATION IN DNA

A. SPECIFIC SEQUENCES

A few examples of methylated cytosines associated with restriction sites are shown in Table I for some species of bacteria. The bacterial

TABLE I

CYTOSINE METHYLASES ISOLATED FROM BACTERIA

Source	Restriction enzyme	Sequence[a]	Reference
H. parainfluenzae	$^m Hpa$II	C↓ ^{m}C G G	Mann and Smith (1978)
H. aegyptius	$^m Hae$III	G G↓ ^{m}C C	Mann and Smith (1978)
H. haemolyticus	$^m Hha$I	G ^{m}C G↓ C	Mann and Smith (1978)
Bacillus brevis S	$^m Bbv$SI	A G ^{m}C T G C	Vanyushin and Dobrita (1975)
E. coli	$^m Eco$RII	A ↓C ^{m}C T G G	Boyer *et al.* (1973)

[a] The conventional way for indicating the sequences for restriction enzymes is to show the sequence necessary for binding the enzyme written in the 5′ → 3′ direction without indicating the flanking nucleotides. In the remainder of the chapter this convention will be followed. An m as a superscript to the left of a letter representing a base indicates the modified or methylated one. The arrow shows the site of cleavage on the chain represented. The other chain which is not shown is cleaved in a symmetrical position.

cells also contain the corresponding site-specific endonuclease in addition to the cytosine methylases specific for each site shown. Isolated enzymes will methylate any native DNA that has the specific sites unmodified. Heat-denatured DNA will also serve as a substrate for the methylases, but the rate of methylation is slower (Mann, 1977). The cleavage by the nuclease, indicated by an arrow, always leaves a 3'-OH end and a 5'-PO_4 end. The methylases, unlike the nucleases, do not require Mg^{2+}, and therefore traces of nuclease activity which may contaminate the methylases can usually be inhibited by using EDTA in the assay mixtures.

The DNAs of the species shown in Table I may also be methylated on the amino groups at the 6-position in adenine, but since the DNAs of vertebrates are methylated almost exclusively at the 5-position on cytosine, we have drawn attention only to the DNAs modified at these sites. Whether the methylated cytosines in vertebrates are regularly in palindromic sequences similar to those in Table I is not known, but the evidence cited below suggests that this may be the situation. HeLa cell DNA has been methylated *in vivo* using L-[methyl-^{3}H]-methionine along with 0.02 M sodium formate to limit the incorporation of the methyl group of methionine into pyrimidines via the C_1 pool (Roy and Weissbach, 1975). The DNA was isolated and digested with DNase I under conditions which gave 40–50% yield of dinucleotides. The dinucleotides were recovered from the digest by chromatography on a DEAE-cellulose–7 M urea column and further digested with alkaline phosphatase to produce the dinucleoside monophosphates. These were then resolved by thin-layer electrophoresis. The two labeled cytosine-containing dinucleoside monophosphates ran identically with two added carrier molecules, the dinucleoside monophosphates GpC and CpC.

To determine the order of the methylated cytosines, the dinucleoside monophosphates were further digested with snake venom phosphodiesterase, which should yield from XpY, the X (nucleoside) and pY (the nucleoside monophosphate from the 3'-end). When the digests were chromatographed to separate the nucleosides from the nucleotides, the dimer containing G and C produced 60% ^{m}C and 40% p^{m}C; therefore, the sequences are 60% ^{m}CpG-3' and 40% Gp^{m}C-3'. On the other hand, the dinucleoside monophosphates containing two cytosines produced 100% of the label in p^{m}C, which indicates the sequence is always Cp^{m}C-3'. The methylated DNA produced by the HeLa cell methylase reacted *in vitro* with DNA from *Micrococcus luteus* gave the same distribution of methyl-labeled cytosine-containing dimers. Therefore the 5'-neighbor of ^{m}C may be either G or C; but if two C's are neighbors, only the 3' neighbor can be methylated. This indicates that the HeLa cells may have C ^{m}C G G or G^{m}C G C, but not G G^{m}C C.

The enzyme from HeLa cells also methylated the synthetic alternating doubled-stranded copolymer $(dG-C)_n \cdot (dG-dC)_n$ and the single-stranded random polymer $(dG,dC)_n$. However, only one methyl group per 1800 bases was added to the double-stranded alternating copolymer in a reaction run to the limit (sequential addition of the methylating enzyme to a small amount of polymer until a plateau of methylation was reached). On the other hand, one methyl group per 200 bases was added to the random polymer in a similar reaction. These experiments indicate a sequence specificity on both single- and double-stranded molecules, but the sequences methylated are not likely to be simply GCGC, since there would be many of these sites in the alternating copolymer. The authors suggest the sites which can be methylated in the alternating copolymer may be produced by a rare and unknown irregularity in the alternating bases.

There is also evidence from the use of restriction enzymes on DNA of the amphibian, *Xenopus*, that the palindromic sequences C ^{m}C G G and G ^{m}C G C may be methylated at the sites indicated. Federoff and Brown (1977) have analyzed the digestion products of the DNA coding for the 5 S RNA. A segment of the DNA cloned into the *E. coli* plasmid, PSC 101, was isolated and cleaved with *Hae*III, *Hpa*II, and *Hha*I. Two *Hae*III sites are in the 5 S coding sequence, one toward the left end and one toward the right of the 120 nucleotides. An *Hpa*II site is near the middle of the sequence. In the 5 S RNA-coding region of the DNA from the frog, the *Hpa*II site, as well as the *Hha*I site, is protected. One of the cytosines in each *Hpa*II and *Hha*I site is methylated, but the *Hae*III sites are not. The methylated cytosine in each instance is known to be in the ^{m}CpG doublet. This fits with the data from the mouse and HeLa cells to indicate that some CCGG and GCGC sites are methylated in cells of vertebrates.

B. DISTRIBUTION OF 5-METHYLCYTOSINE IN CHROMATIN

A recent report indicates that there may be a correlation of methylation patterns and attachment of the histones to the DNA. Razin and Cedar (1977) reported that fractions of DNA protected by the histones in the nucleosomes or "core particles" of chromatin has a significantly higher level of methyl groups than the fraction more available to micrococcal nuclease. Nuclei from calf thymus and chicken tissues were obtained by homogenizing tissues and washing with sucrose–Triton buffers according to a procedure described by Axel *et al.* (1973). The nuclei were washed in buffer and digested in micrococcal nuclease until about 50% of the DNA was solubilized. The ratio of 5-methylcytosine (M^5Cyt) to total cytosine was then determined in nuclei from several tissues of the chicken. All showed the same trend with somewhat more M^5Cyt in the protected or "covered" fraction; the ratio of M^5Cyt in covered DNA to cytosine in the total DNA varied from 1.3 to 1.7 in the different tissues. The variations among tissues might be significant, but the authors did not try to evaluate that variation. The interesting variation was the consistent differences between the covered and open fractions of DNA. Several determinations on chicken erythrocytes in which both the covered and open fractions were measured also support the idea that the covered fraction has a higher level of methylation. Two methods of collecting the open fraction was used. In one the soluble DNA fragments were collected after digestion and in another procedure the open DNA was reacted with poly-D-lysine and then the chromatin was digested with Pronase to which the poly-D-lysine is resistant. The DNA was then digested with nuclease to which the poly-D-lysine bound material was resistant. Separation of the poly-D-lysine from the residual fraction yielded a fraction similar to the soluble fragments which also had a low level of methylation, 2.4 ± 0.2 mole% of M^5Cyt. Covered to total DNA had a ratio of 4.5 ± 0.2 to 3.2 ± 0.1 or 1.4.

When DNA and proteins were dissociated with salt and urea and then reconstituted by gradual dialysis to a low salt concentration, the open and protected fractions both had the same level of methylation, 3.3 mole%. Although the reconstituted chromatin has some structural and biological properties of the native chromatin, the randomization of the major portion of chromatin proteins destroys the differential distribution of methylated cytosines.

The authors noted that the percentage of M^5Cyt was lower in their determinations by more than 30% when compared to determinations by previous workers. They point out that satellite DNA has been reported to have a high level of M^5Cyt and that preferential loss of these fractions in precipitating DNA could have affected the ratios. However, the failure to find a difference in reconstituted chromatin digested with micrococcal nuclease would indicate that such selective precipitation could not account for differences observed. The authors did note that a commercially prepared calf thymus DNA had a M^5Cyt/Cyt molar ratio of 0.065 consistent with previous reports. After ethanol precipitation the ratio was only 0.041. Such variations make reports of variations of levels of M^5Cyt among different tissues in an animal, plant, or different cell lines of a species in culture subject to the same variable unless comparable methods of DNA isolation and purification were used.

IV. PATTERNS OF METHYLATION OF DNA AND THE CELL DIVISION CYCLE

A number of studies have shown that methylation usually follows DNA replication very closely, and that methylation does not occur unless DNA has been recently replicated. If DNA is methylated on a palindromic sequence, such as —N C ^{m}C G G N—, two half-methylated sequences will be produced during replication as

$$5' \cdots N C^m C \ G G N \cdots 3'$$
$$\cdots N G G^m C \ C N \cdots$$

$$5' \cdots N C^m C G G N \cdots 3' \qquad\qquad 5' \cdots N C C \ G G N \cdots 3'$$
$$\cdots N G G C C N \cdots \qquad\qquad\qquad \cdots N G G^m C \ C N \cdots$$

The half-methylated sites are excellent substrates for the methylases and are usually methylated within a few minutes after replication. Lark (1968) and Billens (1968) have studied the methylation in bacteria (*E. coli*). The modification occurs quickly, and there is no lag for a part of the methylation into the next round of replication. However, in the slime mold, *Physarum polycephalum*, there appears to be a considerable lag between synthesis and full methylation. Evans *et al.* (1973) used [^{14}C]deoxycytidine to label the DNA in one cycle in the slime mold where there is a natural synchrony of cell division. They followed methylation by isolating DNA at intervals, and after hydrolysis of the DNA to free bases with 70% $HClO_4$, the components were separated by paper chromatography. When the parental strands were separated from the new progeny chains before hydrolysis, the new strands contained about two times as much radioactivity as the parental chains. Moreover, the labeling of DNA-cytosine with [methyl-^{3}H]-S-adenosylmethionine was stable through at least two cycles, which probably means that there is no significant turnover of the methyl groups except during repair of DNA. Therefore, the methylation of parental chains very likely represents a long delay in part of the methylation following replication. One possible complication in these experiments was the use of the density label, BrdUrd, to allow separation of the strands synthesized in the different cycles. BrdUrd in the DNA could interfere with methylation patterns or increase the repair, especially if cells were exposed to light at any time. However, the incorporation into parental strands was measured in the second S phase when BrdUrd was being incorporated for the first time. At that time the parental chains were methylated, even though no BrdUrd had been previously incorporated in these chains. Any effect on these DNA chains would have to be through a peculiar interaction of the enzymes with the hybrid DNA. Singer *et al.* (1978) have recently reported that they did not find a measurable change in methylation, even with 95% substitution of BrdUrd in rat hepatoma cells in culture. This may be taken as a good indication that one may be able to use BrdUrd in the design of experiments involving methylation. However, Drahovsky *et al.* (1976) mentioned that they had observed an effect of incorporated BrdUrd on methylation and promised a detailed report. Some caution is necessary, especially with *in vivo* experiments where differentiation is occurring since BrdUrd incorporated into DNA is known to prevent differentiation of myoblasts as well as several other types of differentiating cells (Bischoff and Holtzer, 1970; Weintraub *et al.*, 1972).

In bacteria there is no such delay in methylation beyond the replication cycle, but a delay can be induced in appropriate mutant strains. *Escherichia coli* T^-M^- (mutant requiring thymine and methionine) could be grown without methionine for one full cycle if ethionine was supplied to allow cells to make enough protein to reinitiate DNA replication. However, when the replication fork reached the region of the chromosome where methylation was interrupted for lack of S-adenosylmethionine, replication stopped (Lark, 1968). Resupplying methionine resulted in rapid methylation of the old chains and continuation of DNA replication after a short lag period.

Fibroblasts of mammals in culture do not appear to have a delay in methylation which extends beyond the G_2 stage following the S phase in which synthesis occurs (Adams, 1971). Cells were synchronized by subculturing from stationary phase into a medium with one-tenth the normal concentration of methionine and 20 mM sodium formate to prevent equilibration of the methionine methyl groups with the C_1 unit pool in the cells. After 6 hours of growth, and presumably before S phase, aminopterin, adenosine, glycine, and deoxycytidine were added along with L-[methyl-^{3}H]methionine to label the methylatable sites in DNA, and [^{32}P]orthophosphate was added to label all new DNA. Aminopterin, a folic acid analogue, inhibits the synthesis of purines and the methylation of deoxyuridine monophosphate to form thymidylate. The cell cycle stops at early S phase because of the thymidylate deficiency when DNA replication begins. The additional supplements are required because of other pathways requiring the folic acid derivative, tetrahydrofolate. After 16 hours the block was released by adding thymidine. The cells were then pulse labeled with

bromodeoxyuridine for 1 hour at Time 0 and at several intervals thereafter. Incorporation of [³H]methyl groups into DNA during the pulse
was determined by separating the density hybrid DNA from the remainder in a CsCl gradient and determining the ratio of [³H]methyl
groups to the ³²P in the bromodeoxyuridine-containing DNA. A
change in the ratio of the two labels indicates a change in methylation
of the DNA. The DNA replicated during the first hour after release
was fully methylated by 6 hours and did not change over the next 18
hours. However, the rise in methylation after synthesis was low (less
than 5%), and the initial level of methylation was high. On the other
hand, DNA replicated during 4–5 hours after release had the lowest
level of methylation immediately after the pulse, less than one-half
that of the early replicated DNA, but it gained relatively more in the
next 3 hours. Part of this was probably methylated in the G_2 phase.
The level of methylation of this late replicating DNA never rose to
that of the early replicated fraction.

Studies of mammalian cells have generally shown no methylation
during stationary phase. For example, Adams (1971) showed that
horse lymphocytes do not incorporate the methyl group from methionine into their DNA until DNA replication begins after stimulation
with phytohemagglutinin. Methylation can be induced by treating
nonproliferating cells with agents which induce repair synthesis.
Drahovsky et al. (1976) treated human lymphocytes from peripheral
blood with nitrogen mustard and were able to detect incorporation
of [³H]thymidine and ³H from L-[methyl-³H]methionine, but not
[¹⁴C]thymidine, presumably because the amount of synthesis was so
little that the low specific activity ¹⁴C could not be counted. The incorporation was measured over a period of 24 hours with both nitrogen
mustard-treated cells and control cells. These were compared with
leucoagglutinin-stimulated cells which were allowed to incorporate
[¹⁴C]thymidine and to methylate the DNA with [³H]methionine for 96
hours, i.e., time for one or more division cycles. The methylation in
relation to repair synthesis could be measured by comparing ³H in
5-methylcytosine to ³H in thymidine in the DNA of cells which did not
reach S phase, while in the stimulated cells the ratio of ¹⁴C to ³H gave
the ratio of synthesis to methylation. About 30% of the radioactivity
found in the DNA was in rapidly reassociating DNA (C_0t of 0.1) when
fractionated by hydroxylapatite chromatography. Comparing methylation of unique DNA and the repetitive fraction (30%) during repair
synthesis indicated that the levels were the same. However, in mito-
gen-stimulated lympocytes, which are undergoing cell cycle replication, the ratios indicated more extensive methylation of the repetitive

sequences. A comparison of methylation in rat hepatoma cells has also shown a difference in methylation of the highly repetitive DNA (3% of cytosines) compared to unique and moderately repetitive DNA (2.3%) (Singer *et al.*, 1977). The higher methylation of the repetitive sequences during the cell cycle replication compared to repair synthesis could involve some methylation process associated with the mitotic cycle or the G_2 stage which is different from that which is proposed to involve the half-methylated palindromic sequences [see the scheme proposed by Holliday and Pugh (1975) in Section VII].

V. Changes in Methylation during Differentiation and Development

Any indication that modifications of the DNA occur during the early embryonic stages in development provides an attractive idea to explain differentiation. However, to prove that there is a causal relationship between changes in methylation of DNA and regulatory changes in cells is quite a different matter. Sperm cells are reported to have a lower level of methylation than cells such as those of liver, spleen, or kidney, according to Vanyushin *et al.* (1970). For example, they reported that bull sperm has 0.75 mole of 5-methylcytosine per 100 moles of DNA bases (0.75 mole%), while other tissues have higher levels, 1.34 for kidney, 1.07–1.66 for spleen, and 1.40 for liver. Sheep sperm also has 0.76 mole%, while liver has 1.13, kidney 1.15, and spleen 1.07–1.15. For these animals the time during which the changes occur is unknown, but in sea urchin there may be changes in early cleavage stages. Adams (1973) has reported that BrdUrd incorporated into the DNA of embryos of *Paracentrotus lividus* for the first 3 hours after fertilization along with [^{14}C]thymidine as a marker for the heavy chain allows one to follow methylation of those chains over several cell cycles. The BrdUrd hybrid DNA was slowly methylated over a period up to 60 hours postfertilization. Since several cell division cycles occur in the interval, the lag indicates either that some sites would lose their methylated chains by segregation before the late methylation or that completely new sites are being methylated, perhaps by the initiation-type methylases. This experiment has the disadvantage that one is following the methylation of a BrdUrd-containing chain which may be undergoing an abnormal amount of repair or have some other interference with normal methylation. However, no such BrdUrd-induced delay was seen in similar experiments with mouse fibroblasts (Adams, 1971).

On the other hand, our recent measurements indicate that there is no detectable change in the levels of methylation of cytosines from sperm through various embryonic stages to early pluteus in the sea urchin (*Lytechinus variegatus*). Table II shows that sperm and cells from embryos up to 37.5 hours after fertilization had no detectable change in the ratio of cytosine to 5-methylcytosine (Pollock *et al.*, 1978). The ratios of all bases are included in Table II to indicate that the DNA recovery was probably good and that no significant contamination by cytoplasmic DNA occurred. A comparison of DNA from sperm and liver cells of a fish (*Mugil cephalus*, the black mullet) did not reveal any differences in the level of methylation of cytosine in the different tissues of a species. The ratios were obtained by a technique of high-pressure liquid column chromatography on Aminex

TABLE II

BASE RATIOS AND 5-METHYLCYTOSINE IN DNA OF SEA URCHIN AND MULLET

	Sea urchin (*Lytechinus variegatus*)[a]			
	Sperm (2)	Embryos (1) 4 hours	Embryos (3) 5 hours	Embryos (4) 6 hours
M⁵Cyt	0.85 ± 0.05	0.90	0.97 ± 0.13	0.87 ± 0.07
C	18.7 ± 1.7	18.2	16.8 ± 0.5	17.1 ± 0.6
G	16.5 ± 0.5	15.2	17.2 ± 0.2	16.8 ± 0.5
A	31.9 ± 0.9	31.9	32.7 ± 0.1	32.6 ± 0.2
T	32.1 ± 0.2	33.8	32.7 ± 0.8	32.8 ± 0.3
	Embryos (2) 7 hours	Embryos (2) 8 hours	Embryos (2) 13½ hours	Embryos (2) 37½ hours
M⁵Cyt	0.80 ± 0.1	0.80 ± 0.00	0.89 ± 0.00	0.81 ± 0.01
C	18.3 ± 3.3	16.8 ± 0.1	16.6 ± 0.1	16.9 ± 0.1
G	15.5 ± 0.5	17.0 ± 0.1	17.1 ± 0.1	15.4 ± 1.0
A	31.5 ± 0.1	32.7 ± 0.5	32.5 ± 0.1	32.7 ± 0.0
T	34.1 ± 2.7	32.9 ± 0.4	33.0 ± 0.2	34.2 ± 0.9

	Mullet (*Mugil cephalus*)[a]	
	Sperm (4)	Liver (2)
M⁵Cyt	1.87 ± 0.13	1.80 ± 1.0
Cyt	18.5 ± 0.3	19.0 ± 0.1
Gua	19.1 ± 0.8	18.9 ± 0.8
Ade	29.6 ± 1.2	29.8 ± 0.4
Thy	30.8 ± 0.6	30.6 ± 1.3

[a] Number of determinations is given in parentheses; the ± numbers show the range of values, i.e., the largest deviations from the means.

A-10 (Bio-Rad, Inc.) after hydrolysis of the lyophilized DNA to free bases with formic acid sealed under nitrogen. This method should be more accurate than the thin-layer chromatographic method used in most previous determinations, because of the problems of quantitative recovery of the bases from thin-layer plates. In view of these results, perhaps one should be cautious in attributing significance to reports of variations in levels of methylation between tissues of the same species and between different stages of development until more data are available.

The changes of methylation during transformation by virus or other agents varies considerably. The most striking change occurred in the transformation of BHK 21 cells by polyoma virus (Rubery and Newton, 1973). The transformed cells had almost two times as much 5-methylcytosine as the original cell line. The difference in methylation was detected by labeling with [^{14}C]cytidine and then at intervals by determining the ratio of labeled 5-methylcytosine and 5-methylcytosine + cytosine in the DNA. A search for some difference in methylation patterns over the cell cycle, or in the way 5-methylcytosine was produced, did not reveal any differences. There was a small but perhaps significant drop in the level of methylation 20 hours after removal of cells from the labeled precursor. This drop could be due to a delay in methylation of a significant part of the DNA cytosines following synthesis in parasynchronously dividing cells. For example, if a significant amount of the methylation occurred in late S and G_2, the cells at 20 hours could have been mostly in S phase and still incorporating [^{14}C]cytosine from precursors released by the turnover of labeled RNA. Nevertheless, the interesting finding was that a very significant increase in methylated bases occurred during transformation. This change was maintained by the transformed cells as if it were now a permanent feature of the genetic system. Whether there was a causal relationship between transformation and the methylation is not indicated. Certainly some transformed cell lines have smaller changes. Kappler (1971) has examined several mouse cell lines, but failed to include comparisons with normal cells. Two teratomas were reported to have 4.58 ± 0.18 and 4.02 ± 0.14% 5-methylcytosine based on four determinations in each instance (standard deviations shown), while a melanoma and a neuroblastoma had 2.97 ± 0.02 and 3.38 ± 0.02% 5-methylcytosine, respectively. Vanyushin et al. (1970) reported 1.1 mole% for mouse liver and 42.2% G + C + m^{5}Cyt, which indicates about 5.2% of the cytosines methylated for normal cells. Kappler's (1971) examination of normal tissues of chick embryos showed variations from 3.60 ± 0.04 for the chorioallantoic membrane to 3.65 ± 0.09

for kidney and 4.19 ± 0.02 for brain. These differences, which are all based on the conversion of incorporated [^{14}C]deoxycytidine into the 5-methyl derivative, are comparable to several other comparisons of different tissues of animals. The missing element, of course, is the knowledge of which sequences are methylated in the different tissues. Striking differences could exist despite the small changes in overall level of methylation. However, differences that could be important in differentiation would be expected to be very minor changes affecting as little as 1% or even less of the total DNA. Such changes will be very difficult to determine and probably can only be studied in connection with a known sequence around certain genes. A revealing correlation might be found by studying sequences to the 5'-side of the initiation codon for a particular protein, globin, for example, in various tissues, one in a cell where the gene is functional at some stage and in another where it never functions during the life cycle. Are there differences in methylation somewhere in the regulatory sequences of such genes—differences which are set up in the early embryo and propagated by the maintainence type methylase?

Since not all sequences, such as CCGG, are methylated in a cell, there must be some discriminating enzymic differences. It could be due to a pattern set up by an early acting initiation-type methylase with specificity for the flanking base pairs as well as the central CCGG cluster, or possibly by a maintenance-type activity with specificity beyond the known central palindromic sequence.

VI. Methylating Enzymes Isolated from Cells

DNA methylating enzymes were originally isolated from bacteria by Gold and Hurwitz (1963) and by Srinivasan and Borek (1964). S-Adenosyl-L-methionine was shown to be the methyl donor and the only cofactor required to methylate DNA *in vitro*. Since then similar enzyme activities have been isolated from a variety of eukaryotic cells. The sites methylated have been deduced in bacteria in a number of instances, as shown in Table I. This has been possible because of the isolation of the corresponding endonuclease which is inhibited by the methylation of one or both chains of the specific sequence susceptible to the endonuclease. Since no specific endonucleases have been isolated from eukaryotic cells, the identities of the specific sequences methylated are largely unknown. However, the evidence from what is known of the action of methylases in eukaryotes both *in vivo* and *in vitro* would indicate that the methylases have a similar specificity.

Morris and co-workers (Morris and Pih, 1971, Kalousek and Morris,

1969) have obtained methylases from rat spleen and rat liver. The enzymes were partially purified and characterized with respect to substrate. In general, the methylases catalyze the addition of very few methyl groups to homologous DNA, but will use both native or denatured DNA from almost any source as a substrate if it is not fully methylated. The homologous DNA is presumably nearly fully methylated at the sites for which its methylases are specific. Some of the more recent studies have further purified the methylases, but none can be said to be highly purified.

Roy and Weissbach (1975) purified a protein fraction 270-fold from HeLa cells by chromatography on DEAE-cellulose, phosphocellulose, and hydroxylapatite. The product of the reaction with S-adenosylmethionine as the donor was exclusively 5-methylcytosine in DNA. The enzyme was active on either single chains or double-stranded native DNA. A fully methylated substrate (*M. luteus* DNA) had one 5-methylcytosine per 116 bases on native DNA and one per 430 bases on denatured DNA. When denatured HeLa cell DNA was used as substrate, the ratio was the same as for native DNA versus the single strands, but in each assay this was only 9% of the amount added to *M. luteus* DNA. The specific sites methylated were unknown except that an analysis of the dinucleotides indicated G ^{m}C (60%) and the ^{m}C G (40%), but the CC neighbors were always C ^{m}C.

Turnbull and Adams (1976) purified a protein fraction from mouse acites cells 405-fold and tested its methylating activities. This fraction also methylated cytosines in DNA to form 5-methylcytosine, and 90% of the incorporated radioactivity from S-adenosyl-[methyl-^{3}H]methionine could be accounted for in that base. No other specific moeity was identified with the labeled methyl groups. The enzyme was similar to other preparations, but had one peculiar feature not previously reported. Nearly 100% of the radioactivity incorporated from the S-adenosyl-[methyl-^{3}H]methionine into native *E. coli* DNA *in vitro* was made acid soluble by digestion of the product of the reaction with a single-stranded nuclease from *Neurospora crassa*. This was not interpreted by the authors, but it could mean either that the DNA has a few single-stranded regions which serve as the substrate for the enzyme, or the activity digests away one chain as it methylates the other.

One of the most interesting enzymes was found in a preparation which has some of the properties one would look for in a maintenance methylase. Sneider *et al.* (1975) obtained the activity in a 980-fold purification of a protein fraction from Novikoff rat hepatoma. It was 85-fold more active on half-methylated Novikoff cell DNA than on the fully methylated DNA isolated from the same source. However, like most, if not all, preparations, it was more active on denatured than native heterologous DNAs. It methylated denatured $E.$ $coli$ DNA several times as much as the native material. Even denatured homologous DNA was a little better than the native molecule unless the native molecules were the half-methylated type.

Methylating enzymes from bacterial cells should prove very useful in identifying methylated sequences in eukaryotes. For example, the methylases listed in Table I along with corresponding endonucleases could be used to identify specific sequences in eukaryotes. Since half-methylated palindromic sites are reported to be fully protected from the corresponding endonucleases, one could search for such sites with the two activities. If one obtained half-methylated DNA by growing cells with very limited amounts of methionine or with ethionine substituted for methionine, the only sites protected from the endonuclease and yet available for methylation would be the half-methylated ones. By digesting first with a specific endonuclease, all unmethylated sites of a specific sequence would be destroyed. The corresponding methylase would then methylate only the half-methylated sites. The fully methylated sites would still be unrevealed, but if one could produce a DNA with a large percentage of the sites half-methylated, the specific sequences corresponding to a bacterial methylase–endonuclease pair could be identified in the eukaryotic DNA.

VII. Inactivation or Activation of the X Chromosome

Riggs (1975) has proposed that the modification of one X chromosome, which is characteristic of mammalian female cells, somehow involves methylation of cytosines in DNA. The modification of one X chromsome is an event that takes place every time an embryo develops through the blastula stage. Once a cell is set on a pathway with one of its X chromosomes activated or potentially activated, that cell and all of its descendents are maintained that way as if the trait had its basis in some change in the genetic code. DNA methylation has that potential if the enzymes (methylases) operate as we assume they might. A particular pattern is initiated at some stage in development, and thereafter a maintenance methylase keeps those sites and others with a common central palindromic sequence fully methylated on the basis of the specificity provided by the half-methylated sites after each replication.

One difficulty is in imagining a mechanism which would allow one chromosome in the nucleus or one part of a chromosome in some species to become modified to the exclusion of its homologue. Since not only replication pattern but also gene expression of many loci is involved, the problem is even more complicated. Probably methylation is the best model we have, but in this case it may be combined with other initiating events to assure regularity of activation or deactivation of a single homologue. Unfortunately there is no evidence that the active or inactive X chromosome has a different pattern of methylation from any other chromosome. As has been pointed out previously, activation of one chromosome makes more sense than deactivation because one and only one X chromosome is active whether the cell has one, two, or even five X chromosomes (Grumbach *et al.*, 1963). If one of these is activated by a mechanism that shuts off further events of the same type in the cell, one does not have to explain how a cell can count the number of X chromosomes to inactivate in those rare events where the cell has more than two X chromosomes and has no evolutionary history in which it could evolve a mechanism for handling such events with multiple chromosomes. The only model known for exclusion of this type is the episomal model in bacteria, where integration of one episome tends to exclude others of the same type from the cell. Riggs (1975) argues that the episome model has the defect of being rather unappealing as a general model of differentiation. I agree, but it should be pointed out that X chromosome activation is a late evolving special mechanism, and is probably not part of the general mechanism in its initiation process. Methylation could be the general model, but a single event, or one-hit phenomenon, would be more likely for the initiation of the methylation pattern. The details of how such a system would operate are too speculative to elaborate upon at this time. The single event, perhaps an integration, is conceivable, but how that event could lead to the methylation of many potential sites for DNA replication or transcription without affecting the homologue is a more difficult problem.

VIII. Biological Clocks in Cell Division

One of the most elusive concepts for development involves the control of size or order of progression in the formation of an organ which might be determined in part by the number of cell divisions required to form the structure or, more specifically, the number of divisions before another step in development is initiated. Holliday and Pugh (1975) have proposed a novel scheme that involves hypothetical methylases which would be different from the maintenance type we have considered up to now. Such clocks could also determine the aging of cells by a limitation on the number of divisions which cells appear to be capable of until transformed (i.e., until they lose or escape the limitation placed by such a clock) and thereafter could continue division forever.

The idea is that a cell may utilize a segment of DNA with repeating sequences of seven or ten nucleotides, a type of DNA found in many satellites or highly repetitive sequences. Such sequences do have additional variations in sequences superimposed on the short repeats as revealed by digestion with a restriction endonuclease. Sometimes a DNA composed of short repeating sequences yields longer segments, which is due to this larger repeat in sequences. Such a DNA could be a prototype for the clock mentioned above.

There are two methylases in this imaginary scheme, E1 initiates the clock by recognizing a starter sequence (Fig. 1) and methylating one chain of the duplex. The next enzyme in the series, E2, then adds a methyl group to a cytosine in the other chain in the complementary segment of the repeat. It also adds two methyl groups in the next segment of the repeat downstream, so to speak. Both sites are now fully methylated, and the methylation stops until after the next replication occurs. After replication there is a site on each daughter chromosome with only one chain methylated in the last repeat unit modified. The E2 recognizes the site and methylates the complementary chain and again both chains in the next repeat downstream. Since there is no longer a half-methylated site, the methylation stops until the next replication has occurred. The process is repeated each division until a differential segment downstream from the last methylatable site has been altered by a pair of methyl groups adjacent to it. Now another regulatory molecule or RNA polymerase fits the differential segment and binds to the site in a way not possible before the methyl groups appeared. We may suppose it represents a transcribable locus now or initiates some specific event which was impossible before modification.

Evolution could utilize such a scheme to count cell divisions for many processes with the same or similar repeats as long as the differential segment or sequences were specific for the particular process which was being clocked. The model would be useful for programming morphological patterns in development that depend on sequences of cell divisions. It would be a type of methylation which would not cause a turnover of methyl groups, but neither would the methylation pattern necessarily be maintained upstream. Only the last methyl group downstream in the repeat, which is half-methylated, would serve as a signal for the E2 and the subsequent methylation. If many such sequences were being used simultaneously, the level of methylation might rise enough to be detectable, but otherwise these

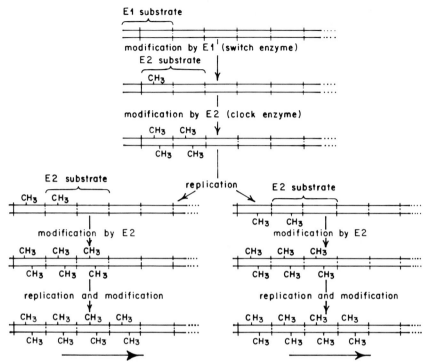

FIG. 1. A mechanism for counting cell divisions based on the methylation of palindromic controlling sequences. The first enzyme, E1, switches on the clock by recognizing a starter sequence, at the extreme left, which is adjacent to the first of the repeated sequences of the clock. One strand of this sequence is methylated by E1, and this provides a substrate for E2, which inserts three more methyl groups in the first two controlling sequences. E2 does not act further once both strands are modified. However, after replication new substrates of E2 are formed, allowing the next sequence to be methylated. All the sequences behind the "growing point" may become modified, but this does not affect the clock mechanism. (From R. Holliday and J. E. Pugh, 1975, *Science* **187**, 226–332. Copyright 1975 by the American Association for the Advancement of Science.)

events would be very difficult to detect. The highly repeated DNA is indeed methylated to a higher level, but the pattern of those methyl groups would have to be known to get any further clues about its significance.

IX. Hypothesis on the Control of Replication by Methylation

It is well known that chromosomes of eukaryotes replicate one time each cell cycle. Mistakes are rare or nonexistent, and one assumes that at the level of the DNA helix the same restriction applies. Yet chromosomes replicate by initiation from many sites, at which chain growth proceeds in both directions. The chains growing in the $5' \rightarrow 3'$ direction could be polymerized continuously, but the other complementary chain would have to grow $3' \rightarrow 5'$ if it moved toward the fork. It is likely that it moves $5' \rightarrow 3'$ in short segments from the fork toward the origin, as has been proposed originally by Okazaki *et al.* (1968) for bacteria.

In bacteria, replication proceeds from a single origin bidirectionally until the two advancing forks meet around the circular chromosome. To reinitiate, the replication begins at the origin and proceeds around the circle a second time. Evidence indicates that the origin may operate a second time in cells growing in rich medium before the first cycle is complete. However, in chromosomes of eukaryotes, the reinitiation does not occur until all segments of all chromosomes are complete. This process must be carefully regulated despite the presence of thousands of initiation sites in each cell. Rarely, initiation is resumed in G_2, but it normally does not occur until the cell has divided and passed through the G_1 stage. When reinitiation occurs in G_2, it affects the whole complement of chromosomes, and the process called endoreduplication results in chromosomes with four chromatids, rather than the usual two, reaching metaphase.

Recent evidence from studies of Chinese hamster ovary cells which were highly synchronized indicate that there are many more potential origins in cells than we had previously suspected for cells from the adult animal (Taylor and Hozier, 1976; Taylor, 1977). By holding cells, synchronized by mitotic selection, at the G_1-S phase interface by completely depriving them of thymidine, it was possible to show that many more of the potential origins in a limited part of the DNA could be activated, or made ready for use when replication was initiated by adding [³H]thymidine to the medium. Fiber autoradiographs, which show the position of individual labeled loops which have been stretched into single fibrils by flow, indicate that many more origins are used in the limited amount of DNA. The origins, estimated by measuring center-to-center distance between labeled segments, were from 4 to over 100 μm apart. When these distances were plotted in a frequency diagram, the most common distances between origins were found at intervals of 4, 8, 12, 16, and perhaps 20 μm. This observation led to the hypothesis that potential sites in these fibroblasts, and probably all cells of the hamster, exist every 4 μm, the common integral for each interval measured. Not all of the potential sites could be made active by the treatment, but the number actually used after holding cells for 20 hours from division, i.e., blocked at the G_1–S phase boundary for 12–14 hours, was about one in three compared to one in 15–20 for cells not interrupted in the cycle. The activated origins accumulated linearly from 6 to 20 hours, but only a small fraction of the total DNA has its origins activated. Such cells proceed through S phase after such a block with only a change in rate, 4 hours instead of the usual 6 hours (Carnevali and Mariotti, 1977).

Not only the used origins but also the unused sites that the fork passes over must be modified so that initiation is excluded during the S phase involved. One might suppose that a protein remains at the site and prevents reinitiation, but it is necessary to imagine a protein with a binding constant that is extremely high and to assume there is enough to cover all sites essentially 100% of the time. All of these criteria are probably possible, but differences in methylation provide a fail-safe mechanism that is extremely simple. In a sense, it has a counterpart in the observation that *E. coli* DNA can replicate one round without methylation of its DNA. However, the process stops when the fork reaches the region where the DNA is half-methylated (Lark, 1968).

For eukaryotes, let us suppose that the replication complex binds at a fairly large palindromic sequence which includes one or more small palindromic sites on each side of the larger palindrome.

If methylation of these sites is delayed after replication until G_2, the site is modified automatically by the fork moving across the site. No other assumptions are necessary except to suppose that the replication complex cannot bind to a half-methylated site. It requires a fully methylated DNA. After S phase is completed, a signal of a yet unknown type causes the activation or synthesis of the specific methylase for these sequences on each side of every potential initiation site. After methylation in G_2, the DNA is available for reinitiation and another cycle if the proper enzymes are available. The operation of such a cycle is purely hypothetical but, as indicated in Section IV of this chapter, there is a delay of a portion of methylation until the G_2 stage in mammalian cells in culture as well as all other cells of eukaryotes studied to date.

X. OTHER MODELS FOR REGULATORY OR DEVELOPMENTAL ROLES OF METHYLATION

As pointed out in the Introduction, DNA methylation has the potential of being a long-term modification of DNA which could affect the affinity of a protein for a certain sequence, perhaps regulatory protein or possibly a "swivelase" or relaxation complex. If we made the assumption that some such activity is necessary not only at replication origins but also at origins for transcription, then the methylation could inhibit or promote transcription. The higher level of methylation of highly repetitive DNA which is not transcribed would be consistent with the methylation being inhibitory for the binding of a positive regulator, whether a "swivelase" or some other regulatory molecule. Without the swivelase, the DNA would remain closed to transcription and the cell could have a very inexpensive way to regulate large blocks of genes. If the regulatory sequences of a gene family were very similar and carried a specific methylatable site, a single initiation-type methylase could suppress the entire family at some stage in development. The suppression could be maintained by the maintenance-type methylation assumed to operate after replication.

Some methylated sites might have a significance immediately, but others may not affect anything one could measure for many cell generations after it was initiated. This type of change is what Hadorn (1965) called determination. Differentiation might be delayed many cell generations until another specific signal (a hormone, for example) reached a certain critical level in the cell's environment. The differentiation of the cells in a imaginal disk of a insect is a case in point.

Perhaps the most difficult question to approach with a feasible model is the one of how certain cells are originally methylated or modified according to one pattern, while others in the immediate vicinity are methylated by a different pattern or simply not modified in the same way as the adjacent cells. The question certainly cannot be answered yet with proved models, but some recent studies on membrane phenomena, such as patching and capping of cell surface receptors, give a clue (Edelman, 1976). Let us take a simple example of the two-layered gastrula with a few free floating cells which will become the mesoderm. Even if all such cells have a promethylase induced, only a few may be capable of activating it by a mechanism that requires clustering of receptors for a protein inducer which is secreted into the fluid between the two layers of cells, ectoderm and endoderm. The cells could all be the same at the stage in question, except that some are free in the fluid layer between the relatively immobilized layers of contact inhibited cells. Those which happen to remain free in the cavity could retain mobile membrane glycoproteins in contrast to their neighbor in the layers. An induced macromolecule (protein) secreted in the fluid would react with all surfaces, but only in those where capping could occur would activation of the protein (methylase) actually occur. Assume the protein is in short supply and as soon as a certain average number of cells have been capped and a methylation pattern set up on the specific sequences which an initiation-type methylase can modify, that particular protein disappears forever in that life cycle. Only when another embryo passes through the same stage will those particular sites be methylatable again by the same mechanism. In that particular individual, the pattern will be maintained by a maintenance methylase with a lower specificity than the original methylase. Any failure or loss of the pattern would be equivalent to a mutation, a somatic mutation which could contribute to aging of the cells and the individual of which it is a part. Such aging would mean loss of regulatory circuits, including those which modulate cell division. These could lead to transformation and the first step in cancer formation.

The modification of tumor cells (teratocarcinoma) in the mouse by recycling them through the blastocyst stage (Mintz and Illmensee, 1975; Illmensee and Mintz, 1976) could involve a modification of this type. Indeed, if cells carry oncogenes, these may consist of integrated and modified virus genomes which are suppressed in the early embryo by methylation of all sites which serve to initiate transcription of many sequences. The constitutive heterochromatin of eukaryotes may represent large clusters of sequences regulated in such a way. Cells may have evolved some such suppression mechanism to inactivate all sequences not specifically protected at some stage in the early embryo. Such regions can replicate but are turned off for transcription. Selective activation could occur by induction of a swivelase that could fit families of the methylated and normally repressed sites. Alternatively, sequences could be permanently activated by the inhibition of the maintenance methylases, specific for those sites during two replication cycles, when one-half the DNA molecules would have completely lost the control imposed by methylation.

A whole new class of regulatory mutations could be attributed to loss of methylation, perhaps by semiconservative replication and repair replication going on near the same locus, if the methylase should be a little slow in acting. If repair occurs downstream from a replication fork and the fork passes before the methylase arrives, one of the newly replicated molecules will have lost its methylcytosine and will have no signal for the maintenance enzyme to restore it on either chain.

An interesting paradox arises in relation to mutation and 5-methylcytosine in DNA. If the 5-methylcytosine is deaminated, it becomes thymine and no enzyme could recognize it as an unusual base unless a special enzyme has evolved to cut such missmatched bases in a palindromic sequences which is half-methylated, a very special enzyme indeed. This failure to recognize a deaminated 5-methylcytosine (thymine) as an odd base could make these sites hot spots for mutation. It is perhaps interesting that the highly radiation-resistant species, *Micrococcus radiodurans*, is reported to be without methylated DNA.

ADDENDUM

Another interesting and potentially useful combination in which one class II restriction endonuclease, *Hpa*II, cleaves DNA at only unmethylated sites, whereas an isoschizomer of the enzyme, from another species of bacteria (*Msp*I reported to be available from New England BioLabs) cleaves either the methylated or unmethylated sites, has been reported by Waalwijk and Flavell (1978). *Hpa*II cleaves CCGG if the second cytosine is unmethylated, but the isoschizomer was reported to cleave the sites with or without the methylated base.

A role for 6-methyladenine has been suggested in a paper by Glickman *et al.* (1978). *Escherichia coli* strains carrying the *dam*-3 and *dam*-4 mutations resulting in reduced levels of 6-methyladenine in the DNA have an increased sensitivity to mutagenesis by base analogues compared to the *dam*⁺ strains. Mutation by ethylmethanesulfonate was also increased by undermethylation, but elimination of dimers produced by uv light did not result in higher mutation rates. These data are consistent with a model proposed by Wagner and Meselson (1976), in which half methylation would allow discrimination between strands in repair of mismatched base pairs in newly replicated DNA so that the new chain (unmethylated) would be preferentially "corrected" without leading to base substitution. The authors also cite unpublished data by Radman, Wagner, and Meselson which demonstrated that transfection of *Escherichia coli* with heteroduplex λ DNA prepared from methylated and undermethylated DNA were always corrected to yield the phenotype coded by the methylated strands. These data suggest another role for methylation that the authors suggest is likely to be ubiquitous in all organisms.

REFERENCES

Adams, R. L. P. (1971). *Biochim. Biophys. Acta* **254**, 205–212.
Adams, R. L. P. (1973). *Nature (London), New Biol.* **244**, 27–29.
Arber, W. (1974). *Prog. Nucleic Acid Res. Mol. Biol.* **14**, 1–37.
Axel, R., Cedar, H., and Felsenfeld, G. (1973). *Proc. Natl. Acad. Sci. U.S.A.* **70**, 2029–2032.
Billens, D. (1968). *J. Mol. Biol.* **31**, 477–486.
Bird, A. P. (1978). *J. Mol. Biol.* **118**, 49–60.
Bird, A. P., and Southern, E. M. (1978). *J. Mol. Biol.* **118**, 27–47.

Bischoff, R., and Holtzer, H. (1970). *J. Cell Biol.* **44**, 134–150.
Boyer, H. W., Chow, L. T., Dugaiczyk, A., Hedgepeth, J., and Goodman, H. M. (1973). *Nature (London), New Biol.* **244**, 40–43.
Carnevali, F., and Mariotti, D. (1977). *Chromosoma* **63**, 33–38.
Drahovsky, D., Lacks, I., and Wacker, A. (1976). *Biochim. Biophys. Acta* **447**, 139–143.
Edelman, G. M. (1976). *Science* **192**, 218–226.
Evans, H. H., Evans, T. E., and Littman, S. (1973). *J. Mol. Biol.* **74**, 563–572.
Federoff, N., and Brown, D. D. (1977). Carnegie Inst. Washington, Washington, D.C. (personal communication).
Glickman, B., van der Elsen, P., and Radman, M. (1978). *Mol. Gen. Genet.* **163**, 307–312.
Gold, M., and Hurwitz, J. (1963). *Cold Spring Harbor Symp. Quant. Biol.* **38**, 149–156.
Gorovsky, M. A., Hattman, S., and Pleger, G. L. (1973). *J. Cell Biol.* **56**, 697–701.
Grumbach, M. M., Morishima, A., and Taylor, J. H. (1963). *Proc. Natl. Acad. Sci. U.S.A.* **69**, 3138–3141.
Hadorn, E. (1965). *Brookhaven Symp. Biol.* **18**, 148–161.
Holliday, R., and Pugh, J. E. (1975). *Science* **187**, 226–232.
Illmensee, K., and Mintz, B. (1976). *Proc. Natl. Acad. Sci. U.S.A.* **73**, 549–553.
Kalousek, F., and Morris, N. R. (1969). *J. Biol. Chem.* **244**, 1157–1163.
Kappler, J. W. (1971). *J. Cell. Physiol.* **78**, 33–76.
Klied, D., Humayun, Z., Jeffery, A., and Ptashne, M. (1976). *Proc. Natl. Acad. Sci. U.S.A.* **73**, 293–297.
Lacks, S., and Greenberg, B. (1975). *J. Biol. Chem.* **250**, 4060–4066.
Lark, C. (1968). *J. Mol. Biol.* **31**, 401–414.
Mann, M. B. (1977). Department of Microbiology, Johns Hopkins University. Baltimore, Maryland (personal communication).
Mann, M. B., and Smith, H. O. (1977). *Nucleic Acids Res.* **4**, 4211–4221.
Meselson, M., Yuan, R., and Heywood, J. (1972). *Annu. Rev. Biochem.* **41**, 447–466.
Mintz, B., and Illmensee, K. (1975). *Proc. Natl. Acad. Sci. U.S.A.* **72**, 3585–3589.
Morris, N. R., and Pih, K. D. (1971). *Cancer Res.* **31**, 433–440.
Okazaki, R., Sugimoto, K., Okazaki, T., Imae, Y., and Sugino, A. (1968). *Proc. Natl. Acad. Sci. U.S.A.* **59**, 598–605.
Pollock, J. M., Swihart, M., and Taylor, J. H. (1978). *Nucleic Acids Res.* **5**, 4855–4863.
Razin, A., and Cedar, H. (1977). *Proc. Natl. Acad. Sci. U.S.A.* **74**, 2725–2728.
Riggs, A. D. (1975). *Cytogenet. Cell Genet.* **14**, 9–25.
Roberts, R. J. (1976). *Crit. Rev. Biochem.* **4**, 123–164.
Roy, P. H., and Weissbach, A. (1975). *Nucleic Acids Res.* **2**, 1669–1684.
Rubery, E. D., and Newton, A. A. (1973). *Biochim. Biophys. Acta* **324**, 24–36.
Singer, J., Stellwagen, R. H., Roberts-Ems, J., and Riggs, A. D. (1977). *J. Biol. Chem.* **252**, 5509–5513.
Sneider, T. W., Teague, W. M., and Rogachevsky, L. M. (1975). *Nucleic Acids Res.* **2**, 1685–1700.
Srinivasan, P., and Borek, E. (1964). *Science* **145**, 548–553.
Taylor, J. H. (1977). *Chromosoma* **62**, 291–300.
Taylor, J. H., and Hozier, J. C. (1976). *Chromosoma* **57**, 341–350.
Turnbull, J. F., and Adams, R. L. P. (1976). *Nucleic Acids Res.* **3**, 677–695.
Vanyushin, B. F., and Dobritsa, A. P. (1975). *Biochim. Biophys. Acta* **407**, 61–72.
Vanyushin, B. F., Tkacheva, S. G., and Belozersky, A. N. (1970). *Nature (London)* **225**, 948–949.
Waalwijk, C., and Flavell, R. A. (1978). *Nucleic Acids Res.* **5**, 3231–3236.
Wagner, R., Jr., and Meselson, M. (1976). *Proc. Natl. Acad. Sci. U.S.A.* **73**, 4135–4139.
Weintraub, H., Campbell, G. L., and Holtzer, H. (1972). *J. Mol. Biol.* **70**, 337–350.

Chapter IV

Transcriptional Units in Eukaryotic Chromosomes

ANN L. BEYER, STEVEN L. McKNIGHT, AND OSCAR L. MILLER, JR.

I. INTRODUCTION

The task of unraveling the structural and functional organization of even the smallest eukaryotic genome is formidable. During the last decade, however, several valuable experimental methods have been devised or adapted for probing genomic organization at the molecular level. Techniques such as nucleic acid hybridization, gene isolation, restriction endonuclease mapping, DNA sequencing, and electron microscopy have allowed the accumulation of considerable knowledge regarding the size, molecular topography, and multiplicity of transcription units of known genetic function.

This review of eukaryotic transcription units is divided into three sections which concentrate on details of the better documented examples. The first deals with the general organization of the eukaryotic genome and of its primary transcription products; the second deals with specific multiple-copy genes that are either repeated tandemly or

MOLECULAR GENETICS, PART III

interspersed within genomes; and the third considers specific single-copy genes.

II. GENOME ORGANIZATION AND TRANSCRIPTION

A. DNA SEQUENCE ORGANIZATION

When DNA sequence complexity is analyzed by monitoring the kinetics of renaturation of sheared denatured DNA, it is found that the typical eukaryotic genome consists of three general categories as defined by reiteration frequency (Britten and Kohne, 1968). Sequences which renature most rapidly consist of highly repetitive elements represented some 10^3–10^6 times per haploid genome equivalent. Sequences which renature at an intermediate rate, termed middle repetitive, are reiterated in frequencies of 50–1000 times, and sequences which renature very slowly consist of either single-copy sequences or sequences present in only a very few copies per haploid genome.

The highly repetitive sequence class includes "satellite" DNA which can be distinguished from "main band" DNA by its buoyant density in neutral CsCl gradients. Satellite DNA, which can constitute a significant proportion of total genomic DNA, consists of tandem repeats of very short segments and does not seem to contain protein encoding information (e.g., Gall and Atherton, 1974; Peacock et al., 1977, for review). Of more interest in this review are the nonrepetitive and moderately repetitive DNA sequences. Single-copy DNA sequences include most of the structural genes coding for messenger RNA (mRNA) (Davidson and Britten, 1973), while moderately repetitive DNA families are represented by sequences coding for ribosomal RNA (rRNA), transfer RNA (tRNA), and a limited number of mRNAs. Other classes of moderately repetitive sequences, as discussed below, are interspersed nonrandomly among single copy sequences in most eukaryotes and may serve a regulatory function in gene expression.

Interspersion of middle repetitive and single-copy sequences is demonstrated by reassociating trace quantities of labeled DNA fragments of known length with short unlabeled DNA fragments present in excess (Davidson et al., 1973). Such an experiment is depicted for a typical metazoan, *Xenopus laevis*, in Fig. 1. The samples are incubated to allow the renaturation of highly and moderately repetitive sequences only, and then the amount of DNA present in duplex form is determined. When the shortest tracer fragments are used, the quantity of DNA which reanneals represents the repetitive fraction of the genome and constitutes about 25% of the total. When the fragment

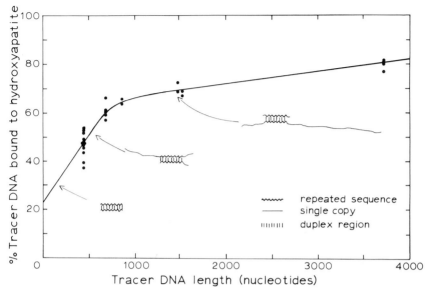

FIG. 1. Demonstration of interspersed repetitive and nonrepetitive sequences in *Xenopus laevis* DNA. Renaturation was carried out between short (450 nucleotides) unlabeled carrier DNA fragments present in excess and tracer-labeled DNA fragments of increasing length as plotted on the abscissa. The DNA fragments for each reaction point were annealed to a middle repetitive C_0t value, (C_0t 50 under these conditions). Duplex formation was detected by hydroxyapatite column binding under conditions in which all nucleic acids with duplex regions will bind. The ordinate shows the fraction of labeled DNA binding to hydroxyapatite as a result of the renaturation of repetitive sequence elements in the labeled DNA strands with the short, excess unlabeled DNA fragments. The ordinate intercept represents the fraction of the DNA which is repetitive sequence DNA, and the increase in binding as fragment length increases from zero is due to nonrepetitive DNA sequences convalently linked to the repetitive sequence elements. Also shown are schematic diagrams of the structures formed by reassociation of the carrier DNA with increasingly longer tracer DNA fragments. (From Davidson, 1977, reproduced with permission.)

length is increased to 3–4 kilobases (kb), it is found that 80% of the labeled DNA renatures. These 3–4 kb molecules must contain at least one moderately repetitive sequence since they reanneal by a middle repetitive C_0t value, but they must be only partially duplexed since only 25% of the DNA is repetitive. Electron microscopic visualization of such reannealed molecules, as shown in Fig. 2, confirms the proposed interspersion pattern and allows an average length of 350 base pairs (bp) for the repetitive sequence regions to be determined (Chamberlin *et al.*, 1975).

Through experiments of this type, it has been found that somewhat

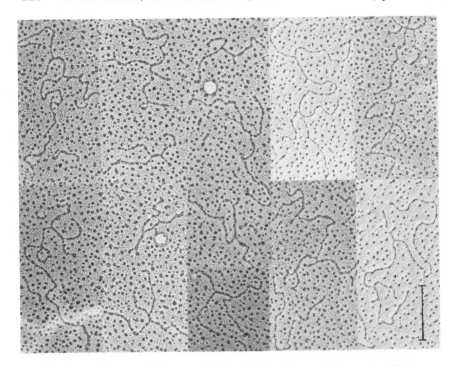

FIG. 2. A montage of electron micrographs showing partially renatured *Xenopus laevis* DNA. The DNA was sheared to a mean length of 2.5 kb, denatured, and renatured to a middle repetitive C_0t value (C_0t 20 under these conditions). Renatured molecules were recovered by hydroxyapatite column chromatography and spread for electron microscopy. Short duplex regions, representing middle repetitive sequences, are attached to four single-stranded tails consisting of nonrepetitive DNA. Analysis of more than 500 such molecules shows the mean length of the renatured segments to be approximately 0.35 kb. Scale bar represents 1.0 kb. (From Davidson, 1977, reprinted with permission.)

more than 50% of the typical animal genome consists of interspersed repetitive and nonrepetitive sequences. The 300–400 bp repetitive sequences are interspersed among single-copy sequences which range from 1000 to several 1000 bp in length (reviewed by Davidson *et al.*, 1975). There is suggestive evidence, as discussed below, that this DNA sequence organization is related to the organization of eukaryotic transcription units.

A different pattern of sequence organization has been found in the *Drosophila melanogaster* and *Apis mellifera* (honeybee) genomes, in which both repetitive and nonrepetitive sequences are greater than 5000 bp in length (Manning *et al.*, 1975; Crain *et al.*, 1976a,b). However, some dipteran genomes have the more typical "*Xenopus* pattern" described above, and it is possible that these exceptions to the general sequence organization pattern reflect an extremely small genome size.

B. EARLY TRANSCRIPTION PRODUCTS:
BIOCHEMICAL CHARACTERIZATION

Analyses of a general class of early transcription products can yield information on the size, complexity, and function of transcriptional units. There has been a considerable amount of research on heterogeneous nuclear RNA (hnRNA) (defined in this review as rapidly labeled nuclear RNA that is not related to rRNA or tRNA) that will be briefly reviewed here [for detailed reviews see Lewin (1975a,b); Greenberg (1975); Perry (1976)]. hnRNA is generally isolated from the nucleus after brief labeling periods, as it incorporates radioactive RNA precursors rapidly and turns over with a 20–30-minute half-life (Brandhorst and McConkey, 1974). It can be distinguished from rRNA precursor molecules (rpRNA), which are also rapidly labeled, by its lower $G + C$ base composition, more heterogeneous size distribution, and relative insensitivity to actinomycin D.

This class of nuclear RNA from a wide variety of metazoans has a heterogeneous size distribution as determined by sucrose gradient sedimentation in conditions that disrupt secondary structure and prevent intermolecular aggregation. Many of the nuclear molecules are in the mRNA size range (0.4–4 kb), but some range upward to 30–40 kb (e.g., Warner *et al.*, 1966; Holmes and Bonner, 1973). Electron microscopic visualization of isolated hnRNA molecules prepared under partially denaturing conditions also indicates that molecules of 15–30 kb exist (Granboulan and Scherrer, 1969). When sucrose gradient distributions of hnRNA mass are replotted to reflect actual molecular size distribution, it is seen that although some molecules are quite large, there is considerable overlap in length between hnRNA and mRNA (see Perry *et al.*, 1976). It was determined, for example, that although 80% of nascent (10-second label) HeLa hnRNA is larger than the average mRNA, 50% is less than 5 kb in length and only 10–20% is derived from large transcription units in the 20–30-kb size range (Derman *et al.*, 1976).

The function of some hnRNA molecules as precursors to cytoplasmic mRNA molecules is supported by the presence of shared chemical modifications in the two populations. These modifications include polyadenylation at the 3′-OH terminus, the "capping" of the 5′-terminus by a guanylate residue in a 5′–5′ triphosphate linkage, and the methylation of two or three 5'-terminal nucleotides and a few internal residues (reviewed by Perry, 1976). These modifications are found in lesser proportions in the hnRNA population, presumably occurring [although not necessarily in the case of poly(A)] on pre-mRNA molecules (Perry *et al.*, 1974, 1975; Herman *et al.*, 1976). hnRNA and mRNA have similar sensitivities to drug inhibition (Penman *et al.*, 1968; Perry and Kelley, 1970) and share common nucleotide sequences (Hough *et al.*, 1975; Perry *et al.*, 1976; Herman *et al.*, 1976). In mouse L cells, for example, about 25% of hnRNA sequences are complementary to mRNA (Perry *et al.*, 1976). In the case of *D. melanogaster* tissue culture cells, kinetic relationships are consistent with a chase of about 40% of hnRNA radioactivity into mRNA (Levis and Penman, 1977).

A considerable proportion of hnRNA turns over rapidly in the nucleus, however, and generally only 2–10%, but up to 40% in some cases, reaches the cytoplasm (Soeiro *et al.*, 1968; Perry *et al.*, 1974; Levis and Penman, 1977). Using techniques of DNA–RNA hybridization, it has been determined that hnRNA is representative in base composition of 5–10 times more of the genome than is mRNA (Galau *et al.*, 1974; Hough *et al.*, 1975; Getz *et al.*, 1975). The function of the 60–98% of the hnRNA that is not converted to mRNA is not known. It is known, however, that whereas over 95% of mRNA is derived from single-copy DNA (reviewed by Davidson and Britten, 1973), a significant proportion of hnRNA (e.g., 75% of the 15–30 kb molecules in rat ascites cells) reflects the typical DNA sequence organization pattern of interspersed repetitive and nonrepetitive sequences (Darnell and Balint, 1970; Molloy *et al.*, 1974; Holmes and Bonner, 1974; Smith *et al.*, 1974). Because of these findings, and because hnRNA is generally larger than mRNA and appears to represent more sequences than are used as structural genes (Davidson *et al.*, 1977), it is concluded that there are both quantitative and qualitative selection mechanisms in the hnRNA–mRNA transition.

C. EARLY TRANSCRIPTION PRODUCTS: ELECTRON MICROSCOPIC CHARACTERIZATION

An independent approach to the study of nuclear transcription was devised by Miller and colleagues, who found that chromatin can be spread for electron microscopic visualization by hypotonic dispersion (Miller and Beatty, 1969a,b). Under such conditions, nascent ribonucleoprotein (RNP) fibrils remain attached to chromatin fibers, thereby defining individual transcription units. Ribosomal RNA transcription units have been characterized by this method (see Section III,A,1), and recently quantitative analyses have been made on active nonnucleolar transcription units (Foe et al., 1976; Laird and Chooi, 1976; McKnight and Miller, 1976). As reviewed below, these studies have shown that the initial products of transcription viewed in the electron microscope are, on the average, two to three times longer than the average hnRNA molecule isolated by biochemical techniques.

Figure 3 shows nonribosomal transcription units, identified by the presence of nascent RNP fibrils on chromatin, from embryos of D. melanogaster and Oncopeltus fasciatus. In this type of electron microscopic preparation, rpRNA transcription units typically are tandemly repeated, are densely packed with RNA polymerase molecules, and have a length close to the B-conformation DNA gene length required for the synthesis of rpRNA molecules (Hamkalo and Miller, 1973; McKnight and Miller, 1976). In comparison to rpRNA genes, the fibrillar arrays shown in Fig. 3 are longer, have a lower density of nascent RNP fibrils, and are not linked in tandem with similar units. In Oncopeltus, the unique unbeaded morphology of rpRNA gene chromatin can also be used to distinguish the two types of transcription units (Foe et al., 1976).

There is considerable variation in length and RNA polymerase density among nonribosomal transcription units, as might be expected of putative hnRNA coding regions. RNP fibril frequency is sometimes quite high and approaches the dense packing seen in rpRNA transcription units of about $52/\mu m$ of chromatin (McKnight and Miller, 1976). However, the fibril packing density averages $21/\mu m$ for nonribosomal transcription units of cellular blastoderm Drosophila embryos (McKnight and Miller, 1976), and $6/\mu m$ for nonribosomal transcription units of Oncopeltus embryos (Foe et al., 1976). This parameter reflects RNA polymerase initiation frequency and presumably also, in a simplistic way, the nuclear abundance of the species. In fact, the majority of nascent RNP fibrils seen in these preparations occur singly rather than in fibrillar gradients (Laird and Chooi, 1976; McKnight and Miller, 1976), and may correspond to the large number of low abundance transcripts present in nuclear RNA (e.g., Smith et al., 1974). That the activity exhibited in these transcription units does indeed correspond in part to mRNA production is supported by the finding that a marked increase in fibrillar arrays (McKnight and Miller, 1976) occurs concomitantly with a two- to threefold increase in polyadenylated RNA synthesis (Lamb and Laird, 1976a) at cellular blastulation in D. melanogaster embryos.

The length of nonnucleolar transcription units also varies con-

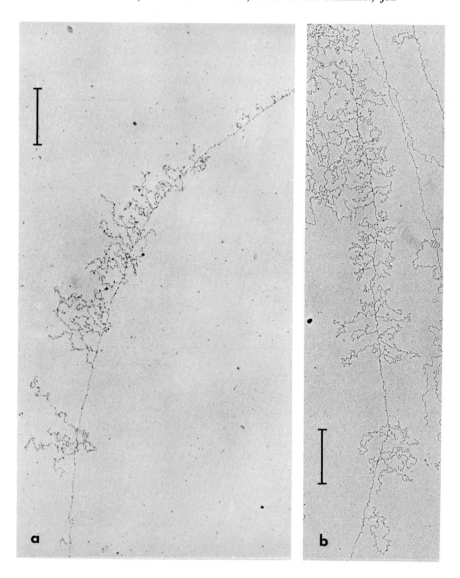

FIG. 3. Electron micrographs showing nonribosomal RNA transcriptional units observed on the chromatin of two insect species. (a) shows a configuration spread from an embryo of *Oncopeltus fasciatus*. Approximate sites for the initiation and termination of transcription can be assigned from the distinct gradient of short-to-long ribonucleoprotein fibrils. (From V. E. Foe *et al.*, 1976, *Cell* **9**, 131–146, reproduced with permission. Copyright © 1976 by MIT Press.) (b) shows a similar type of configuration observed on chromatin spread from a *Drosophila melanogaster* embryo. Scale bars represent 1 μm. (From McKnight *et al.*, 1977, reprinted with permission.)

siderably. On those fibrillar arrays containing several nascent RNA molecules, it is possible to approximate transcription initiation and termination sites. In fact, Laird *et al.* (1976) have shown that if the length of an RNP fibril is plotted against its position of attachment to the chromatin template in relation to other fibrils of the same array, a simple linear pattern is usually obtained, supporting the visual impression that transcription proceeds from a specific origin to a specific terminus. Since initiation and termination sites can be defined on such units, quantitative information can be obtained on transcription unit length. The initial data, expressed in micrometers of chromatin, can be converted to kilobase units to facilitate comparison with biochemical information on chromosome organization. Chromatin exists in a nucleosomal configuration in transcriptionally active nonnucleolar regions (Foe *et al.*, 1976; McKnight *et al.*, 1977). Since the length of DNA per nucleosome and the number of nucleosomes per micrometer of chromatin can be directly measured, it can be calculated that transcribed regions of nonribosomal *Drosophila* chromatin contain approximately 4.5 kb/μm (Laird *et al.*, 1976; McKnight *et al.*, 1977). Laird and Chooi (1976) carefully analyzed eight *Drosophila* nonribosomal transcription units and found a mean DNA content of 18 kb. McKnight and Miller (1976) surveyed the spread chromatin of many cellular blastoderm stage *Drosophila* embryos and determined the mean DNA length of over 150 transcription units to be 13 kb, with a range from 1.5 to 40 kb.

Both of these determinations, i.e., 18 and 13 kb, for the average length of *Drosophila* nonribosomal transcription units, are larger than estimates of the average size of *Drosophila* hnRNA. Lengyel and Penman (1975) labeled cultured *Drosophila* cells with radioactive uridine under conditions where rRNA synthesis was inhibited and found the rapidly labeled RNA (2–15 minutes of incorporation) to average 26 S, or about 4 kb, on denaturing sucrose gradients. Lamb and Laird (1976b) determined the median size of steady-state, polyadenylated nuclear RNA of *Drosophila* embryos to be 6 kb. Both of these communications report a small proportion of hnRNA that sediments on sucrose gradients at 55–60 S (equivalent to RNA molecules at least 20 kb in length), which may correspond to intact transcripts similar in size to those visualized by electron microscopy.

It should be noted that estimates of lengths of transcription units based on electron microscopic observations preferentially reflect units which exhibit numerous RNP fibrils. As mentioned earlier, single attached RNP fibrils are more common than multiple fibrillar arrays, and, although it is not possible to determine accurately the length of the transcription units from which such fibrils are derived, Laird and Chooi (1976) suggest that they may be somewhat shorter. That is, assuming, on the average, these isolated fibrils are halfway through transcription, it is estimated that upon release from the chromatin template they average 10 kb. Although inclusion of this value would lower the median transcription unit length as determined by microscopic visualization, there is still a discrepancy in the modal values which indicates that isolated hnRNA is one-half to one-third the size expected for primary transcription products of most transcription units visualized in the electron microscope. Lamb and Laird (1976b) point out that their 6 kb length estimate was obtained on polyadenylated nuclear RNA molecules and that some nucleolytic processing may have already occurred. The 4 kb median estimate of Lengyel and Penman (1975), however, was independent of polyadenylation, and was obtained after brief labeling periods. If, as indicated by microscopy, nonribosomal transcription units average 13–18 kb, it may be that their primary transcription products are short-lived as very large molecules, being subjected to endonucleolytic processing either before or immediately after transcriptional termination. In fact, Laird and Chooi (1976) have occasionally observed *Drosophila* transcription units on which some of the RNP fibrils appear to have been cleaved at a specific point prior to completion of transcription. Another consideration is the well-known observation that exposure of hnRNA to denaturing conditions effects a moderate to extreme reduction in its size (e.g., Mayo and deKloet, 1971). It is interesting that the size of the large, denaturable hnRNA complexes (> 15 kb) overlaps with the estimated size of nascent RNP fibrils observed in the electron microscope. Since rapid intranuclear cleavage of hnRNA is occurring, it is possible that some of these large molecular complexes on nondenaturing sucrose gradients represent nicked or processed primary transcripts held together by short duplex regions rather than nonspecific aggregates (Derman and Darnell, 1974; Federoff *et al.*, 1977).

D. SIZE AND FUNCTION OF PRIMARY TRANSCRIPTION PRODUCTS

Both biochemical and microscopic methods of analysis indicate that the average transcription unit length in *Drosophila* is much larger than necessary to encode the average mRNA molecule, which is about 2 kb for polyadenylated mRNA (Lamb and Laird, 1976b; Levis and Penman, 1977). As previously mentioned, it is a general finding that hnRNA is significantly larger than mRNA. It will be necessary to determine if this size discrepancy reflects the simultaneous transcription of contiguous genomic coding and noncoding sequences, such that mRNA molecules are derived from large transcripts in a series of enzymatic processing steps (Darnell *et al.*, 1973; Georgiev *et al.*, 1973), or whether mRNA molecules are derived from relatively small transcription units (Davidson *et al.*, 1977). It has been shown in the sea urchin that mRNA coding sequences are located contiguous to ~300 bp repetitive sequence elements (Davidson *et al.*, 1975), and it appears that the repetitive sequences which are transcribed in a particular tissue represent a restricted subset of repetitive sequence families (Costantini *et al.*, 1978; Scheller *et al.*, 1978). These observations are quite interesting, but at the present time sufficient information is not available concerning the exact function of the nonmessenger sequences in the hnRNA population and the sequence arrangement of most structural gene transcripts to postulate a "typical" organization for eukaryotic mRNA-coding transcription units.

An approach which has yielded information in this area has been to examine the size and sequence organization of specific transcriptional units in relation to their functional RNA products. For example, one can ask if specific mRNA molecules are transcribed as part of larger hnRNA molecules. This question will be considered in some detail in the following sections on functionally defined genes.

III. MULTIPLE-COPY GENES

A. TANDEMLY REPEATED GENES

1. *Ribosomal RNA Genes*

The rRNAs are the nucleic acid constituents with which ribosomal proteins interact in highly specific modes to form the large and small ribosomal subunits of the protein synthesizing apparatus of cells [for reviews, see articles in Nomura *et al.* (1974); Weissbach and Pestka (1977)]. The large ribosomal subunit in mammals and other higher eukaryotes contains three rRNA molecules with sedimentation coefficients of 28 S, 5.8 S, and 5 S, whereas the small subunit contains a single 18 S rRNA molecule. In lower eukaryotes, the counterparts to the 28 S and 18 S molecules occur as slightly smaller molecules of approximately 25 S and 17 S, respectively. One each of the 28 S, 18 S, and 5.8 S rRNA molecules are derived from a single large rRNA precursor molecule (rpRNA), and 5 S rRNA molecules are synthesized independently.

a. rpRNA Genes. The size of rpRNA molecules, and thus their respective genes, differs between classes of eukaryotes, ranging from a 45 S (~13 kb) molecule in mammals and birds to 36–40 S (7.5–8.1 kb) molecules in lower vertebrates and invertebrates (Perry, 1976; Tartof, 1975). The processing steps involved in the cleavage of the three rRNA molecules from rpRNA molecules have recently been reviewed (Perry, 1976; Hadjiolov and Nikolaev, 1976). The number of chromosomal rpRNA genes per haploid genome equivalent, as estimated by RNA/DNA hybridization, varies considerably between eukaryotic species, ranging from a single gene in the protozoan, *Tetrahymena pyriformis* (Yao and Gall, 1977), to many thousands of genes in some plants (Cullis and Davies, 1974). Three examples between these extremes are *D. melanogaster*, ~250 (Tartof, 1971), *X. laevis* ~600 (Birnstiel *et al.*, 1972), and *Homo sapiens*, ~300 rpRNA genes per haploid genome equivalent (Jeanteur and Attardi, 1969). In many species, the rpRNA genes are located at only one specific chromosomal region per haploid set, designated the nucleolus organizer (NO); familiar examples of this arrangement occur in the genomes of *D. melanogaster*, *X. laevis*, and *Zea mays*. In other instances, multiple NOs occur per haploid genome; examples of this condition are *Mus musculus*, which has three NOs on separate chromosomes (Elsevier and Ruddle, 1975), and *H. sapiens*, which has NOs on the short arms of five of the acrocentric D and G chromosomes (Evans *et al.*, 1974).

Although chromosomal NOs are inherited as Mendelian units (Miller and Gurdon, 1970), rpRNA genes have been shown to be differentially amplified as extrachromosomal genes in certain primitive eukaryotes (Prescott *et al.*, 1973; Engberg *et al.*, 1974; Gall, 1974; Vogt and Braun, 1976) and in the oocytes of amphibia and many insects (reviewed by Tobler, 1975). The amplification occurring in oocytes of *X. laevis* has been studied extensively. The process begins prior to meiosis and is completed by the end of pachytene, at which time several hundreds of thousands of extrachromosomal rpRNA genes have accumulated in each oocyte nucleus (Bird and Birnstiel, 1971; Watson-Coggins and Gall, 1972; Kalt and Gall, 1974). The amplified genes apparently are derived initially from chromosomal NOs, but the molecular aspects of this process are not known (Brown and Blackler, 1972). The subsequent cascade of extrachromosomal amplification, however, appears to involve a rolling circle mechanism of DNA synthesis (Hourcade *et al.*, 1973; Rochaix *et al.*, 1974).

The amplification process in amphibia and insects gives rise predominantly to circular molecules containing rpRNA genes (Gall and Rochaix, 1974; Tobler, 1975; Trendelenburg *et al.*, 1977). On the other hand, it has been shown that amplification in three primitive eukaryotes results primarily in linear molecules, which are large palindromes containing two rpRNA genes each [*T. pyriformis* (Engberg *et al.*, 1976; Karrer and Gall, 1976); *Physarum polycephalum* (Vogt and

Braun, 1976; Grainger and Ogle, 1978); and *Dictyostelium discoideum* (N. Maizels and R. Grainger, personal communication)]. In all three of the latter cases, the two rpRNA genes exhibit divergent transcriptional polarity and are located toward the distal ends of the inverted repeats (Grainger and Ogle, 1978; N. Maizels and R. Grainger, personal communications). Yao and Gall (1977) have proposed a tentative model for the origin of extrachromosomal rDNA palindromes which involves replication via branch migration of a single integrated rpRNA gene and associated nontranscribed sequences to form an extrachromosomal molecule which unfolds into a linear palindrome via semiconservative replication. These authors also suggest that the single integrated rpRNA gene present in *T. pyriformis* may represent an evolutionary link between the low multiplicity of rpRNA genes found in prokaryotes and the high multiplicity of chromosomal rpRNA genes typical of most eukaryotes.

In all cases in which multiple chromosomal rpRNA genes have been examined in detail, the genes have been found to exist in tandemly repeated units, with each repeat unit consisting of an rpRNA gene and a nontranscribed spacer (NTS) segment of DNA. As illustrations of this arrangement, the rDNA repeat unit of *X. laevis* is presented schematically in Fig. 4, and an electron micrograph showing

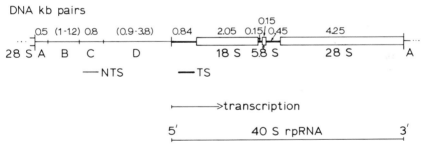

FIG. 4. The ribosomal DNA repeat unit of *Xenopus laevis*. A schematic representation of one rDNA repeat unit is shown at the top, and the 40 S primary transcript of the rpRNA gene is drawn at the bottom. The transcriptional polarity is that determined by Dawid and Wellauer (1976) and Reeder *et al.* (1976a). The DNA regions which code for the 18 S, 5.8 S, and 28 S rRNAs are depicted by open bars, and the nontranscribed spacer (NTS) and transcribed spacer (TS) regions are drawn as light and heavy lines, respectively. The divisions and length heterogeneity within the NTS shown in this figure were derived from Botchan *et al.* (1977) and are discussed in the text. Wellauer *et al.* (1976a,b), however, have reported NTSs as short as 1.8 kb. (From Speirs and Birnstiel, 1974, adapted with permission.) This and all the following gene schemata have been drawn to scale, and are graduated in base pairs (bp) or kilobase pairs (kb). In each case the endpoints of the unit depict one complete repeat unit and do not reflect specific restriction endonuclease cuts.

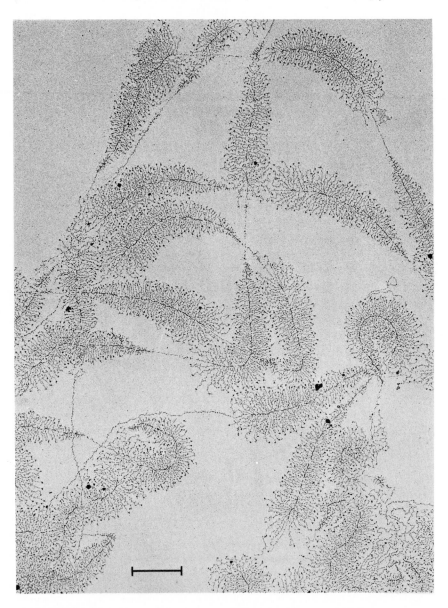

FIG. 5. Electron micrograph of transcriptionally active, amplified ribosomal DNA isolated from an oocyte of the spotted newt, *Triturus viridescens*. Gradients of short to long nascent ribonucleoprotein fibrils delimit the rpRNA genes within tandem rDNA repeats. Spacer chromatin segments which are not transcribed separate neighboring rpRNA genes. Scale bar represents 1 μm. (From Miller and Beatty, 1969b, reproduced with permission.)

transcriptionally active rpRNA genes of *Triturus viridescens* is shown in Fig. 5. The order and polarity of the 28 S, 18 S, and 5.8 S rRNA coding elements within rpRNA genes have been studied by several experimental approaches. It is now well established that the arrangement of —initiation–18 S–5.8 S–28 S–termination— is essentially universal for eukaryotic rpRNA gene transcription (Hackett and Sauerbier, 1975; Dawid and Wellauer, 1976; Reeder *et al.*, 1976a; Schibler *et al.*, 1976; Carlson *et al.*, 1977).

Observations of chromosomal NOs by chromatin spreading procedures suggest that all of the rDNA repeats within a single NO have the same transcriptional polarity. With the exceptions of the amplified palindromes described earlier, which exhibit divergent transcriptional polarity, this arrangement generally holds also for amplified rDNA. [It should be noted that occasional examples of convergent or divergent polarity have been demonstrated in extrachromosomal rDNA, but such exceptions occur infrequently (Miller and Beatty, 1969b; Trendelenburg *et al.*, 1974).] As shown in Fig. 6, however, a unique arrangement of rpRNA genes, in which rDNA repeats exhibit a consistent pattern of alternating polarity, has recently been found in the plant species, *Acetabularia exigua* (Berger *et al.*, 1978). It is not known, however, whether these genes are integrated within chromosomal DNA or are amplified rDNA copies.

The NTSs of various organisms can be quite different in average length. One of the shortest NTS lengths occurs in the amplified rDNA of the giant unicellular alga, *Batophora oerstedii*, where the NTS is estimated to be less than 0.5 kb long (Berger and Schweiger, 1975). Similar very short NTSs also occur in amplified rDNA of *Acetabularia mediterranea* (Spring *et al.*, 1976). At the other extreme are the amplified rDNA repeats (gene plus NTS) of *D. discoideum* (Maizels, 1976) and the chromosomal rDNA repeats in the mouse (Arnheim and Southern, 1977; Cory and Adams, 1977), both of which are nearly 40 kb long. In certain instances, most of the NTSs within a species appear to be homogeneous in length, such as in the amplified palindromes of *D. discoideum*. On the other hand, the chromosomal NTSs of *D. melanogaster* and *X. laevis* can differ significantly in size (Wellauer and Dawid, 1977; Wellauer *et al.*, 1974). In general, it has been observed that repeat units of amplified rDNA for a given individual are more homogeneous in size than are chromosomal repeats.

The arrangement of NTS length heterogeneity has been examined in detail for *X. laevis*, in which the NTSs of rDNA vary in length between 1.8 and 6.3 kb (Wellauer *et al.*, 1976a,b; Botchan *et al.*, 1977). Reeder *et al.* (1976b) found that the pattern of spacer lengths within

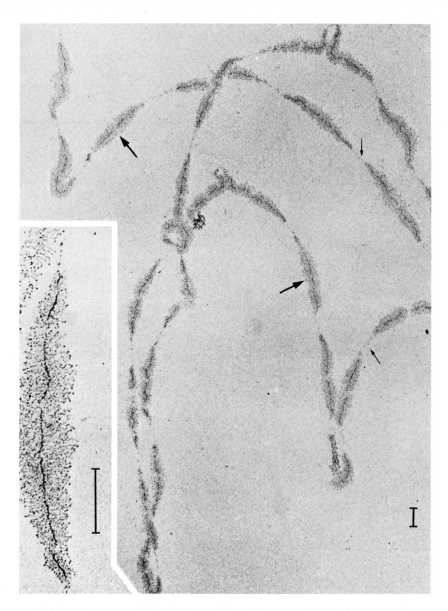

FIG. 6. Electron micrographs of transcriptionally active ribosomal DNA repeats isolated from *Acetabularia exigua*. Each rpRNA gene is localized by a gradient of short to
long nascent ribonucleoprotein fibrils. Neighboring rpRNA genes alternate in polarity,
with the termination ends of convergent pairs occurring extremely close together (large
arrows). Short nontranscribed spacers separating adjacent initiation sites are indicated
by small arrows. The inset shows a higher magnification of one pair of converging
rpRNA genes. Scale bars represent 1 μm. (Micrographs kindly provided by S. Berger
and H. Schweiger, Max-Planck-Institut für Zellbiologie, Ladenburg bei Heidelberg.)

individual frogs varied sufficiently to allow their use as genetic markers. Using controlled matings, these authors showed that the spacer patterns of rDNA are inherited in a Mendelian manner without detectable change. Wellauer *et al.* (1976b) compared the arrangement of NTS length heterogeneity within chromosomal and amplified rDNA of *X. laevis.* Amplified rDNA contains repeats of the same size classes as are found in the chromosomal rDNA of the same animal, but some animals selectively amplify repeat lengths rarely present in their chromosomal rDNA complement, whereas others amplify their most abundant size classes. Furthermore, the preference for size-class-specific amplification is inherited. For chromosomal rDNA, between 50 and 68% of adjacent repeats of a given molecule were found to differ in NTS contour length. Alternatively, most, if not all, tandem repeats along single molecules of amplified rDNA exhibit homogeneous contour lengths when analyzed in the electron microscope by heteroduplex mapping. These observations lend support to the proposal that amplified rDNA arises primarily by a rolling circle mechanism since a template circle containing a single repeat would be expected to replicate linear molecules containing homogeneous repeats. However, observations of different lengths of adjacent NTSs in spread chromatin preparations of amplified rDNA from *X. laevis* (Scheer *et al.*, 1977), *Dytiscus marginalis, Acheta domesticus* (Trendelenburg *et al.*, 1976), and *T. viridescens* (Miller and Beatty, 1969a) are compatible with a rolling circle mechanism of replication only if the initial templates contain multiple heterogeneous repeats within the same circle or if recombination occurs relatively frequently within extrachromosomal rDNA.

The molecular basis for variable NTS lengths in *X. laevis* has been studied by heteroduplex mapping and restriction enzyme digestion of cloned rDNA molecules (Wellauer *et al.*, 1976a,b; Botchan *et al.*, 1977). The spacers can be divided into four distinguishable regions based on nucleotide sequence (see Fig. 4). Immediately adjacent to the termination end of the rpRNA gene is a 0.5 kb sequence (A in Fig. 4) that has no obvious internal repetition and is conserved in spacers of different lengths. Adjacent to this region is a 1.0–1.2 kb segment (B in Fig. 4) that consists of multiples of a 15 bp repeat. Proximal to the initiation end of each rpRNA gene is a region (D in Fig. 4) that also contains 15 bp repeats but which is highly variable in length and accounts for most of the observed heterogeneity in spacer lengths. The repeats within the two variable regions differ but may have arisen by divergence from a common ancestral 15 bp sequence. Botchan *et al.* (1977) speculate that such a divergence may have occurred after the formation of a conserved, nonrepetitive segment (C in Fig. 4), which lies between the two variable repetitive regions. If so, this separation may have resulted from the evolution of an essential function for the sequence of the middle segment, thus prohibiting recombination events from spanning the boundary and eliminating the conserved sequence. One potential function suggested for the conserved element of the X. *laevis* spacer is that it may contain another type of gene. However, transcription typically is not observed in such regions by electron microscopic procedures nor have biochemical experiments detected transcriptionally functional sequences within the NTS. Another possible function might be that a site for the initiation of rDNA replication is present in the conserved element of the NTS. In syncytial embryos of *D. melanogaster*, the brief S phase and the rapid rate of DNA chain elongation require that at least one replication initiation event occur per rDNA repeat at that developmental stage (Tartof, 1975). The possibility of a replication initiation site localized in the NTS is supported by evidence from chromatin spreads of *D. melanogaster* embryos in which replicon structures appear to have initiated within NTS regions (McKnight *et al.*, 1977).

In addition to some NTS length differences, another basis for length heterogeneity in the rDNA repeat has been discovered in *D. melanogaster*. It has been found that nonribosomal DNA segments are present in approximately 60% of the rpRNA gene sequences of the common fruit fly (Wellauer and Dawid, 1977; White and Hogness, 1977; Pellegrini *et al.*, 1977). Such segments are found inserted within the 28 S coding sequence, predominantly in the NO of the X chromosome rDNA repeats (see Fig. 7); genes with and without insertions appear to be randomly distributed within the NO. The insertions occur at a reproducible point in the 28 S region, and range in size from 0.5 to 6.0 kb, with 5.0 kb inserts being the predominant size class. There are at least two types of insertions (Wellauer *et al.*, 1978). Type 1 includes the abundant 5 kb class and homologous small (0.5 and 1 kb) insertions. These interrupt about 49% of all X chromosome rDNA genes, and are probably not present in Y chromosome rDNA genes. Type 2 insertions are generally between 1.5 and 4 kb long, have no homology to type 1, and interrupt about 16% of both X and Y chromosome rDNA repeats (Wellauer *et al.*, 1978). Sequences homologous to the rDNA inserts have been found to occur elsewhere in the *D. melanogaster* genome, comprising some 0.2% of the haploid genome (Dawid and Botchan, 1977).

The functional significance of rDNA insertion sequences in *D. melanogaster* remains obscure. The lengths of rpRNA transcription units

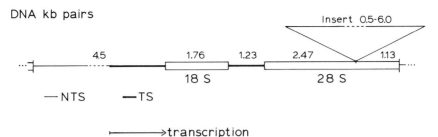

FIG. 7. Ribosomal DNA repeat unit of *Drosophila melanogaster*. This schematic representation shows the location of the variable-length sequence which does not encode any part of 28 S rRNA but which is inserted within the 28 S cistron of about 60% of the X chromosome rDNA repeat units. As discussed in the text, it is not yet clear whether or not these insertion sequences are transcribed. Note the general similarity to the *Xenopus laevis* rDNA repeat unit in Fig. 4. Although not shown here, the 5.8 S rRNA cistron occurs in the transcribed spacer segment between the 18 S and 28 S coding regions (Pellegrini *et al.*, 1977). The indicated transcription initiation site is an estimate based on the molecular weight of *Drosophila* rpRNA, ~2.8 × 10⁶ daltons (Perry *et al.*, 1970). As indicated by dashed lines, there is also a degree of length heterogeneity in rDNA repeats contributed by variation in length of the nontranscribed spacer region (Wellauer and Dawid, 1977). (From Wellauer and Dawid, 1977, adapted with permission.)

observed from chromatin spreads of *D. melanogaster* embryos suggest that the inserts are not transcribed as parts of larger than normal rpRNA molecules (Laird and Chooi, 1976; McKnight and Miller, 1976). One possible mechanism for obtaining a normal rpRNA transcript from an insert-containing gene would be for the insert to form an untranscribed loop, allowing RNA polymerase molecules to transcribe only the normal sequences (Pellegrini *et al.*, 1977; Wellauer and Dawid, 1977). However, electron microscopic analysis provides no evidence of such loops in rpRNA transcription units of *D. melanogaster*. It is possible, therefore, that if insert-containing rDNA repeats are ever active, transcription terminates close to the beginning of the insert sequence, mimicking transcription of a normal rpRNA gene.

Chromatin spreading studies provide some insight into the control of rpRNA synthesis. McKnight and Miller (1976) observed that when rpRNA genes first activate at the cellular blastoderm stage during *D. melanogaster* embryogenesis, each rDNA repeat is regulated independently of other nucleolar genes in the same NO. Evidence for the existence of an independent promoter for each rpRNA gene also has been obtained from the kinetics of ultraviolet inactivation of rpRNA synthesis in mouse L cells (Hackett and Sauerbier, 1975) and from biochemical experiments on the nature of the 5'-sequence of *X. laevis* rpRNA (Reeder *et al.*, 1977). As shown in Fig. 8, maximal packing of

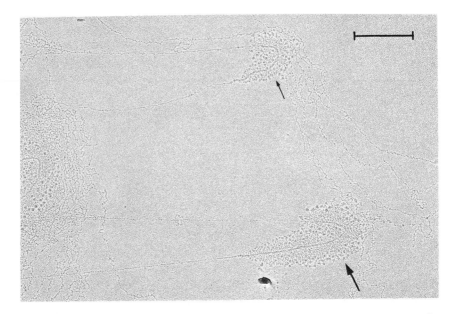

FIG. 8. Electron micrograph of transcriptionally active *D. melanogaster* rDNA. The sample was prepared from an early cellular blastoderm embryo at which stage rpRNA synthesis is first initiated. Two active rpRNA genes are shown bracketed by extensive regions of transcriptionally inactive chromatin. The large arrow indicates a transcriptionally mature rpRNA gene, whereas the small arrow indicates a newly activated rpRNA gene which was not fully loaded with RNA polymerases at the time of specimen preparation. Note that the density of nascent ribonucleoprotein fibrils per unit length is the same on both genes. Scale bar represents 1 μm. (From S. L. McKnight and O. L. Miller, Jr., 1976, *Cell* **8**, 305–319, reproduced with permission. Copyright © 1976 by MIT Press.)

RNA polymerase molecules occurs on both newly activated and fully transcribed rpRNA genes of *D. melanogaster* embryonic chromatin. Such evidence indicates that the rate of transcription rather than the frequency of polymerase-binding events regulates rpRNA production per individual gene (McKnight and Miller, 1976). As cellularization proceeds in the *Drosophila* embryo, the number of active rpRNA genes increases, yet fewer than 50% of the rpRNA genes of each NO are activated at that stage of development. A similar situation has been reported for primary spermatocytes of *D. hydei,* where the number of active rpRNA genes increases during spermiogenesis, but, again, only some 50% of the total number of rpRNA genes are ever activated (Meyer and Hennig, 1974). Thus, in *Drosophila,* it appears that total rpRNA synthesis is regulated by a combination of the rate of RNA polymerase movement and the total number of individual rpRNA genes activated at any specific developmental stage.

Alternative to the scheme for regulation of rpRNA synthesis in *Drosophila*, two other chromatin spreading studies have implicated a degree of control modulated at the level of RNA polymerase initiation. Foe *et al.* (1976), studying embryogenesis of the milkweed bug *O. fasciatus*, observed that newly activated rpRNA genes typically are sparsely loaded with RNA polymerase molecules. Similarly, Scheer *et al.* (1976) observed that amplified rpRNA genes from young oocytes of *Triturus alpestris* contain reduced RNA polymerase packing ratios as compared to the rDNA of more mature oocyte nuclei. Since RNA polymerase classes I, II, and III occur in excess during all stages of amphibian oogenesis (Roeder, 1974), it appears likely that the modulation of RNA polymerase initiation events is due to regulatory factors that interact directly with the RNA polymerases or with the rDNA chromatin template (Biswas *et al.*, 1975). However, the important question of what mechanisms control the total number of active rDNA genes within a cell cycle or during development remains unanswered.

b. 5 S rRNA Genes. 5 S rRNA genes (5 S genes) are present in multiple copies in the genomes of all eukaryotes so far examined. However, as for rpRNA genes, the haploid number and chromosomal location of 5 S DNA repeats vary considerably between different organisms. For example, the genome of *D. melanogaster* contains about 160 5 S genes which have been localized by *in situ* hybridization to bands 56 e–f on chromosome 2R (Procunier and Tartof, 1975; Prensky *et al.*, 1973). *Homo sapiens* contains about 2000 5 S genes located on several chromosomes, with a predominant locus present on chromosome 1 (Johnson *et al.*, 1974; Steffensen *et al.*, 1974). *Xenopus laevis* has over 20,000 5 S genes, which, as Fig. 9 demonstrates, are distributed among the ends of the long arms of most, if not all, of the 18-chromosome complement (Pardue *et al.*, 1973; Pardue, 1973).

Although 5 S rRNA is present in the large ribosomal subunit in an equimolar ratio with 28 S and 5.8 S rRNA, 5 S genes show no linkage with rpRNA genes in higher eukaryotes. Linkage has been reported, however, for two primitive eukaryotes, *D. discoideum* (Cockburn *et al.*, 1976; Maizels, 1976) and *Saccharomyces cerevisiae* (Maxam *et al.*, 1977), in which 5 S genes are present in the spacer intercepts between rpRNA genes. This latter arrangement in which the 5 S and rpRNA genes are topographically linked but under the control of separate promoters (Batts-Young and Lodish, 1978) may represent an intermediate evolutionary state of divergence from the bacterial organization, where the 16 S, 23 S, and 5 S rRNA cistrons have a common promotor,

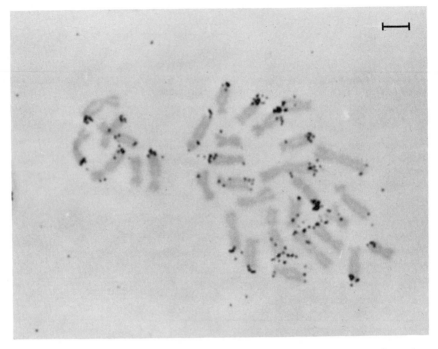

FIG. 9. Autoradiograph showing localization of 5 S rRNA genes on metaphase chromosomes of *X. laevis*. Labeled 5 S cRNA was hybridized *in situ* to denatured chromosomal DNA. The silver grains show hybridization to be localized at the telomeric regions of the long arms of most chromosomes. Scale bar represents 5 μm. (From Pardue *et al.*, 1973, reproduced with permission.)

to higher eukaryotes, where 5 S genes are transcriptionally and topographically unlinked from rpRNA genes (Maizels, 1976).

The organization of repeat units containing 5 S genes (5 S DNA) has been extensively studied in two species of the genus, *Xenopus*. A dual 5 S rRNA system occurs in *Xenopus* and other amphibia in that somatic cells normally synthesize predominantly only one type of 5 S rRNA, whereas oocytes synthesize both the somatic type and several oocyte-specific types (Ford and Southern, 1973; Wegnez *et al.*, 1972; Ford and Brown, 1976). In *X. laevis* oocytes, about 10% of the 5 S rRNA synthesized is somatic type, which differs by six nucleotides from the major oocyte-type 5 S rRNA. The predominant oocyte-type accounts for over 50% of oocyte 5 S rRNA.

The DNA coding for the major oocyte 5 S rRNA has been isolated from total genomic DNA by density banding and also by molecular cloning in a bacterial plasmid (Brown and Sugimoto, 1973b; Brownlee

et al., 1974; Carroll and Brown, 1976a; Jacq *et al.*, 1977). The nucleotide sequence of oocyte 5 S DNA has also been obtained (Federoff
and Brown, 1978; Miller *et al.*, 1978). The oocyte 5 S DNA repeat unit
of *X. laevis* consists of three regions: the G + C-rich 5 S gene, a G + C-
rich segment containing sequences essentially homologous to the first
101 bp of the 120 bp 5 S genes, and an A + T-rich spacer segment (see
Figs. 10 and 11b). The gene and G + C-rich segment containing the
"pseudogene" are essentially constant in size within families of 5 S
DNA repeats, but the A + T-rich spacer segments exhibit length heterogeneity. The spacers are primarily composed of repeating multiples
of closely related 15 bp sequences, and the length heterogeneity is attributed to the presence of differing numbers of these subrepeats per
spacer segment. There is a G + C-rich region of 49 nucleotides immediately adjacent to the 5′-terminus of the 5 S rRNA sequence. It is less
repetitive in sequence than the rest of the spacer, contains several palindromes, and is identical in the six cloned repeating units of 5 S DNA
which were analyzed (Federoff and Brown, 1978). A possible explanation for the presence of the pseudogene sequence distal to the termination end of each 5 S gene is that it is a relic of a 5 S gene duplication
which was followed by the mutational inactivation of one of the genes
(Jacq *et al.*, 1977).

A minor oocyte-type 5 S DNA has also been isolated in which the

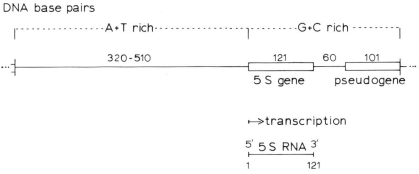

FIG. 10. The 5 S rDNA repeat unit of *X. laevis*. The 5 S gene lies within the G + C-
rich region of the repeat which is essentially constant in size from one repeat to the
next. A "pseudogene," which is an almost perfect repeat of the first 101 residues of the
gene and is not transcribed, also is present in the G + C-rich segment. Data on the variability in length of the A + T-rich region of the repeat were obtained from Carroll and
Brown (1976a). The length heterogeneity of this segment is based primarily on the presence of different numbers of closely related 15 bp subrepeats per 5 S DNA repeat
(Brownlee *et al.*, 1974; Carroll and Brown, 1976a). (From Jacq *et al.*, 1977, adapted with
permission.)

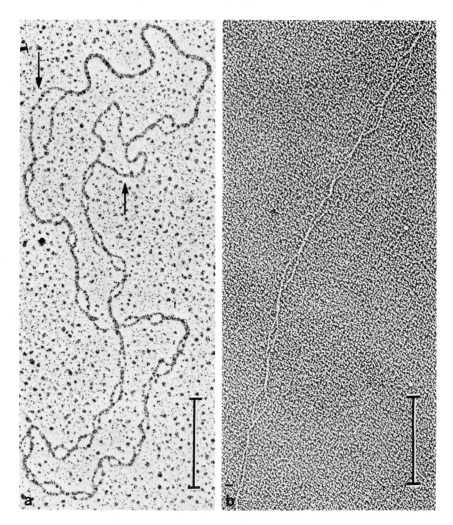

FIG. 11. Electron micrographs of partially denatured 5 S DNA demonstrating the alternating A + T-rich and G + C-rich segments of the repeat units. The DNA molecule in (a) contains a cloned segment of 5 S DNA of *D. melanogaster*. The large bubbles (arrows) are the poly(AT) regions used to insert the 5 S DNA fragment into the plasmid. The largely duplex DNA segment at the top is the plasmid DNA, whereas the repeating bubbles in the bottom two-thirds of the molecule are the A + T-rich segments which separate neighboring G + C-rich duplex regions containing the 5 S genes. (From N. D. Hershey *et al.*, 1977, reproduced with permission. Copyright © 1977 by MIT Press.) The partially shown molecule in (b) is 5 S DNA isolated by buoyant density centrifugation from *X. laevis*. The bubble regions again are denatured A + T-rich regions separating the tandemly repeated 5 S genes. Scale bar represents 1.0 kb. (Kindly supplied by D. D. Brown, Department of Embryology, Carnegie Institution of Washington in Baltimore, Maryland.)

gene has both somatic and major oocyte-type specific residues plus its own specific residues. The latter consists of about 2000 homogeneous repeats with a repeat length one-half that of the average, major oocyte-type repeat of 720 bp (Brown *et al.*, 1977a).

The 5 S DNA repeat unit of *D. melanogaster* has been partially characterized by restriction enzyme mapping and partial denaturation mapping in the electron microscope. In this species, the repeat units consist of approximately 375 bp, contain no pseudogene structures, and, within limits of detection, exhibit only slight length heterogeneity (Artavanis-Tsakonas *et al.*, 1977; Hershey *et al.*, 1977). As the electron micrograph in Fig. 11a shows, adjacent 5 S genes of *D. melanogaster*, similar to those of *X*. laevis, are separated by A + T-rich spacer segments.

Numerous studies have shown that 5 S RNA synthesis is not coordinated with that of rpRNA production [*X. laevis* ovaries (Ford, 1971; Mairy and Denis, 1971), *X. laevis* embryos (Miller, 1974), HeLa cell cultures (Zylber and Penman, 1971; Leibowitz *et al.*, 1973), and mouse L cells (Perry and Kelley, 1968)]. Noncoordinated synthesis might well be expected since rpRNA genes and 5 S genes are transcribed by different RNA polymerase types (Weinmann and Roeder, 1974). The differential regulation of somatic and oocyte-type 5 S transcription remains an intriguing problem. Regarding this, Ford and Mathieson (1976) report that oocyte-type 5 S rRNA is synthesized in a transformed *X. laevis* tissue culture cell line in which oocyte-type 5 S genes are translocated to a site adjacent to the nucleolus. It was suggested that oocyte-type genes may normally be repressed in somatic cells by being located in genomic regions not available for transcription (e.g., heterochromatin) and that the translocation of such genes into a euchromatic region may cause their activation.

2. Transfer RNA Genes

Transfer RNAs provide the molecular link between the genetic code in messenger RNAs and the amino acids which form nascent polypeptide chains (for review, see Ofengand, 1977). Because of the degeneracy of the genetic code, more than one type of tRNA molecule exists for many of the 20 known amino acids. Some estimates of the average number of genes per haploid genome for each tRNA species are ~9 in yeast (Feldman, 1976), ~10 in *D. melanogaster* (Weber and Berger, 1976), ~20 in HeLa cells (Hatlen and Attardi, 1971), and ~200 in *X. laevis* (Clarkson *et al.*, 1973a). For *X. laevis*, it has been shown that the actual number of genes varies by as much as a factor of 2 for certain different aminoacyl-tRNAs and even for different isoaccepting tRNAs for the same amino acid (Clarkson and Birnstiel, 1973).

DNA base pairs

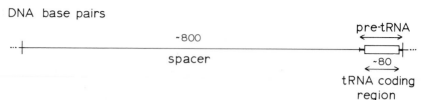

FIG. 12. Repeating unit of *X. laevis* tDNA. The tRNA precursor genes are clustered, probably in isogene families (Clarkson and Birnstiel, 1973), and a spacer segment is present between neighboring genes which is lower in G + C content and ~10 times longer than the tRNA precursor (pre-tRNA) coding region. (From Clarkson and Birnstiel, 1973, adapted with permission.)

In situ hybridization studies of tRNA genes using total tRNA probes have revealed their location at many sites on the chromosome set of HeLa cells (Aloni *et al.*, 1971) and *D. melanogaster* (Steffensen and Wimber, 1971). *In situ* hybridization with a few purified tRNAs suggest that at least some tRNA gene families are partially, if not totally, present in clusters in the *D. melanogaster* genome (Grigliatti *et al.*, 1974; Delaney *et al.*, 1976). Experiments on the DNA-containing tRNA genes (tDNA) of *X. laevis* (Clarkson *et al.*, 1973b; Clarkson and Birnstiel, 1973) and *D. melanogaster* (Yen *et al.*, 1977) provide further evidence that in these two species tRNA gene families are arranged in clusters and, furthermore, show that the tRNA genes are interdigitated with noncoding spacer sequences. Clarkson and co-workers have studied the arrangement of tRNA genes of *X. laevis* by determining the relative buoyant density position of different native tDNAs and by analyzing the buoyant density shift of denatured DNA prehybridized with specific tRNAs. Figure 12 shows the proposed arrangement of the basic repeating unit of *X. laevis* tDNA as derived from these studies. In *X. laevis*, the spacer adjoining tRNA genes is about ten times the length of the tRNA precursor coding sequence. Within the limits of detection of the buoyant density shift experiments, tDNA repeats of several different isocoding tRNA gene clusters appear to be homogeneous in size. However, the spacer sequences, although homologous within tRNA gene families, probably differ between these families. Evidence for identical repeating units and long-range periodicity in the DNA coding for tRNA$_1^{Met}$, and for possible interdigitation of two tDNA families, has been reported in a study of *X. laevis* tDNA (Clarkson and Kurer, 1976).

Figure 13 shows a schematic representation of a tRNA coding DNA fragment isolated by molecular cloning from the *D. melanogaster* genome which contains four tRNA genes (Yen *et al.*, 1977). Hybridization analyses of restriction fragments show that genes *1*, *2*, and

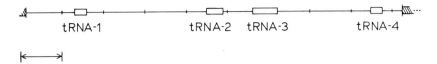

tRNA-1   tRNA-2   tRNA-3   tRNA-4

1 kb

FIG. 13. Map of a cloned *D. melanogaster* tDNA fragment which contains four tRNA genes. The locations of the four genes were determined by restriction endonuclease mapping and electron microscopic observations of tRNA binding regions. The lengths of the bars represent standard deviations of measurements rather than exact gene lengths. Genes *1*, *2*, and *3* code for the same (although unidentified) tRNA, while gene *4* codes for a different tRNA. Unlike *X. laevis,* the intergene spacers appear to be variable in length. (From Yen *et al.,* 1977, adapted with permission.)

3 are identical, and different from the terminal number *4* gene. The position of the genes, secondary structural features, and restriction sites all indicate that, unlike *X. laevis,* the intergene spacers are not homogeneous in size. cRNA derived from the cloned fragment for *in situ* hybridization to polytene chromosomes has been used to localize these tDNAs in a puffed locus within region 42A of chromosome 2R. Careful analysis of the pattern of silver grains suggests that hybridization occurs at two slightly separated regions in the puff. However, it is not known whether tRNA gene *4* is interspersed within the *1–2–3* class cluster, or if the cloned DNA was fortuitously derived from the junction between two different tRNA gene clusters.

For yeast, there is conflicting evidence regarding whether homologous tRNA genes are clustered (Feldmann, 1976) or unlinked (Olson *et al.,* 1977). Members of two isogenic tRNA families in yeast have been obtained as plasmid inserts and cloned in bacterial hosts. Four of the eight tRNA^{Tyr} genes, including one *ochre* suppressor mutant, have been sequenced (Goodman *et al.,* 1977), and three tRNA^{Phe} genes have been sequenced (Valenzuela *et al.,* 1978). In both cases, it was found that, although the isogenic coding sequences are essentially identical, there is often no sequence homology between clones in the region preceding the 5′-end of the mature tRNA sequence. This region, however, is AT-rich. In the coding strand, at or near the 3′-end of the tDNA molecule, an oligo(dA) sequence of at least seven bases can be found in each case, but the rest of the region is nonhomologous between clones. This nonhomology in the flanking sequences is reminiscent of the situation in the cloned fragment containing tRNA genes of *D. melanogaster* (Yen *et al.,* 1977), where several lines of evidence indicate that the intergenic spacers between homologous tRNA genes are dissimilar.

A very interesting finding from the yeast tDNA sequencing studies is the discovery of a short intervening DNA segment within the region coding for the tRNA, which is not found in the mature tRNA molecule. In the cloned tRNA^{Tyr} genes, the insert is 14 bp long and invariant except in one residue among the four genes (Goodman et al., 1977). There is an 18 bp insertion in the three tRNA^{Phe} genes which, again, is essentially identical for these three genes (Valenzuela et al., 1978), but shows no homology to the tRNA^{Tyr} insertion sequence. In both families however, the intervening sequence, which contains no inverted repeats, is located immediately to the 3'-side of the anticodon triplet. Two other yeast tRNA genes that have been sequenced (tRNA^{Arg} and tRNA^{Asp}) do not contain intervening sequences (Knapp et al., 1978).

Transfer RNA precursor molecules are transcribed by RNA polymerase IIIa (Weinmann and Roeder, 1974), and synthesis, similar to that of 5 S rRNA, is not coordinately regulated with that of rpRNA molecules (Zylber and Penman, 1971; Ford, 1971). Even though the genes are clustered in some eukaryotes, there is no evidence for multigene transcription units (Perry, 1976; Feldmann, 1977). Mature cytoplasmic tRNAs consist of 80 nucleotides and are derived from precursor molecules which, depending on the particular tRNA species, range from 15–35 bases larger than the mature tRNAs (reviewed by Perry, 1976). In the case of yeast tRNA genes containing inserts (tRNA^{Phe} and tRNA^{Tyr}), it has been determined that the intervening sequences are transcribed as part of the tRNA precursor (Knapp et al., 1978). A yeast mutant is available which accumulates the tRNA precursors containing intervening sequences. The presence of the intervening sequence could entirely account for the observed size of the precursors. An enzymatic activity obtained from the ribosomal wash fraction of wild-type yeast specifically removes the intervening sequences (Knapp et al., 1978).

In X. laevis, tRNA$_1^{Met}$ (330-fold reiterated gene) and tRNA$_2^{Met}$ (170-fold reiterated gene) molecules are present in about equal amounts in kidney cells, whereas tRNA$_2^{Met}$ appears to be absent in liver and ovary (Clarkson et al., 1973a). Such observations suggest that the concentration of a particular type of tRNA is not based simply on its genomic redundancy, but rather is controlled by a differential synthesis (Clarkson and Birnstiel, 1973).

Additional evidence for the regulation of tRNA synthesis comes from studies of the relative amounts of tRNAs in the posterior portion of the silk gland of Bombyx mori (Siddiqui and Chen, 1975). During developmental periods in which silk fibroin is the predominant translational product, the proportions of total tRNA corresponding to tRNA^{Gly}, tRNA^{Ala}, and tRNA^{Ser} approximate the relative abundance of the three most common amino acids in fibroin: glycine, 44% of the total residues; alanine, 29%; and serine, 12% (Sprague, 1975). There is evidence in this system that differential degradation of mature tRNAs does not occur, suggesting perhaps that a control at the transcriptional or pre-tRNA maturation level operates to adapt the tRNA population to the biosynthetic requirements of the cell (Fournier et al., 1976).

Valenzuela et al. (1978) and Knapp et al. (1978) point out that the recently discovered intragenic insertion sequences may introduce another level of tRNA processing in yeast that could be regulatory in nature. These authors show that when the intervening sequence is transcribed, a secondary structure might be assumed by the putative tRNA precursor which would eliminate anticodon accessibility (by specific base pairing with a triplet in the intervening sequence) and prevent the tRNA from functioning in protein synthesis until the intervening sequence is excised.

3. Histone Genes

Histones are small basic proteins that complex with DNA in a near 1:1 mass ratio to form nucleosomes, which are the basic structural units of eukaryotic chromosomes (reviewed by Kornberg, 1977). Chromatin fibers visualized by electron microscopy appear as periodic arrays of closely spaced nucleosomes which are approximately 100 Å in diameter and roughly spherical in shape. Each nucleosome consists of approximately 140 bp of DNA coiled on the surface of an octomeric histone core which contains two molecules each of histones H2A, H2B, H3, and H4. Neighboring nucleosomes are connected by a DNA strand about 50 bp long, which has associated with it a single molecule of histone H1.

Details regarding the molecular organization of histone genes in sea urchins and Drosophila have recently been obtained. During early sea urchin embryogenesis, the supply of histone proteins is met, in part, by an intense synthesis of histone mRNA (Kedes and Gross, 1969), thus allowing histone-specific nucleic acid probes to be isolated. DNA–mRNA hybridization experiments show that the histone genes of various echinoderms are repeated from 300 to 1000 times per haploid genome (Kedes and Birnstiel, 1971; Weinberg et al., 1972; Skoultchi and Gross, 1973). Similar studies indicate that the Drosophila genome contains about 100 copies of each histone gene (Lifton et al., 1977) and that H. sapiens has some 30–40 repeats of each histone gene per haploid genome (Wilson and Melli, 1976).

Kedes and Birnstiel (1971) reported that echinoderm DNA com-

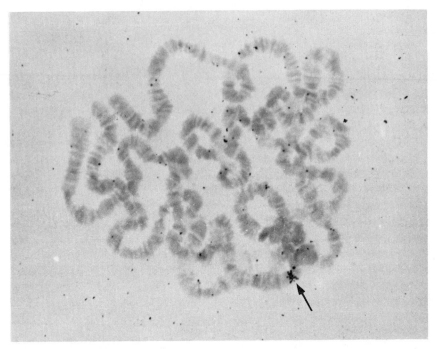

FIG. 14. Autoradiograph of polytene chromosomes from *D. melanogaster* hybridized *in situ* with labeled histone mRNA isolated from embryos of the echinoderm, *Psammechinus miliaris*. The cluster of silver grains (arrow) shows the localization of the histone genes in region 39D–E on the left arm of chromosome 2. (From Birnstiel *et al.*, 1973, reproduced with permission.)

plementary to histone mRNA forms a distinct buoyant density band in CsCl, suggesting a tandem repetition of histone genes. Further evidence for the clustering of histone genes in eukaryotes resulted from *in situ* hybridization experiments (Pardue *et al.*, 1977) in which histone mRNA from sea urchins was hybridized to the salivary gland chromosomes of *D. melanogaster*. As Fig. 14 shows, histone sequences are localized in the 39D–E region of chromosome 2. When individual mRNA fractions, specific for one histone only, were used to probe gene organization, hybridization occurred throughout the 12-band region, indicating that the various histone genes of *Drosophila* are interspersed throughout the region.

More recently, the histone DNAs of various sea urchin species and *D. melanogaster* have been cloned in plasmids, and definitive evidence for the interdigitated clustering of repetitive histone genes has been provided by restriction enzyme mapping. Such studies, combined with electron microscopic techniques, have also established the 3'–5' orientation of the five primary histone genes in both *Psammechinus miliaris* and *Strongylocentrotus purpuratus* (Kedes, 1976; Schaffner *et al.*, 1976; Gross *et al.*, 1976; Cohn *et al.*, 1976; Wu *et al.*, 1976). As Fig. 15 illustrates, the five individual genes of sea urchin histone DNA are clustered within repeating units of 6–7 kb, and the coding sequences for each of the genes are present on the same DNA strand. Similar to the tandemly repeated genes discussed previously, the histone coding sequences are separated by spacer segments, and partial denaturation mapping of histone DNA by electron microscopy (Portmann *et al.*, 1976) demonstrates that the spacers have relatively high A + T content (Fig. 16). The structural genes are included within the G + C-rich stretches, but these latter stretches are somewhat longer than necessary to code for histone proteins (Portmann *et al.*, 1976). DNA sequencing of more than half of a 6 kb histone repeat unit from *P. miliaris* has shown that the spacers are 32–44% GC, the structural gene sequences are 46–61% GC, and the "prelude" sequences of 100–200 bp in front of the structural genes are 50–71% GC (Schaffner *et al.*, 1978). The spacers that have been sequenced have a relatively simple nucleotide arrangement but no extensive homologies are obvious between them. They do not encode proteins, but are

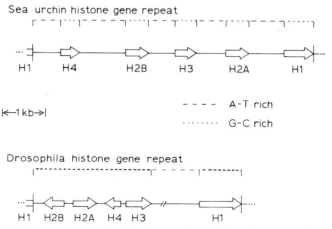

FIG. 15. Histone gene repeat units of sea urchin and *D. melanogaster*. In both cases, one each of the five main histone genes is contained within the repeat unit. However, the five genes are located on the same DNA sense strand in the sea urchin, whereas two of the genes are on one strand and the other three are on the opposite strand in *Drosophila*. (Sea urchin repeat adapted with permission from Cohn *et al.*, 1976. *Drosophila* and sea urchin repeats adapted with permission from Lifton *et al.*, 1977.)

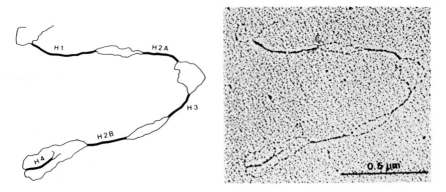

FIG. 16. Electron micrograph and interpretive drawing of partially denatured histone DNA of the sea urchin, *P. miliaris* (for details, see Portmann *et al.*, 1976). The bubblelike, denatured regions are A + T-rich regions separating the histone mRNA coding sequences. (Photograph and drawing kindly supplied by R. Portmann and M. Birnstiel, Institut für Molekularbiologie II der Universität Zürich.)

also not a simple satellite sequence. Slight-to-moderate heterogeneities in length and restriction enzyme sites, which most likely map to the spacer regions, exist between different cloned histone gene repeats of sea urchins (Overton and Weinberg, 1978; Schaffner *et al.*, 1978). No intervening, non-mRNA sequences are found within individual histone genes.

Lifton *et al.* (1977) have isolated and characterized the histone DNA of *D. melanogaster*. As expected from the *in situ* hybridization pattern reported by Pardue *et al.* (1977), the individual histone genes of *Drosophila* are interdigitated in tandem repeats. However, as Fig. 15 shows, the organization of the genes within the repeats is quite different from the sea urchin in that the coding strands alternate with respect to transcriptional polarity. Since the echinoderms and *Drosophila* diverged some 600 million years ago (Dickerson, 1971), selection pressure appears to have maintained the clustered interdigitated arrangement of histone genes. Lifton *et al.* (1977) suggest two possible functional advantages of such an arrangement. First, the dosage of each individual histone gene would be maintained by a clustered interdigitated arrangement in the event of unequal crossovers, and, second, linkage of the five histone sequences may facilitate the coordinate expression of the histone genes.

A transcriptional mechanism for coordinate synthesis of histone mRNAs is attractive since it is known that four of the five major histones are present in equimolar amounts in eukaryotic chromatin, with H1 histone being present in one-half that amount (Kornberg and Thomas, 1974). The arrangement of the histone genes in sea urchins would allow a transcriptional scheme in which a single RNA molecule, spanning the entire histone gene repeat, is the primary transcriptional product. However, no such transcripts have been detected (Kunkel *et al.*, 1977). In *Drosophila*, it is clear that more than one initiation must occur for the production of histone mRNAs because of the opposite strand polarities. Identical sequences have been shown to exist in an inverted repetitive array at the boundaries between the H3–H4 and H2A–H2B genes in *Drosophila*. Lifton *et al.* (1977) speculate that such sequences may be bidirectional promotors which, when derepressed, effect an equimolar production of the mRNAs for those four histones. The inability to detect transcripts complementary to histone DNA that are appreciably larger than mature histone mRNAs in either sea urchins (Kunkel *et al.*, 1977) or *Drosophila* (Lifton *et al.*, 1977) supports the concept of individual promotors for each histone coding sequence in these organisms. The largest transcription products of histone genes in a sea urchin *in vitro* system are discrete in length (650–1100 nucleotides) and are somewhat larger than the relatively small histone mRNAs (Levy *et al.*, 1978). Each of the larger transcripts seems to contain sequences complementary to only one of the histone genes and is about 50% larger than the respective mRNA. On the other hand, studies by Melli *et al.* (1977) suggest that a large precursor to histone mRNAs (up to 4×10^6 daltons) may be the primary transcript in HeLa cells.

Using procedures developed by Pukkila (1975), in which labeled DNA is hybridized to nascent chromosomal RNA *in situ*, Old *et al.* (1977) studied the transcription of histone mRNA on the lampbrush chromosomes of *Triturus carnifex*. Using various sea urchin histone DNA probes, it was shown that hybridization occurs on some seven loop pairs of the newt chromosomes. Furthermore, the pattern of hybridization suggests that histone mRNAs are components of very large primary transcripts (Fig. 17). It was concluded that, in some cases, transcription of histone DNA in the amphibian oocyte can occur following initiation upstream from the histone DNA sequences and that termination may occur downstream from such sequences, with the histone RNA sequences being cut from the transcripts before termination of transcription. Whether this pattern of histone RNA synthesis reflects the regulatory mechanisms existing in somatic cells of the same species remains to be determined.

4. *Immunoglobulin Genes*

The structure of all functional antibody molecules (immunoglobulins) is based on a protomer containing one light (L) polypeptide chain (214 amino acids) and one heavy (H) chain (446 amino acids) linked

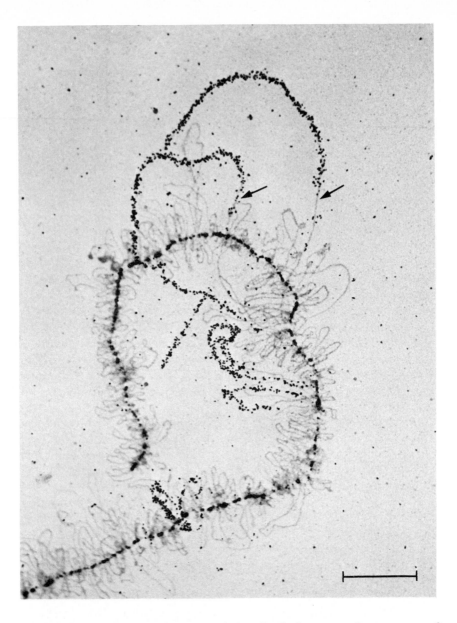

FIG. 17. Autoradiograph of a portion of a lampbrush chromosome from an oocyte of the newt, *Triturus cristatus carnifex*, hybridized *in situ* with labeled histone DNA of the echinoderm *S. purpuratus*. Conditions were used such that hybridization occurred exclusively to nascent chromosomal RNA. Silver grains indicate localization of histone RNA sequences on three pairs of loops on this portion of chromosome 1. Note that labeling of the largest pair of loops stops abruptly at intermediate positions (arrows), indicating the absence of histone coding sequences in the RNA further along the loops. One loop of this pair was broken during specimen preparation. Scale bar represents 20 μm. (From Old *et al.*, 1977, reproduced with permission.)

through a disulfide bond. Two L–H protomors held together by specific interchain disulfide bonds and noncovalent forces constitute the basic four-chain antibody (7 S antibody), while some classes of antibodies are present predominantly as multimers of the basic four-chain structure (Edelman, 1973; Williamson, 1976). Each polypeptide chain contains variable (V) and constant (C) regions relative to amino acid sequences. The differences in the constant region sequences are the primary bases for classification of the various immunoglobulins (Igs). The V regions of the H and L chains allow for many permutations of amino acid sequences and thereby antigenic-binding specificity, whereas the C regions of the H chains are sites of antibody effector activities, such as complement fixation, cell membrane receptor interactions, and transplacental transfer.

Genetic and protein structural evidence indicates that two discrete DNA segments, coding for a V region and a C region, respectively, combine to specify each H and each L polypeptide chain (the so-called two gene–one polypeptide hypothesis for Ig synthesis). V and C sequences fall into three families of antibody genes, as depicted in Fig. 18. Each family is thought to consist of a tandem array of V genes linked to related C genes. The genes of the different families do not exhibit genetic linkage. The reiteration frequency and molecular organization of various Ig genes have been intensively studied using mRNA probes (or cDNA copies thereof) purified from mouse myeloma cell lines (Kuehl, 1977). Such studies, combined with amino acid sequence data, suggest that the total number of V_κ L chain genes in germ line cells is in the 50–100 range, whereas the number of C_κ genes is in the 1–5 range (Cohn et al., 1974; Tonegawa et al., 1977). A more recent study indicates that there may be 200 or more V_κ L-chain genes. There are at least 26 V_κ subgroups distinguished by amino acid sequence, and Seidman et al. (1978) have shown that each of several subgroups consists of six to eight discrete V region genes. In addition, each subgroup has a characteristic flanking sequence that extends over a thousand nucleotides on each side of the gene and that may promote recombination within subgroups.

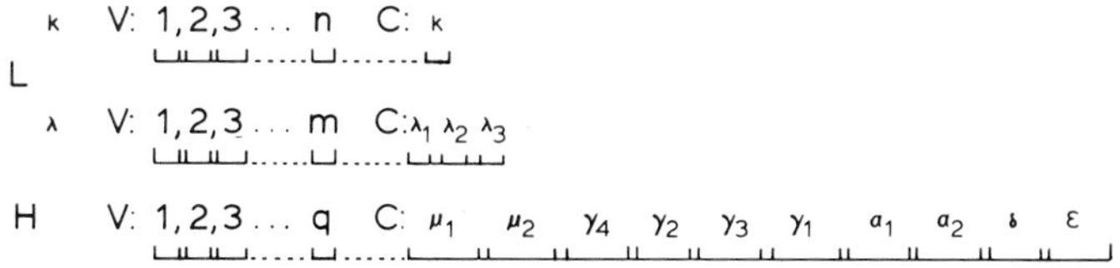

FIG. 18. The arrangement of three families of genes coding for immunoglobulins in humans. (Adapted, with permission, from A. R. Williamson, 1976, Annu. Rev. Biochem. 45, 467–500. Copyright © 1976 by Annual Reviews, Inc.)

On the other hand, hybridization studies of V_λ and C_λ genes indicate that both of these sequences are represented only one to three times in the mouse germ line genome (Leder *et al.*, 1975; Tonegawa *et al.*, 1977). Since there are already nine different known V_λ amino acid sequences, the low V_λ gene number estimate suggests that a mechanism exists by which multiple V_λ region sequences are derived during differentiation from a very few germ line genes (Rabbitts and Milstein, 1977; Tonegawa *et al.*, 1977).

V and C sequences are covalently linked in Ig mRNA molecules. The joining of such sequences theoretically is possible at either the genome or transcript level. However, direct evidence for the translocation of DNA sequences to form closely linked V and C sequences during lymphocyte differentiation has been provided by experiments in which κ light chain mRNAs were hybridized to restriction fragments of embryonic and myeloma DNAs (Tonegawa *et al.*, 1977). *Bam*HI restriction endonuclease was used to produce fragments of embryonic DNA and DNA of the myeloma line, MOPC321. When the restricted embryonic DNA was probed with κ mRNA purified from the MOPC321 line, two hybridization components were found, one having sequences homologous with the C portion and the other with the V portion of the mRNA. Alternatively, when restricted MOPC321 DNA was probed, a single new DNA fragment was found which hybridized with both the C and V portions of the mRNA. The V sequences of a κ light chain mRNA purified from a myeloma line (MOPC21) belonging to another subgroup were shown to be unlinked to the C_κ sequences in the MOPC321 DNA, indicating that a specific joining to the C_κ sequences of a single V_κ segment had occurred in the derivation of the MOPC321 line. As an additional control, the hybridization pattern of the MOPC321 mRNA to restriction fragments from a λ chain myeloma was found to be indistinguishable from the pattern obtained with embryonic DNA, showing that mutual exclusiveness for V–C sequence joining exists for κ and λ chains, as well as for κ chains of different subgroups.

As shown by restriction enzyme mapping, translocation of V_λ and/or C_λ DNA sequences occurs during lymphocyte differentiation into a λ-light chain-producing cell (Tonegawa *et al.*, 1976) and in a different κ chain-producing myeloma line, MOPC21 (Rabbits and Forster, 1978). The latter authors showed that the DNA translocation is probably specific to lymphocytic cells, because restriction enzyme-cleaved liver and kidney DNA exhibited similar "germ line" patterns of hybridization with cDNA probes to V and C regions.

Recent experiments have indicated, however, that even after translocation of V and C genes to a neighboring location, they are still not immediately contiguous on the chromosome (Brack and Tonegawa, 1977; Rabbitts and Forster, 1978). A DNA fragment has been cloned which contains both the V_λ and C_λ regions, presumably after translocation, from the light chain-producing plasmacytoma, HOPC2020, and the molecular topography of the V and C regions examined by electron microscopy (Brack and Tonegawa, 1977). R-loop mapping of hybrids between purified λ chain mRNA and the cloned DNA fragment carrying both the V and C regions for that mRNA shows that the V_λ and C_λ DNA sequences are separated by a 1250-base segment of DNA whose sequences are not present in the λ light chain mRNA. Rabbitts and Forster (1978) also have evidence for a gap between V_κ and C_κ genes in the MOPC21 cell line. They were able to show that the hybrid formed between cDNA (specific to the κ chain mRNA) and the myeloma DNA fragment containing V and C regions was not fully protected from a single strand-specific S_1 nuclease, suggesting that the mRNA and DNA sequences are not colinear.

Brack *et al.* (1978) subsequently explained the origin of this intervening sequence by identifying the site of the "homology switch" on the λ chain gene, where embryonic DNA segments are joined in the myeloma cell. The homology switch occurs near the boundary of the 3′-end of the V sequence and the 1.2 kb intervening sequence. Actually, it appears that the nucleotides coding for the final 15 amino acids of the V region comprise this joining region and are separated in embryonic DNA from the C region coding segment by the same 1.2 kb intervening sequence that is conserved in the myeloma gene. It is the remaining 5′ V region coding sequence that is more distantly removed from the joining region and the C gene region before the translocation event (Brack *et al.*, 1978)

Gilmore-Hebert and Wall (1978) have recently found that a mouse κ light chain mRNA is derived from large nuclear RNA in P3 myeloma cells. A DNA probe produced by molecular cloning was used to find three discrete size classes of nuclear RNA that contain the light chain mRNA sequence. The largest class is ~11 kb long (40 S), or ten times the length of the mature mRNA, and appears after a 5-minute pulse with [³H]uridine. After ~10 minutes of labeling, an RNA molecule ~4 kb long (24 S) which hybridizes to the probe becomes apparent. This is soon followed by the appearance of a nuclear RNA species that is about the same size (13 S) as the cytoplasmic message, which can be detected in the cytoplasm only after 20–30 minutes of labeling. Thus, this κ light chain is apparently generated by the stepwise cleavage and processing of a large nuclear transcript. Further information re-

garding the nature of the nuclear precursor molecules comes from
Rabbitts (1978), who repeated the cDNA hybridization–S1 nuclease
digestion experiment with a 27 S nuclear RNA complementary to light
chain mRNA. He found that the cDNA hybrid formed with 27 S nuclear RNA was not completely protected from nuclease digestion.
Only the 13 S hybrid provided complete protection, suggesting that
the intervening sequence was still present in the 27 S precursor. Comparison of the VC junction sequence in the 27 and 13 S molecules supported this conclusion. Thus, a light chain mRNA is apparently generated first by DNA translocation followed by transcription of a large
nuclear precursor molecule that is sequentially cleaved and processed.

5. *Evolution of Tandemly Repeated Genes*

Excluding the special case of Ig genes from consideration, the repeat units of the tandemly repeated multigene families discussed in
the previous sections exhibit striking similarities. Each gene is separated from neighboring genes by spacer segments of DNA that are not
typically transcribed and, at least for rDNA and 5 S DNA, consist of
repetitive sequences. Comparisons of the families of rDNA and 5 S
DNA repeats in the closely related species of X. *laevis* and X. *borealis*
(formerly misidentified as X. *mulleri;* Brown *et al.,* 1977b) raise several questions regarding the evolution of redundant gene families.
Within each species, the rDNA repeats are similar with respect to the
nucleotide sequences of both the rpRNA genes and the NTSs. However, whereas the sequences coding for the 18 S and 28 S rRNAs are
the same for both species, the spacer region sequences have diverged
markedly (Brown *et al.,* 1972; Forsheit *et al.,* 1974). Thus, while the
rRNA cistrons have remained essentially unchanged, the repetitious
spacer segments within each species have evolved together. Similarly, the 5 S gene sequences of the two species have not diverged significantly, while the spacer regions exhibit no detectable molecular
hybridization between the species. The 5 S DNA spacer region sequences within each species, however, are quite similar (Brown and
Sugimoto, 1973a,b). Furthermore, the X. *borealis* spacer is about
twice the size of typical X. *laevis* spacers. It is clear, therefore, that a
mechanism must exist either to propagate or to eliminate mutations
relatively rapidly throughout repetitive gene families within species
so that members of each family evolve together. This phenomenon
has been termed "horizontal evolution" (Brown *et al.,* 1972).

Several hypothetical sequence-change mechanisms, which can be
subclassified as "sudden" or "gradual," have been proposed to ac-

count for horizontal evolution (Brown and Sugimoto, 1973b). Examples of sudden mechanisms are the master–slave concept (Callan, 1967; Thomas, 1970) and the contraction–expansion model (Brown *et al.*, 1972), both of which, in their strictest form, require that all members of a repetitive gene cluster be identical to one another. The facts that adjacent repeating units of chromosomal rDNA (Wellauer *et al.*, 1976b), 5 S DNA (Carroll and Brown, 1976b), and tDNA (Yen *et al.*, 1977) can differ in size, and that patterns of rDNA spacer lengths within different NOs are inherited with no detectable changes (Reeder *et al.*, 1976b) argue against sudden mechanisms as explanations for horizontal evolution within tandem multigene families.

The most plausible mechanism available at present for gradual horizontal evolution is the unequal crossover model advanced by Smith (1973, 1976). In this model, repeated unequal crossing-over events between members of a family of repeats can lead to either the elimination or the spread and fixation of variants within the cluster. It has been proposed that the simple repetitive sequence regions within spacers may be involved in this mechanism, since their internal homologies would allow for out-of-register pairing that could generate unequal crossovers both with respect to the number of subrepeats within spacers as well as to one or more entire repeats (Carroll and Brown, 1976b; Botchan *et al.*, 1977). The function of such internal repetition, then, might be to increase the frequency of recombination and thereby to promote sequence homogeneity within gene families. If internal repetition is a general property of spacers of tandem gene clusters which promotes recombination, then a DNA segment can have a function that is not dependent on a specific nucleotide sequence, but, rather, on the presence of homologous sequence repeats (Carroll and Brown, 1976b). In strong support of this hypothesis is the intraspecies homogeneity of spacer sequence within multigene families as opposed to the highly diverged sequence characteristics observed for spacers of functionally homologous multigene families of different species.

In both X. *laevis* and X. *borealis,* the thousands of 5 S genes per haploid genome are distributed at the ends of the long arms of most, if not all, of the chromosomes within the set. In both species, there are clusters of genes that are either somatic cell specific or oocyte specific in their expression. The analysis of variants within the 5 S gene sequences of both species suggests that there is a mechanism that allows occasional genetic exchange between somatic- and oocyte-specific genes (Ford and Brown, 1976). The simplest explanation for the tissue-specific control of 5 S genes is that different gene sequence families are clustered on different chromosomes which are independently regulated with respect to gene expression. If this is the case, one possible mechanism for exchange could be infrequent recombinations between 5 S gene clusters on nonhomologous chromosomes during the bouquet stage of meiosis when the telomeres of the chromosomes are in close proximity (Pardue *et al.*, 1973). Such nonhomologous exchanges could be limited to 5 S genes because of their telomeric location, but would also have to be limited to even-numbered double-stranded exchanges, so that the integrity of individual chromosomes would be maintained.

B. INTERSPERSED MULTIPLE-COPY GENES

In Section III,A, of this chapter, the organizational patterns of five tandemly repeated gene families have been described. Four of the multigene families considered to this point, namely, rDNA, 5 S DNA, tDNA, and histone DNA, share a distinct similarity in the organization of their repeating units. This observation might be taken to indicate that a tandem arrangement is generally present for the members of all repeated gene families. It has recently been found, however, that there are moderately repetitive gene families in *D. melanogaster* whose members are widely dispersed throughout the genome.

Hogness and co-workers have characterized the sequences of two repeated gene families of *D. melanogaster*, designated *copia* and *412*, which are complementary to abundant, cytoplasmic poly(A) + mRNA (Rubin *et al.*, 1976; Finnegan *et al.*, 1977). DNA fragments complementary to abundant, poly(A) + mRNA were isolated by challenging a random selection of cloned *Drosophila* DNA fragments with radioactively labeled mRNA. Restriction mapping procedures were employed to define the structural gene boundaries within a number of independently cloned DNA fragments. Similar restriction enzyme maps were obtained for the *412* coding sequences in six independently cloned fragments. That these six chromosomal segments are members of a dispersed multigene family was established by three criteria. First, the frequency at which these DNA sequences were picked up from random clones was approximately 30-fold greater than the frequency expected of single copy sequences. Second, it was found that the sequences flanking the mRNA coding regions of the DNA segments were not homologous for each of the six clones which otherwise shared common coding sequences. Third, when the coding segment of one of the six cloned fragments was used as a probe for *in situ* hybridization with polytene chromosomes, hybridization was

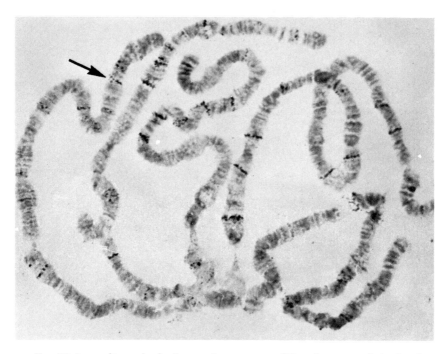

FIG. 19. Autoradiograph of polytene chromosomes of *D. melanogaster* hybridized *in situ* with labeled cRNA homologous to coding sequences of *412* DNA. The arrow points to the 3C2-7 region of the X chromosome, which is one of the approximately 30 sites labeled. Note the wide dispersion of labeled sites. (From Rubin *et al.*, 1976, reproduced with permission.)

found to occur with equal intensity at approximately 30 sites. As Fig. 19 shows, the gene family designated as *412* contains members at interspersed locations on each of the chromosome arms of the *D. melanogaster* genome. Although the noncoding sequences were found to differ in the *412* cloned fragments, an identical repetitive DNA sequence was discovered at both boundaries of the coding regions of each of three *412* clones analyzed in detail. Electron microscopic techniques were used to show that the repetitive elements, which measure approximately 0.5 kb, are aligned with the same polarity rather than in an inverted manner and that they occur at the approximate boundaries between mRNA-coding and noncoding sequences.

The organization of the multigene family designated as *copia* was also studied and found to be similar to that of the *412* family (Finnegan *et al.*, 1977). Ten different clones containing the *copia* structural gene sequence were analyzed. In each, the 4.2–4.9 kb coding sequence was found to exhibit the same restriction digestion pattern.

However, as with the *412* family, the restriction fragments containing the endpoints of the structural gene sequences were observed to vary considerably among the 10 *copia*-containing clones. The structural sequence of *copia* produces an *in situ* hybridization pattern on polytene chromosomes which is characterized by nearly equivalent labeling at approximately 35 locations interspersed throughout the genome. Perhaps the most striking similarity between the *copia* and *412* gene families is the presence of a repetitive sequence at both ends of each structural gene region. The repetitive *copia* segments were found to be approximately 0.3 kb long compared to the 0.5 kb repetitive sequences of the *412* gene family. No appreciable homology was found to occur between the gene-bracketing repetitive sequences of *copia* and *412* DNA.

Although the functions of these two families of genes have not been firmly established, compelling evidence indicates that both families code for mRNA. A better understanding of this type of multigene family will come with the identification of the gene products. However, these experiments clearly establish the existence of multigene families whose members are interspersed throughout the genome, and whose function most likely is the production of mRNA. How such dispersed multigene families originate and maintain homologous mRNA-coding and noncoding sequences remain intriguing questions. It seems probable, however, that significant differences will be found between the evolutionary mechanisms involved in the organization of tandemly repeated gene families versus those for dispersed multigene families.

IV. SINGLE-COPY GENES

Unlike histone DNA and the interspersed multigene families of *D. melanogaster*, most mRNA-coding gene sequences appear to exist only once per haploid genome equivalent in eukaryotic chromosome sets. Early evidence in support of this was provided by kinetic analysis of mRNA–DNA hybridization (Davidson and Britten, 1973). More recently, it has been established that a number of specific structural genes are represented at, or very near, single-copy gene proportions within eukaryotic genomes. Using either purified mRNA or cDNA probes, various workers have established the single-copy nature of the duck β-globin gene (Bishop and Rosbash, 1973), the mouse β-globin gene (Harrison *et al.*, 1974), the chicken ovalbumin gene (Harris *et al.*, 1973), and the fibroin gene of the silkworm (Suzuki *et al.*, 1972). It is important to note that the single-copy reiteration frequency of such structural gene sequences has been determined both for the genomes of differentiated tissues which synthesize large quantities of the specific gene product as well as for tissues in which the sequences are either transcriptionally silent or expressed at a very low level. Therefore, a gene amplification mechanism, which allows an extreme accumulation of rRNA in some types of oocytes, does not appear to operate in these specialized somatic cells that produce relatively large quantities of a single gene product.

The genomic organization and primary transcription products of four single-copy genes which have been studied in detail will be considered here. These are the β-globin gene of mouse and rabbit tissue, the ovalbumin gene of the chicken, the silk fibroin gene of *Bombyx mori*, and the Balbiani ring 2 locus of *Chironomus tentans*.

A. β-GLOBIN GENE

Hemoglobin functions to transport oxygen in the bloodstream. It is a tetramer composed of two pairs of similar globin polypeptide chains, which in the major mammalian adult type are designated α- and β-globin. There are generally several other β-globin-like chains that interact with α-globin in minor adult and fetal hemoglobins, and whose genes, in humans, are present in single-copy proportion and are genetically linked with the β-globin gene (reviewed by Weatherall and Clegg, 1976).

In terms of genomic and transcriptional unit organization, the single-copy β-globin gene has been most extensively studied. A physical map has been obtained of the restriction endonuclease cleavage sites in the DNA sequences within and surrounding the rabbit β-globin gene (Jeffreys and Flavell, 1977a,b). There is a repetitive DNA sequence (10^3–10^4 copies/haploid genome) within 2.5 kb of the 3'-end of the gene, but the 2 kb contiguous with the 5'-end are present in less than 100 copies, and perhaps one copy, per genome (Flavell *et al.*, 1977). It was also discovered that the gene contains a 600 bp segment in the middle of the structural gene sequence which is not complementary to sequences in the β-globin mRNA molecule. Specifically, a *Hae*III restriction fragment of the globin DNA probe, which had been obtained by reverse transcription of the mRNA, hybridized to a *Hae*III restriction fragment of genomic DNA that was more than twice as large as itself (Jeffreys and Flavell, 1977b). The 600 bp insert occurs somewhere within the coding sequence for amino acid residues 101–120 of the 146 residue β-globin chain, and it is found in the globin gene obtained from several different tissues, including erythropoietic spleen which actively synthesizes globin.

Restriction enzyme and EM R-loop mapping have shown that

mouse (Leder *et al.*, 1977; Tilghman *et al.*, 1978a) and human (Flavell *et al.*, 1978; Mears *et al.*, 1978) β-globin genes also contain an intervening sequence which maps at a similar position in the coding region. In humans, the β-globin gene has an 800–1000 bp intervening sequence which occurs within the sequence coding for amino acids 101–120. The mouse insert is about 550 bp long and occurs immediately following the codon corresponding to amino acid 104 (Tilghman *et al.*, 1978a). Another small intervening sequence is probably also present near the transcription initiation site in the mouse globin gene. Two cloned β-globin genes have been obtained from one mouse strain which coordinately expresses the two nonallelic genes (Tiemeier *et al.*, 1978). Both have inserts of similar size (550 bp) at similar or identical locations. Heteroduplex mapping between the two reveals homology only between the mRNA-coding regions (as expected) and between a few hundred bp immediately adjacent to the structural gene sequences, and not between the central portion of the large intervening sequences (Tiemeier *et al.*, 1978).

The size and composition of the transcription unit have been analyzed by investigating the initial transcription products of the gene. The existence of intervening sequences within structural sequences suggests three possible mechanisms for mRNA biogenesis: (1) the intervening DNA may be transcribed as part of primary transcripts also containing the coding sequences, which are then enzymatically processed by the excision of intervening sequences and the splicing of the structural gene sequences to form the mature mRNA; (2) the intervening DNA segment may not be transcribed but somehow "loop out" during transcription, allowing the production of primary transcripts in which the coding regions of the mRNAs are contiguous; or (3) the coding elements may be transcribed by multiple initiation events, producing small pre-mRNAs containing no intervening sequences, which are spliced for mRNA maturation. As reviewed below, mouse β-globin mRNA synthesis proceeds by the first mechanism.

β-globin mRNA is transcribed in a molecule which is at least twice (Curtis *et al.*, 1977a; Ross and Knecht, 1978) and perhaps seven times (Bastos and Aviv, 1977; Strair *et al.*, 1977) its coding length. The larger precursor (~5 kb) is not detected by all research groups (cf. Curtis *et al.*, 1977a; Bastos and Aviv, 1977), and has not been assigned specifically to the α or β gene. It may be that the largest primary transcript is present in very small steady state amounts and is rapidly processed, perhaps even before release from the DNA template.

The smaller β-globin mRNA precursor has been carefully analyzed. Ross and Knecht (1978) have shown that it is 1860 nucleotides long as compared to the 780 nucleotide mature mRNA. R-loop formation between the mouse mRNA precursor and the cloned β-globin gene results in an uninterrupted RNA-DNA hybrid, indicating that the intervening sequence(s) in the gene are transcribed in the mRNA precursor (Tilghman *et al.*, 1978b; Kinniburgh *et al.*, 1978). Kinniburgh *et al.* (1978) also have biochemical evidence that the initial transcript in the mouse includes the large insert (estimated in this study to be 780 bp long) and a smaller insert (less than or equal to 125 bp) near the 5' mRNA terminus. The combined lengths of the gene-coding regions plus the intervening sequences is probably enough to account for the 1860 nucleotide precursor. The remainder of the non-coding sequences at the 5'- and 3'-termini of the precursor are relatively short and probably correspond to those which are conserved in the mature mRNA (Efstratiadis *et al.*, 1977; Kinniburgh *et al.*, 1978). Indeed, the 5'-cap (Curtis *et al.*, 1977b) and the 3'-poly(A) terminus (Bastos and Aviv, 1977) of the mRNA precursor are indistinguishable from that of globin mRNA.

B. OVALBUMIN GENE

Ovalbumin is the major secretory protein of the hen oviduct. The gene for ovalbumin in chicken chromosomes exists in single-copy proportion (Harris *et al.*, 1973). Several genomic fragments containing all or part of the ovalbumin gene have been cloned and analyzed by restriction enzyme and R-loop mapping (Doel *et al.*, 1977; Breathnach *et al.*, 1977; Weinstock *et al.*, 1978; Lai *et al.*, 1978; Garapin *et al.*, 1978; Mandel *et al.*, 1978; Dugaiczyk *et al.*, 1978). The sequence of ovalbumin mRNA, including 5'- and 3'-nontranslated regions, is also available for comparison to the genomic sequence (McReynolds *et al.*, 1978). It has been found that the sequences that code for ovalbumin mRNA (1859 bp) are not contiguous in genomic DNA but occur in at least eight segments, separated by seven intervening sequences in a 7 kb stretch of DNA (Breathnach *et al.*, 1978; Dugaiczyk *et al.*, 1978; Mandel *et al.*, 1978). The noncontiguous structural gene sequences occur in the same order and relative orientation as the mRNA (Garapin *et al.*, 1978; Dugaiczyk *et al.*, 1978). The intervening sequences are clustered in the 5'-half of the structural gene sequence and comprise more than 80% of the DNA in this region. The nontranslated 5'-end of the mRNA is encoded in a segment of DNA separated from those coding for the protein (i.e., a leader sequence), but the long 3'-nontranslated region (637 nucleotides) is uninterrupted and contiguous with the 3'-end of the structural gene (Breathnach *et al.*, 1978; Dugaiczyk *et al.*, 1978; Mandel *et al.*, 1978).

This interrupted genomic organization does not represent a somatic translocation event, as it is the same in other chicken tissues, including gametes, and in chick oviduct before and after mRNA synthesis is stimulated by steroid hormone administration (Breathnach *et al.*, 1978; Weinstock *et al.*, 1978; Woo *et al.*, 1978). Like the ovalbumin gene sequence, the ovalbumin inserts are present in single-copy proportion in the chicken genome (Woo *et al.*, 1978).

The ovalbumin transcription unit is apparently comprised of both mRNA-coding and intervening squares. The intervening sequences are coordinately expressed with the gene sequences upon estrogen stimulation, but are present in fewer copies per nucleus, probably because of preferential and rapid degradation (Woo *et al.*, 1978). Roop *et al.* (1978) were able to demonstrate nuclear RNA species containing ovalbumin mRNA sequences that were up to four times larger than the mature message. They separated nuclear RNA on agarose gels prior to hybridization of an ovalbumin probe and identified six to seven discrete molecular weight species that were 1.3 to 4 times larger. The intervening sequences also hybridized to these RNA species in a manner which suggests a processing order exists for the removal of the inserts and biogenesis of the mature mRNA. Based on the size of the largest detected precursor, and the size of the genomic segment containing gene and intervening sequences, it is probable that the non-mRNA coding DNA in the transcription unit is primarily or completely intragenic.

Sequencing of the cloned genomic ovalbumin DNA at the junction of the structural gene and insert sequences reveals a similar sequence for all the 5'- and 3'-extremities of the intervening sequences (Breathnach *et al.*, 1978). A common excision–ligation point of two to four bases can be defined for both boundaries, consistent with a common enzymatic recognition site. In any particular case, however, it appears that splicing could take place in one of several different reading frames and still form the correct mRNA because of direct sequence repeats at either end of the intervening sequence. It is interesting from the sequence data that no obvious base-paired structures can be formed, such as would be expected if the intervening sequence RNA were looped out and excised.

C. SILK FIBROIN GENE

Silk fibroin comprises most of the protein synthesized in the *Bombyx mori* posterior silk gland during the latter stages of the fifth larval instar (reviewed by Suzuki, 1975). The fibroin gene is present in one copy per haploid genome in the posterior silk gland, as well as in other tissues (Suzuki *et al.*, 1972; Gage and Manning; 1976).

Several clones containing all or part of the fibroin gene and flanking sequences have been obtained (Ohshima and Suzuki; 1977; Suzuki and Ohshima; 1977). Restriction enzyme mapping indicates that the entire 16 kb coding region of the gene is present on a 21 kb plasmid insert, and is apparently not interrupted by any large intervening sequences. The initiation site of transcription has been mapped to a 0.9 kb fragment that includes the 5'-end of the mature mRNA. It is probable that the 5'-ends of the primary transcript and of the mRNA are identical or very close to each other for this gene (Suzuki and Ohshima, 1977). Two other lines of evidence, as reviewed below, have shown that the silk fibroin transcription unit is only slightly larger than necessary to encode the 16 kb mRNA molecule.

Silk fibroin mRNA (which is identifiable by its large size, high $G + C$ content, and internally repetitious sequence) comprises about 3% of the cellular RNA in posterior silk glands of mature larvae (Suzuki and Brown, 1972; Suzuki, 1975). Lizardi (1976a,b) has examined the size distribution of pulse-labeled fibroin RNA to determine if a precursor exists which is larger than the 16 kb mRNA. Using affinity column chromatography to purify the message and polyacrylamide gel electrophoresis for molecular sizing, a larger molecule can be detected after 6- to 10-minute labeling periods as a shoulder to the heavy side of the mature message. The material in this shoulder is somewhat more pronounced during shorter pulse-labeling times (Lizardi, 1967a) and during labeling at subnormal temperatures which may slow down processing steps (Lizardi, 1976b), consistent with the interpretation that it may be a slightly larger precursor to the polysomal mRNA. This early transcript, containing about 18.2 kb, is about 13% larger than fibroin mRNA.

An independent approach to determine the length of the fibroin transcription unit closely confirms the estimate of about 18 kb for the silk fibroin mRNA precursor size. McKnight *et al.* (1976) used chromatin spreading techniques for electron microscopic visualization of the nuclear contents of fifth larval instar posterior silk gland cells. Figure 20 shows a putative silk fibroin transcription unit from this tissue. The transcription unit is not tandemly repeated, but rather occurs singly, which is consistent with data showing the fibroin gene to exist only once per haploid genome equivalent. In addition, transcription units of this sort have not been seen in chromatin spreads of the middle portion of the silk gland which synthesizes little or no fibroin mRNA or protein. These transcription units are characterized by their length and the tight packing of nascent RNP fibers. They are essentially filled with RNA polymerases, as would be necessary to account for the large number of fibroin messages synthesized per unit time during

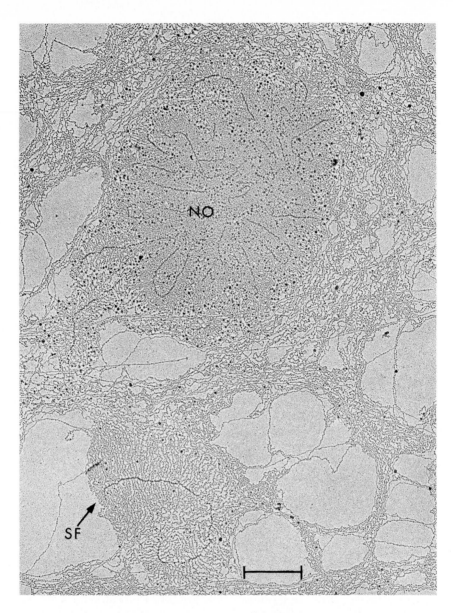

FIG. 20. Electron micrograph of transcriptionally active rDNA and putative silk fibroin DNA from the silkworm *Bombyx mori*. The chromatin was prepared from posterior silk gland tissue of a fifth instar larva. The nucleolus organizer region (NO) containing repeated rpRNA genes is in an intermediate state of dispersion. The silk fibroin transcription unit (SF) measures approximately twice the length of a rpRNA gene. Note the high density of nascent transcripts associated with the fibroin transcription unit. Scale bar represents 1 μm. (Photograph supplied by S. L. McKnight.)

this stage of development. The length of the transcription unit, taking into account the ~ 12% foreshortening of the DNA from its B-conformation length that is characteristic of genes with such high RNA polymerase density (McKnight *et al.*, 1976), corresponds to slightly more than 18 kb, in agreement with the size of the template required for the synthesis of the primary transcript characterized by Lizardi (1976b). The visualization of this transcription unit strongly supports the conclusion that the 16 kb fibroin mRNA is derived from a short-lived precursor which is about 2 kb longer. Both biochemical (Lizardi, 1976b) and electron microscopic evidence (McKnight *et al.*, 1976) suggest that processing may occur on the primary transcript before transcription is complete.

D. BALBIANI RING 2 LOCUS

An example of a gene in which the largest detectable transcription product is essentially the same size as the polysomal mRNA is the Balbiani ring 2 (*BR2*) locus of *Chironomus tentans* polytene chromosomes (reviewed by Daneholt, 1974, 1975; Daneholt *et al.*, 1976). The *BR2* puff is one of three giant puffs on chromosome IV in the salivary glands of *Chironomus* larvae. It is salivary gland specific, and evidence suggests that the *BR2* locus contains information for one or more of the large, structural, secretion polypeptides which constitute 80% of the protein synthesized at this stage (Daneholt, 1974).

Because of their extremely large size, the *BR2* chromosomal puffs can be isolated by microdissection and the associated RNA can be subsequently analyzed. Electrophoretic analysis of the RNA closely associated with the *BR2* locus has shown that there is one main component, a 75 S RNA, and a continuous size spectrum of smaller molecules representing the nascent transcription products. Although no larger molecules can be detected, it cannot be excluded that a primary transcript is cleaved to yield the 75 S molecule before transcription of the unit is complete. This extremely large RNA of 75 S corresponds to approximately 12×10^6 daltons and thus is transcribed from about 37 kb of DNA (Daneholt *et al.*, 1977). Daneholt and co-workers have shown that the BR2 RNA is suitable for analysis of nuclear–cytoplasmic transport as it constitutes a major cellular RNA species and can be recognized by its ability to hybridize *in situ* to Balbiani ring 2. In electron micrographs, large characteristic ribonucleoprotein granules containing the 75 S molecule can be seen at the puff, in the nucleoplasm, closely associated with nuclear pore complexes, and in the ribosome-rich regions of the cytoplasm near the nucleus. A 75 S RNA species, which hybridizes to the *BR2* locus, can be recovered from the large polysome fraction, apparently in the process of translation (Daneholt *et al.*, 1976). The latter authors conclude that the 75 S RNA molecules are transported to the cytoplasm as mature mRNAs without a size reduction that can be detected on polyacrylamide gels.

The four single-copy loci discussed here have included two examples of genes whose primary transcription products are either the same size as, or only slightly larger than, the mature mRNAs, as well as two examples of split genes whose primary transcription products contain the intervening sequences and are significantly larger than the respective mRNAs. Which of these represent the more "typical" structural gene transcription unit will become obvious as more information accumulates on other mRNA-coding genes.

Prior to the recent advances in recombinant DNA technology, much of the information on the transcriptional organization of single-copy genes came from analyses of early transcription products rather than from direct analyses of genomic sequences. Thus, it is not surprising that the best-characterized single-copy genes are those whose mRNA products represent a significant proportion of the mRNA population in the cell types studied. With gene cloning methods, it is now quite feasible to obtain any genomic segment for which an mRNA or cDNA probe is available in quantities sufficient for fine structure analysis. This advance should allow a rapid increase in our knowledge of the organization and regulation of single-copy genes in eukaryotic cells.

V. PERSPECTIVES

Current research emphasis on isolating and sequencing specific genes and their contiguous DNA segments underscores the widely held concept that primary sequence organization of DNA is directly involved in the regulation of gene expression. There is much information consistent with this concept, such as the interspersion of repetitive sequences with single-copy structural genes and the homology between flanking sequences of multiple-copy genes, whether tandemly repeated (as in rDNA or 5 S DNA) or dispersed throughout the genome (as in *D. melanogaster copia* and *412* DNA). The direct relationship of DNA sequence to gene control is, however, far from clear.

DNA sequences nearby coding segments are considered good candidates for possible controlling regions in gene expression. The characterization of specific eukaryotic transcription units has revealed, in most cases, the presence of "extra" DNA that does not encode any of the information present in mature rRNA, mRNA, or tRNA molecules. The most intriguing recent finding is that these sequences are often located as inserts within the regions coding for the mature products. The incidence of the "split-gene" phenomenon is not as yet known, but, as discussed in this review, insertion sequences have already been discovered in *Drosophila* 28 S rRNA cistrons, yeast tRNA genes, mouse myeloma light-chain Ig genes, the mouse β-globin gene, and the chicken ovalbumin gene. A similar phenomenon has also been reported for the genes for adenovirus late mRNAs (Chow *et al.*, 1977; Berget *et al.*, 1977) and late mRNAs of SV40 (Aloni *et al.*, 1977; Lavi and Groner, 1977). Thus, it is quite possible that such insertions may be shown to be very common in the genes of eukaryotes and their viruses. The possible functions of these intervening segments remain the subject of speculation at present. The recent findings that mouse β-globin mRNA and chicken ovalbumin mRNA are derived from larger precursor molecules containing sequences coded by the insertions within the gene suggests the possibility that the excision and ligation steps necessary for the transition from precursor to final-product molecules may play some role in the regulation of the amounts of mature RNA molecules coded for by insert-containing genes.

It is obvious that present knowledge of the sequence organization of specific individual transcription units does not yet reveal how the expression of such genes is regulated. The nature of genomic regulatory elements should become clearer as more information accumulates, particularly on the sequences that neighbor coordinately expressed genes. Such studies, combined with efforts to characterize specific promotor–RNA polymerase interactions and to assess the role of histone and nonhistone proteins in gene expression, should ultimately allow the development of *in vitro* transcription and processing systems that respond correctly to physiological cues and provide definitive information regarding the *in vivo* regulation of gene expression.

ACKNOWLEDGMENTS

We thank Dr. R. M. Grainger, Yvonne Osheim, and Kathy Martin for critical reading of the manuscript, and Lorraine Blanks for technical assistance. During the preparation of this article, A.L.B. was supported by a postdoctoral fellowship from the Jane Coffin Childs Memorial Fund for Medical Research, S.L.M. by a postdoctoral fellowship from the Helen Hay Whitney Foundation for Medical Research, and O.L.M. by Grant PCM 73-01134 from the National Science Foundation and by Grant NIGMS2-R01-GM21020 from the Public Health Service.

REFERENCES

Aloni, Y., Hatlan, L. E., and Attardi, G. (1971). *J. Mol. Biol.* **56**, 555–563.

Aloni, Y., Dhar, R., Laub, O., Horowitz, M., and Khoury, G. (1977). *Proc. Natl. Acad. Sci. U.S.A.* **74**, 3686–3690.

Arnheim, N., and Southern, E. M. (1977). *Cell* **11**, 363–370.

Artavanis-Tsakonas, S., Schedl, P., Tschudi, C., Pirrotta, V., Steward, R., and Gehring, W. J. (1977). *Cell* **12**, 1057–1067.

Bastos, R. N., and Aviv, H. (1977). *Cell* **11**, 641–650.

Batts-Young, B., and Lodish, H. F. (1978). *Proc. Natl. Acad. Sci. U.S.A.* **75**, 740–744.

Berger, S., and Schweiger, H. G. (1975). *Mol. Gen. Genet.* **139**, 269–275.

Berger, S., Zellmer, D. M., Kloppstech, K., Richter, G., Dillard, W. L., and Schweiger, H. G. (1978). *Cell Biol. Int. Rep.* **2**, 41–50.

Berget, S. M., Moore, C., and Sharp, P. A. (1977). *Proc. Natl. Acad. Sci. U.S.A.* **74**, 3171–3175.

Bird, A. P., and Birnstiel, M. L. (1971). *Chromosoma* **35**, 300–309.

Birnstiel, M. L., Sells, B. H., and Purdom, I. F. (1972). *J. Mol. Biol.* **63**, 21–39.

Birnstiel, M. L., Weinberg, E. S., and Pardue, M. L. (1973). *In* "Molecular Cytogenetics" (B. A. Hamkalo and J. Papaconstantinou, eds.), p. 75. Plenum, New York.

Bishop, J. O., and Rosbash, M. (1973). *Nature (London), New Biol.* **241**, 204–207.

Biswas, B. B., Ganguly, A., and Das, A. (1975). *Prog. Nucleic Acid Res. Mol. Biol.* **15**, 145–184.

Botchan, P., Reeder, R. H., and Dawid, I. B. (1977). *Cell* **11**, 599–607.

Brack, C., and Tonegawa, S. (1977). *Proc. Natl. Acad. Sci. U.S.A.* **74**, 5652–5656.

Brack, C., Hirama, M., Lenhard-Schuller, R., and Tonegawa, S. (1978). *Cell* **15**, 1–14.

Brandhorst, B. P., and McConkey, E. H. (1974). *J. Mol. Biol.* **85**, 451–463.

Breathnach, R., Mandel, J. L., and Chambon, P. (1977). *Nature (London)* **270**, 314–319.

Breathnach, R., Benoist, C., O'Hare, K., Gannon, F., and Chambon, R. (1978). *Proc. Natl. Acad. Sci. U.S.A.* **75**, 4853–4857.

Britten, R. J., and Kohne, D. E. (1968). *Science* **161**, 529–540.

Brown, D. D., and Blackler, A. W. (1972). *J. Mol. Biol.* **63**, 75–83.

Brown, D. D., and Sugimoto, K. (1973a). *J. Mol. Biol.* **78**, 397–415.

Brown, D. D., and Sugimoto, K. (1973b). *Cold Spring Harbor Symp. Quant. Biol.* **38**, 501–505.

Brown, D. D., Wensink, P. C., and Jordan, E. (1972). *J. Mol. Biol.* **63**, 57–73.

Brown, D. D., Carroll, D., and Brown, R. D. (1977a). *Cell* **12**, 1045–1056.

Brown, D. D., Dawid, I. B., and Reeder, R. H. (1977b). *Dev. Biol.* **59**, 266–267.

Brownlee, G. G., Cartwright, E. M., and Brown, D. D. (1974). *J. Mol. Biol.* **89**, 703–718.

Callan, H. G. (1967). *J. Cell Sci.* **2**, 1–7.

Carlson, J., Ott, G., and Sauerbier, W. (1977). *J. Mol. Biol.* **112**, 353–357.

Carroll, D., and Brown, D. D. (1976a). *Cell* **7**, 467–475.

Carroll, D., and Brown, D. D. (1976b). *Cell* **7**, 477–486.

Chamberlin, M. E., Britten, R. J., and Davidson, E. H. (1975). *J. Mol. Biol.* **96**, 317–333.

Chow, L. T., Gelinas, R. E., Broker, T. R., and Roberts, R. J. (1977). *Cell* **12**, 1–8.

Clarkson, S. G., and Birnstiel, M. L. (1973). *Cold Spring Harbor Symp. Quant. Biol.* **38**, 451–459.

Clarkson, S. G., and Kurer, V. (1976). *Cell* **8**, 183–195.

Clarkson, S. G., Birnstiel, M. L., and Serra, V. (1973a). *J. Mol. Biol.* **79**, 391–410.

Clarkson, S. G., Birnstiel, M. L., and Purdom, I. F. (1973b). *J. Mol. Biol.* **79**, 411–429.

Cockburn, A. F., Newkirk, M. J., and Firtel, R. A. (1976). *Cell* **9**, 605–613.

Cohn, M., Blomberg, B., Geckeler, W., Raschke, W., Riblet, R., and Weigert, M. (1974). *ICN-UCLA Symp. Mol. & Cell. Biol.* **3**, 89–117.

Cohn, R. H., Lowry, J. C., and Kedes, L. H. (1976). *Cell* **9**, 147–161.

Cory, S., and Adams, J. M. (1977). *Cell* **11**, 795–805.

Costantini, F. D., Scheller, R. H., Britten, R. J., and Davidson, E. H. (1978). *Cell* **15**, 173–187.

Crain, W. R., Eden, F. C., Pearson, W. R., Davidson, E. H., and Britten, R. J. (1976a). *Chromosoma* **56**, 309–326.

Crain, W. R., Davidson, E. H., and Britten, R. J. (1976b). *Chromosoma* **59**, 1–12.

Cullis, C., and Davies, D. R. (1974). *Chromosoma* **46**, 23–28.

Curtis, P. J., Mantei, N., van den Berg, J., and Weissmann, C. (1977a). *Proc. Natl. Acad. Sci. U.S.A.* **74**, 3184–3188.

Curtis, P. J., Mantei, N., and Weissmann, C. (1977b). *Cold Spring Harbor Symp. Quant. Biol.* **42**, 971–984.

Daneholt, B. (1974). *Int. Rev. Cytol., Suppl.* **4**, 417–462.

Daneholt, B. (1975). *Cell* **4**, 1–9.

Daneholt, B., Case, S. T., Hyde, J., Nelson, L., and Wieslander, L. (1976). *Prog. Nucleic Acid Res. Mol. Biol.* **19**, 319–334.

Daneholt, B., Case, S. T., Derksen, J., Lamb, M. M., Nelson, L., and Wieslander, L. (1977). *Cold Spring Harbor Symp. Quant. Biol.* **42**, 867–876.

Darnell, J. E., and Balint, R. (1970). *J. Cell. Physiol.* **76**, 349–356.

Darnell, J. E., Jelinek, W. R., and Molloy, G. R. (1973). *Science* **181**, 1215–1221.

Davidson, E. H. (1977). "Gene Activity in Early Development," 2nd ed. Academic Press, New York.

Davidson, E. H., and Britten, R. J. (1973). *Q. Rev. Biol.* **48**, 565–613.

Davidson, E. H., Hough, B. R., Amenson, C. S., and Britten, R. J. (1973). *J. Mol. Biol.* **77**, 1–23.

Davidson, E. H., Galau, G. A., Angerer, R. C., and Britten, R. J. (1975). *Chromosoma* **51**, 253–259.

Davidson, E. H., Klein, W. H., and Britten, R. J. (1977). *Dev. Biol.* **55**, 69–84.

Dawid, I. B., and Botchan, P. (1977). *Proc. Natl. Acad. Sci. U.S.A.* **74**, 4233–4237.

Dawid, I. B., and Wellauer, P. K. (1976). *Cell* **8**, 443–448.

Delaney, A., Dunn, R., Grigliatti, T. A., Tener, G. M., Kaufman, T. C., and Suzuki, D. T. (1976). *Fed. Proc., Fed. Am. Soc. Exp. Biol.* **35**, 1676.

Derman, E., and Darnell, J. E. (1974). *Cell* **3**, 255–264.

Derman, E., Goldberg, S., and Darnell, J. E. (1976). *Cell* **9**, 465–472.

Dickerson, R. E. (1971). *J. Mol. Evol.* **1**, 26–45.

Doel, M. T., Houghton, M., Cook, E. A. and Carey, N. H. (1977). *Nucleic Acids Res.* **4**, 3701–3713.

Dugaiczyk, A., Woo, S. L. C., Lai, E. C., Mace, M. L., McReynolds, L., and O'Malley, B. W. (1978). *Nature (London)* **274**, 328–333.

Edelman, G. M. (1973). *Science* **180**, 830–840.

Efstratiadis, A., Kafatos, F. C., and Maniatis, T. (1977). *Cell* **10**, 571–585.

Elsevier, S. M., and Ruddle, F. H. (1975). *Chromosoma* **52**, 219–228.

Engberg, J., Christiansen, G., and Leick, V. (1974). *Biochem. Biophys. Res. Commun.* **59**, 1356–1365.

Engberg, J., Andersson, P., Leick, V., and Collins, J. (1976). *J. Mol. Biol.* **104**, 455–470.

Evans, H. J., Buckland, R. A., and Pardue, M. L. (1974). *Chromosoma* **48**, 405–426.

Federoff, N., and Brown, D. D. (1978). *Cell* **13**, 701–716.

Federoff, N., Wellauer, P. K., and Wall, R. (1977). *Cell* **10**, 597–610.

Feldmann, H. (1976). *Nucleic Acids Res.* **3**, 2379–2386.

Feldmann, H. (1977). *Nucleic Acids Res.* **4**, 2831–2841.

Finnegan, D. J., Rubin, G. M., Young, M. W., and Hogness, D. S. (1977). *Cold Spring Harbor Symp. Quant. Biol.* **42**, 1053–1063.

Flavell, R., Jeffreys, A. J., and Grosveld, G. C. (1977). *Cold Spring Harbor Symp. Quant. Biol.* **42**, 1003–1010.

Flavell, R., Kooter, J. M., DeBoer, E., Little, P. F. R., and Williamson, R. (1978). *Cell* **15**, 25–42.

Foe, V. E., Wilkinson, L. E., and Laird, C. D. (1976). *Cell* **9**, 131–146.

Ford, P. J. (1971). *Nature (London)* **233**, 561–564.

Ford, P. J., and Brown, R. D. (1976). *Cell* **8**, 485–493.

Ford, P. J., and Mathieson, T. (1976). *Nature (London)* **261**, 433–435.

Ford, P. J., and Southern, E. M. (1973). *Nature (London), New Biol.* **241**, 7–12.

Forsheit, A. B., Davidson, N., and Brown, D. D. (1974). *J. Mol. Biol.* **90**, 301–314.

Fournier, A., Chavancy, G., and Garel, J. (1976). *Biochem. Biophys. Res. Commun.* **72**, 1187–1194.

Gage, L. P., and Manning, R. F. (1976). *J. Mol. Biol.* **101**, 327–348.

Galau, G. A., Britten, R. J., and Davidson, E. H. (1974). *Cell* **2**, 9–20.

Gall, J. G. (1974). *Proc. Natl. Acad. Sci. U.S.A.* **71**, 3078–3081.

Gall, J. G., and Atherton, D. D. (1974). *J. Mol. Biol.* **85**, 633–664.

Gall, J. G., and Rochaix, J. D. (1974). *Proc. Natl. Acad. Sci. U.S.A.* **71**, 1819–1823.

Garapin, A. C., Lepennec, J. P., Roskam, W., Perrin, F., Cami, B., Krust, A., Breathnach, R., Chambon, P., and Kourilsky, P. (1978). *Nature (London)* **273**, 349–354.

Georgiev, G. P., Varshavsky, A. J., Ryskov, A. P., and Church, R. B. (1973). *Cold Spring Harbor Symp. Quant. Biol.* **38**, 869–884.

Getz, M. J., Birnie, G. D., Young, B. D., MacPhail, E., and Paul, J. (1975). *Cell* **4**, 121–129.

Gilmore-Hebert, M., and Wall, R. (1978). *Proc. Natl. Acad. Sci. U.S.A.* **75**, 342–345.

Goodman, H. M., Olson, M. V., and Hall, B. D. (1977). *Proc. Natl. Acad. Sci. U.S.A.* **74**, 5453–5457.

Grainger, R. M., and Ogle, R. C. (1978). *Chromosoma* **65**, 115–126.

Granboulan, N., and Scherrer, K. (1969). *Eur. J. Biochem.* **9**, 1–20.

Greenberg, J. R. (1975). *J. Cell Biol.* **64**, 269–288.

Grigliatti, T. A., White, B. N., Tener, G. M., Kaufman, T. C., and Suzuki, D. T. (1974). *Proc. Natl. Acad. Sci. U.S.A.* **71**, 3527–3531.

Gross, K., Schaffner, W., Telford, J., and Birnstiel, M. L. (1976). *Cell* **8**, 479–484.

Hackett, P. B., and Sauerbier, W. (1975). *J. Mol. Biol.* **91**, 235–256.

Hadjiolov, A. A., and Nikolaev, N. (1976). *Prog. Biophys. Mol. Biol.* **31**, 95–144.

Hamkalo, B. A., and Miller, O. L., Jr. (1973). *Annu. Rev. Biochem.* **42**, 379–396.

Harris, S. E., Means, A. R., Mitchell, W. M., and O'Malley, B. W. (1973). *Proc. Natl. Acad. Sci. U.S.A.* **70**, 3776–3780.

Harrison, P. R., Birnie, G. D., Hell, A., Humphries, S., Young, B. D., and Paul, J. (1974). *J. Mol. Biol.* **84**, 539–554.

Hatlen, L., and Attardi, G. (1971). *J. Mol. Biol.* **56**, 535–553.

Herman, R. C., Williams, J. G., and Penman, S. (1976). *Cell* **7**, 429–437.

Hershey, N. D., Conrad, S. E., Sodja, A., Yen, P. H., Cohen, M., Jr., Davidson, N., Ilgen, C., and Carbon, J. (1977). *Cell* **11**, 585–598.

Holmes, D. S., and Bonner, J. (1973). *Biochemistry* **12**, 2330–2338.

Holmes, D. S., and Bonner, J. (1974). *Proc. Natl. Acad. Sci. U.S.A.* **71**, 1108–1112.
Hough, B. R., Smith, M. J., Britten, R. J., and Davidson, E. H. (1975). *Cell* **5**, 291–299.
Hourcade, D., Dressler, D., and Wolfson, J. (1973). *Proc. Natl. Acad. Sci. U.S.A.* **70**, 2926–2930.
Jacq, C., Miller, J. R., and Brownlee, G. C. (1977). *Cell* **12**, 109–120.
Jeanteur, P., and Attardi, G. (1969). *J. Mol. Biol.* **45**, 304–324.
Jeffreys, A. J., and Flavell, R. A. (1977a). *Cell* **12**, 429–439.
Jeffreys, A. J., and Flavell, R. A. (1977b). *Cell* **12**, 1097–1108.
Johnson, L. D., Henderson, A. S., and Atwood, K. C. (1974). *Cytogenet. Cell Genet.* **13**, 103–105.
Kalt, M. R., and Gall, J. G. (1974). *J. Cell Biol.* **62**, 460–472.
Karrer, K. M., and Gall, J. G. (1976). *J. Mol. Biol.* **104**, 421–453.
Kedes, L. H. (1976). *Cell* **8**, 321–331.
Kedes, L. H., and Birnstiel, M. L. (1971). *Nature (London), New Biol.* **230**, 165–169.
Kedes, L. H., and Gross, P. R. (1969). *J. Mol. Biol.* **42**, 559–575.
Kinniburgh, A. J., Mertz, J. E., and Ross, J. (1978). *Cell* **14**, 681–693.
Knapp, G., Beckmann, J. S., Johnson, P. F., Fuhrman, S. A., and Abelson, J. (1978). *Cell* **14**, 221–236.
Kornberg, R. D. (1977). *Annu. Rev. Biochem.* **40**, 931–954.
Kornberg, R. D., and Thomas, J. O. (1974). *Science* **184**, 865–871.
Kuehl, W. M. (1977). *Curr. Top. Microbiol. Immunol.* **76**, 1–47.
Kunkel, N. S., Hemminki, K., and Weinberg, E. S. (1977). *J. Cell Biol.* **75**, 339a.
Lai, E. C., Woo, S. L. C., Dugaiczyk, A., Catterall, J. F., and O'Malley, B. W. (1978). *Proc. Natl. Acad. Sci. U.S.A.* **75**, 2205–2209.
Laird, C. D., and Chooi, W. Y. (1976). *Chromosoma* **58**, 198–218.
Laird, C. D., Wilkinson, L. E., Foe, V. E., and Chooi, W. Y. (1976). *Chromosoma* **58**, 169–192.
Lamb, M. M., and Laird, C. D. (1976a). *Dev. Biol.* **52**, 31–42.
Lamb, M. M., and Laird, C. D. (1976b). *Biochem. Genet.* **14**, 357–371.
Lavi, S., and Groner, Y. (1977). *Proc. Natl. Acad. Sci. U.S.A.* **74**, 5323–5327.
Leder, P., Honjo, T., Swan, D., Packman, S., Nau, M., and Norman, B. (1975). In "Molecular Approaches to Immunology" (E. E. Smith and D. W. Ribbons, eds.), pp. 173–188. Academic Press, New York.
Leder, P., Tilghman, S. M., Tiemeier, D. C., Polsky, F. I., Seidman, J. G., Edgell, M. H., Enquist, L. W., Leder, A., and Norman, B. (1977). *Cold Spring Harbor Symp. Quant. Biol.* **42**, 915–920.
Leibowitz, R. D., Weinberg, R. A., and Penman, S. (1973). *J. Mol. Biol.* **73**, 139–144.
Lengyel, J., and Penman, S. (1975). *Cell* **5**, 281–290.
Levis, R., and Penman, S. (1977). *Cell* **11**, 105–113.
Levy, S., Childs, G., and Kedes, L. (1978). *Cell* **15**, 151–162.
Lewin, B. (1975a). *Cell* **4**, 11–20.
Lewin, B. (1975b). *Cell* **4**, 77–93.
Lifton, R. P., Goldberg, M. L., Karp, R. W., and Hogness, D. S. (1977). *Cold Spring Harbor Symp. Quant. Biol.* **42**, 1047–1051.
Lizardi, P. M. (1976a). *Cell* **7**, 239–245.
Lizardi, P. M. (1976b). *Prog. Nucleic Acid Res. Mol. Biol.* **19**, 301–312.
McKnight, S. L., and Miller, O. L., Jr. (1976). *Cell* **8**, 305–319.
McKnight, S. L., Sullivan, N. L., and Miller, O. L., Jr. (1976). *Prog. Nucleic Acid Res. Mol. Biol.* **19**, 313–318.

McKnight, S. L., Bustin, M., and Miller, O. L., Jr. (1977). *Cold Spring Harbor Symp. Quant. Biol.* **42**, 741–754.

McReynolds, L., O'Malley, B. W., Nisbet, A. D., Fothergill, J. E., Givol, D., Fields, S., Robertson, M., and Brownlee, G. G. (1978). *Nature (London)* **273**, 723–728.

Mairy, M., and Denis, H. (1971). *Dev. Biol.* **24**, 143–165.

Maizels, N. (1976). *Cell* **9**, 431–438.

Mandel, J. L., Breathnach, R., Gerlinger, P., LeMeur, M., Gannon, F., and Chambon, P. (1978). *Cell* **14**, 641–653.

Manning, J. E., Schmid, C. W., and Davidson, N. (1975). *Cell* **4**, 141–155.

Mayo, V. S., and deKloet, S. R. (1971). *Biochim. Biophys. Acta* **247**, 74–79.

Maxam, A. W., Tizard, R., Skryabin, K. G., and Gilbert, W. (1977). *Nature (London)* **267**, 643–645.

Mears, J. G., Ramirez, F., Leibowitz, D., and Bank, A. (1978). *Cell* **15**, 15–24.

Melli, M., Spinelli, G., Wyssling, H., and Arnold, E. (1977). *Cell* **11**, 651–661.

Meyer, G. F., and Hennig, W. (1974). *Chromosoma* **46**, 121–144.

Miller, J. R., Cartwright, E. M., Brownlee, J. G., Federoff, N. V., and Brown, D. D. (1978). *Cell* **13**, 717–725.

Miller, L. (1974). *Cell* **3**, 275–281.

Miller, L., and Gurdon, J. B. (1970). *Nature (London)* **227**, 1108–1110.

Miller, O. L., Jr., and Beatty, B. R. (1969a). *Science* **164**, 955–957.

Miller, O. L., Jr., and Beatty, B. R. (1969b). *J. Cell. Physiol.* **74**, Suppl. 1, 225–232.

Molloy, G. R., Jelinek, W., Salditt, M., and Darnell, J. E. (1974). *Cell* **1**, 43–53.

Nomura, M., Tissières, A., and Lengyel, P., eds. (1974). "Ribosomes." Cold Spring Harbor Lab., Cold Spring Harbor, New York.

Ofengand, J. (1977). *In* "Molecular Mechanisms of Protein Biosynthesis" (H. Weissbach and S. Pestka, eds.), pp. 7–79. Academic Press, New York.

Ohshima, Y., and Suzuki, Y., (1977). *Proc. Natl. Acad. Sci. U.S.A.* **74**, 5363–5367.

Old, R. W., Callan, H. G., and Gross, K. W. (1977). *J. Cell Sci.* **27**, 57–79.

Olson, M. V., Montgomery, D. L., Hopper, A. K., Page, G. S., Horodyski, F., and Hall, B. D. (1977). *Nature (London)* **267**, 639–641.

Overton, G. C., and Weinberg, E. S. (1978). *Cell* **14**, 247–257.

Pardue, M. L. (1973). *Cold Spring Harbor Symp. Quant. Biol.* **38**, 475–482.

Pardue, M. L., Brown, D. D., and Birnstiel, M. L. (1973). *Chromosoma* **42**, 191–203.

Pardue, M. L., Kedes, L. H., Weinberg, E. S., and Birnstiel, M. L. (1977). *Chromosoma* **63**, 135–151.

Peacock, W. J., Appels, R., Dunsmuir, P., Lohe, A. R., and Gerlach, W. L. (1977). *In* "International Cell Biology 1976–1977" (B. R. Brinkley and K. R. Porter, eds.), pp. 494–506. Rockefeller Univ. Press, New York.

Pellegrini, M., Manning, J., and Davidson, N. (1977). *Cell* **10**, 213–224.

Penman, S., Vesco, C., and Penman, M. (1968). *J. Mol. Biol.* **34**, 49–69.

Perry, R. P. (1976). *Annu. Rev. Biochem.* **45**, 605–629.

Perry, R. P., and Kelley, D. E. (1968). *J. Cell. Physiol.* **72**, 235–246.

Perry, R. P., and Kelley, D. E. (1970). *J. Cell. Physiol.* **76**, 127–140.

Perry, R. P., Cheng, T. Y., Freed, J. J., Greenberg, J. R., Kelley, D. E., and Tartof, K. D. (1970). *Proc. Natl. Acad. Sci. U.S.A.* **65**, 609–619.

Perry, R. P., Kelley, D. E., and LaTorre, J. (1974). *J. Mol. Biol.* **82**, 315–331.

Perry, R. P., Kelley, D. E., Friderici, K. H., and Rottman, F. M. (1975). *Cell* **6**, 13–19.

Perry, R. P., Bard, E., Hames, B. D., Kelley, D. E., and Schibler, U. (1976). *Prog. Nucleic Acid Res. Mol. Biol.* **19**, 275–292.

Portmann, R., Schaffner, W., and Birnstiel, M. (1976). *Nature (London)* **264**, 31–34.

Prensky, W., Steffensen, D. M., and Hughes, W. L. (1973). *Proc. Natl. Acad. Sci. U.S.A.* **70**, 1860-1864.

Prescott, D. M., Murti, K. G., and Bostock, C. J. (1973). *Nature (London)* **242**, 576, 597-600.

Procunier, J. D., and Tartof, K. D. (1975). *Genetics* **81**, 515-523.

Pukkila, P. J. (1975). *Chromosoma* **53**, 71-89.

Rabbitts, T. H. (1978). *Nature (London)* **275**, 291-296.

Rabbitts, T. H., and Forster, A. (1978). *Cell* **13**, 319-327.

Rabbitts, T. H., and Milstein, C. (1977). *Contemp. Top. Mol. Immunol.* **6**, 117-143.

Reeder, R. H., Higashinakagawa, T., and Miller, O. L., Jr. (1976a). *Cell* **8**, 449-454.

Reeder, R. H., Brown, D. D., Wellauer, P. K., and Dawid, J. B. (1976b). *J. Mol. Biol.* **105**, 507-516.

Reeder, R. H., Sollner-Webb, B., and Wahn, H. L. (1977). *Proc. Natl. Acad. Sci. U.S.A.* **74**, 5402-5406.

Rochaix, J. D., Bird, A., and Bakken, A. (1974). *J. Mol. Biol.* **87**, 473-487.

Roeder, R. G. (1974). *J. Biol. Chem.* **249**, 249-256.

Roop, D. R., Nordstrom, J. L., Tsai, S. Y., Tsai, M. J., and O'Malley, B. W. (1978). *Cell* **15**, 671-685.

Ross, J., and Knecht, D. A. (1978). *J. Mol. Biol.* **119**, 1-20.

Rubin, G. M., Finnegan, D. J., and Hogness, D. S. (1976). *Prog. Nucleic Acid Res. Mol. Biol.* **19**, 221-226.

Schaffner, W., Gross, K., Telford, J., and Birnstiel, M. L. (1976). *Cell* **8**, 471-478.

Schaffner, W., Kunz, G., Daetwyler, H., Telford, J., Smith, H. O., and Birnstiel, M. L. (1978). *Cell* **14**, 655-671.

Scheer, U., Trendelenburg, M. F., and Franke, W. W. (1976). *J. Cell Biol.* **69**, 465-489.

Scheer, U., Trendelenburg, M. F., Krohne, G., and Franke, W. W. (1977). *Chromosoma* **60**, 147-167.

Scheller, R. H., Costantini, F. D., Kozlowski, M. R., Britten, R. J., and Davidson, E. H. (1978). *Cell* **15**, 189-203.

Schibler, U., Hagenbüchle, O., Wyler, T., Weber, R., Boseley, P., Telford, J., and Birnstiel, M. L. (1976). *Eur. J. Biochem.* **68**, 471-480.

Seidman, J. G., Leder, A., Nav, M., Norman, B., and Leder, P. (1978). *Science* **202**, 11-17.

Siddiqui, M. A. Q., and Chen, G. S. (1975). *Brookhaven Symp. Biol.* **26**, 154-164.

Skoultchi, A., and Gross, P. R. (1973). *Proc. Natl. Acad. Sci. U.S.A.* **70**, 2840-2844.

Smith, G. (1973). *Cold Spring Harbor Symp. Quant. Biol.* **38**, 507-513.

Smith, G. P. (1976). *Science* **191**, 528-535.

Smith, M. J., Hough, B. R., Chamberlin, M. E., and Davidson, E. H. (1974). *J. Mol. Biol.* **85**, 103-126.

Soeiro, R., Vaughan, M. H., Warner, J. R., and Darnell, J. E., Jr. (1968). *J. Cell Biol.* **39**, 112-118.

Speirs, J., and Birnstiel, M. (1974). *J. Mol. Biol.* **87**, 237.

Sprague, K. U. (1975). *Biochemistry* **14**, 925-931.

Spring, H., Krohne, G., Franke, W. W., Scheer, U., and Trendelenburg, M. F. (1976). *J. Microsc. Biol. Cell.* **25**, 107-116.

Steffensen, D. M., and Wimber, D. E. (1971). *Genetics* **69**, 163-178.

Steffensen, D. M., Duffey, P., and Prensky, W. (1974). *Nature (London)* **252**, 741-743.

Strair, R. K., Skoultchi, A. I., and Shafritz, D. A. (1977). *Cell* **12**, 133-141.

Suzuki, Y. (1975). *Adv. Biophys.* **8**, 83-114.

Suzuki, Y., and Brown, D. D. (1972). *J. Mol. Biol.* **63**, 409-429.

Suzuki, Y., and Ohshima, Y. (1977). *Cold Spring Harbor Symp. Quant. Biol.* **42**, 947–957.

Suzuki, Y., Gage, L. P., and Brown, D. D. (1972). *J. Mol. Biol.* **70**, 637–649.

Tartof, K. D. (1971). *Science* **171**, 294–297.

Tartof, K. D. (1975). *Annu. Rev. Genet.* **9**, 355–385.

Thomas, C. A., Jr. (1970). In "The Neurosciences: Second Study Program" (F. O. Schmitt, ed.), pp. 973–998. Rockefeller Univ. Press, New York.

Tiemeier, D. C., Tilghman, S. M., Polsky, F. I., Seidman, J. G., Leder, A., Edgell, M. H., and Leder, P. (1978). *Cell* **14**, 237–245.

Tilghman, S. M., Tiemeier, D. C., Seidman, J. G., Peterlin, B. M., Sullivan, M., Maizel, J. V., and Leder, P. (1978a). *Proc. Natl. Acad. Sci. U.S.A.* **75**, 725–729.

Tilghman, S. M., Curtis, P. J., Tiemeier, D. C., Leder, P., and Weissmann, C. (1978b). *Proc. Natl. Acad. Sci. U.S.A.* **75**, 1309–1313.

Tobler, H. (1975). *Biochem. Anim. Dev.* **3**, 91–143.

Tonegawa, A., Hozumi, N., Matthyssens, G., and Schuller, R. (1976). *Cold Spring Harbor Symp. Quant. Biol.* **41**, 877–889.

Tonegawa, S., Brack, C., Hozumi, N., Matthyssens, G., and Schuller, R. (1977). *Immunol. Rev.* **36**, 73–94.

Trendelenburg, M. F., Spring, H., Scheer, U., and Franke, W. W. (1974). *Proc. Natl. Acad. Sci. U.S.A.* **71**, 3626–3630.

Trendelenburg, M. F., Scheer, U., Zentgraf, H., and Franke, W. W. (1976). *J. Mol. Biol.* **108**, 453–470.

Trendelenburg, M. F., Franke, W. W., and Scheer, U. (1977). *Differentiation* **7**, 133–158.

Valenzuela, P., Venegas, A., Weinberg, F., Bishop, R., and Rutter, W. J. (1978). *Proc. Natl. Acad. Sci. U.S.A.* **75**, 190–194.

Vogt, V. M., and Braun, R. (1976). *J. Mol. Biol.* **106**, 567–587.

Warner, J. R., Soeiro, R., Birnboim, H. C., Girard, M., and Darnell, J. E. (1966). *J. Mol. Biol.* **19**, 349–361.

Watson-Coggins, L., and Gall, J. G. (1972). *J. Cell Biol.* **52**, 569–576.

Weatherall, D. J., and Clegg, J. B. (1976). *Annu. Rev. Genet.* **10**, 157–178.

Weber, L., and Berger, E. (1976). *Biochemistry* **15**, 5511–5519.

Wegnez, M., Monier, R., and Denis, H. (1972). *FEBS Lett.* **25**, 13–20.

Weinberg, E. S., Birnstiel, M. L., Purdom, I. F., and Williamson, R. (1972). *Nature (London)* **240**, 225–228.

Weinmann, R., and Roeder, R. G. (1974). *Proc. Natl. Acad. Sci. U.S.A.* **71**, 1790–1794.

Weinstock, R., Sweet, R., Weiss, M., Cedar, H. and Axel, R. (1978). *Proc. Natl. Acad. Sci. U.S.A.* **75**, 1299–1303.

Weisbach, H., and Pestka, S. (eds.). (1977). "Molecular Mechanisms of Protein Biosynthesis." Academic Press, New York.

Wellauer, P. K., and Dawid, I. B. (1977). *Cell* **10**, 193–212.

Wellauer, P. K., Reeder, R. H., Carroll, D., Brown, D. D., Deutch, A., Higashinakagawa, T., and Dawid, I. B. (1974). *Proc. Natl. Acad. Sci. U.S.A.* **71**, 2823–2827.

Wellauer, P. K., Dawid, I. B., Brown, D. D., and Reeder, R. H. (1967a). *J. Mol. Biol.* **105**, 461–486.

Wellauer, P. K., Reeder, R. H., Dawid, I. B., and Brown, D. D. (1976b). *J. Mol. Biol.* **105**, 487–505.

Wellauer, P. K., Dawid, I. B., and Tartof, K. D. (1978). *Cell* **14**, 269–278.

White, R. L., and Hogness, D. S. (1977). *Cell* **10**, 177–192.

Williamson, A. R. (1976). *Annu. Rev. Biochem.* **45**, 467–500.

Wilson, M. C., and Melli, M. (1976). *J. Mol. Biol.* **110,** 511–535.
Woo, S. L. C., Dugaiczyk, A., Tsai, M. J., Lai, E. C., Catterall, J. F., and O'Malley, B. W. (1978). *Proc. Natl. Acad. Sci. U.S.A.* **75,** 3688–3692.
Wu, M., Holmes, D. S., Davidson, N., Cohn, R. H., and Kedes, L. H. (1976). *Cell* **9,** 163–169.
Yao, M. C., and Gall, J. G. (1977). *Cell* **12,** 121–132.
Yen, P. H., Sodja, A., Cohen, M., Jr., Conrad, S. E., Wu, M., and Davidson, N. (1977). *Cell* **11,** 763–777.
Zylber, E. A., and Penman, S. (1971). *Science* **172,** 947–949.

Chapter V

Recognition and Control Sequences in Nucleic Acids

P. ANDREW BIRO AND
SHERMAN M. WEISSMAN

I. INTRODUCTION

Modern methods of microbial genetic manipulation and *in vitro* recombinant DNA technology have made available highly purified segments of genetic material from almost any desired source (1). The proliferation of the discovery of new restriction endonucleases (2), their

177

MOLECULAR GENETICS, PART III

applications to DNA analysis, and the subsequent development of rapid methods of DNA sequence analysis (3–6) make it entirely practical to determine the nucleotide sequence of any gene in a reasonably short period of time in a small laboratory. In the future, as a result of these advances, it is likely that knowledge of the primary structure of genes will be available before detailed functional analysis of the regulation of their expression. If the regulatory functions of the nucleotide sequences could be interpreted, this would bypass a great deal of genetic analysis. This is particularly important for animal cells, where genetic methods are so much more difficult than in prokaryotes that functional interpretation of gene structure at present might provide the only route toward a detailed understanding of the control of gene expression.

The elucidation of the triplet code by which mRNA specifies incorporation of amino acids (Fig. 1) was spectacularly successful in this regard (7). The code is universal in all living forms, except for cells bearing certain mutations of the protein-synthesizing apparatus. At

UUU	Phe	UCU	Ser	UAU	Tyr	UGU	Cys
UUC	Phe	UCC	Ser	UAC	Tyr	UGC	Cys
UUA	Leu	UCA	Ser	UAA	Termination	UGA	Termination
UUG	Leu	UCG	Ser	UAG	Termination	UGG	Trp
CUU	Leu	CCU	Pro	CAU	His	CGU	Arg
CUC	Leu	CCC	Pro	CAC	His	CGA	Arg
CUA	Leu	CCA	Pro	CAA	Gln	CGA	Arg
CUG	Leu	CCG	Pro	CAG	Gln	CGG	Arg
AUU	Ile	ACU	Thr	AAU	Asn	AGU	Ser
AUC	Ile	ACC	Thr	AAC	Asn	AGC	Ser
AUA	Ile	ACA	Thr	AAA	Lys	AGA	Arg
AUG	Met	ACG	Thr	AAG	Lys	AGG	Arg
GUU	Val	GCU	Ala	GAU	Asp	GGU	Gly
GUC	Val	GCC	Ala	GAC	Asp	GGC	Gly
GUA	Val	GCA	Ala	GAA	Glu	GGA	Gly
GUG	Val	GCG	Ala	GAG	Glu	GGG	Gly

FIG. 1. The genetic code. Triplets of capital letters indicate three bases in codons; for example, UUU indicates that a particular codon is made up of three successive uridylic acids. Capital letter followed by two small letters signifies the amino acid corresponding to the codon to its left. "Termination" designates the three triplets UAA, UAG, and UGA that do not encode incorporation of amino acids into protein and polypeptide chains but rather signal termination of the polypeptide chain.

first the code appeared to be entirely independent of syntactical influence; that is, a triplet of bases, if it were translated, would insert the same amino acid regardless of what the bases preceding or following it were. This view has required some minor modification; for example, the codon GUG might specify methionine if it acts as initiator codon (8–10), but specifies valine if it occurs internally in the messenger RNA (mRNA). Also, the efficiency with which suppressors can overcome chain termination mutants appears to be quite variable, depending on the syntax in which the termination triplet is found (11–15). Nevertheless, the genetic code can be used with almost total confidence to predict amino acid sequences from mRNA nucleotide sequences. Initially, it was hoped that a similar simple lexicon would be available for understanding the problem of elementary events in gene regulation, such as the initiation and termination of transcription, and there was considerable thrust to determine the first nucleotide sequences in DNA and RNA about sites where regulation might occur. This expectation has been only incompletely borne out. In some cases it is possible to recognize or at least strongly suspect the function of specific sequences in DNA; however, as new examples emerge, it is apparent that the base sequences specifying these aspects of genetic function are less rigidly determined than are the base sequences coding for individual amino acids. Crystallographic resolution of specific complexes between nucleic acids and proteins would provide a very strong structural basis for prediction of the range of base sequences which permit effective recognition by sequence-specific proteins such as RNA polymerase, cyclic AMP binding protein, etc. So far, however, crystallography of specific recognition complexes between double-stranded DNA and their recognition proteins has not advanced to the level of fine structure resolution. Therefore, deductions about the requirements for recognition are presently based upon comparison of large numbers of functional nucleotide sequences and the effect of mutations on the function of the sequences by relatively indirect methods of studying protein nucleic acid complexes, such as chemical modification of the complex or of either of its components, or by model building based on the known stereochemistry of peptides and nucleic acids.

This chapter will survey the current status of information about sequences which are recognized for initiation or termination of transcription, of . . . iation of DNA replication, and for RNA processing. . . . n mRNA and the newer methods of sequence analy- . . . In most cases the available sequence data permit making only limited generalizations. Numerous recent reviews are available on relevant parts of this subject (16–33), and will be referred to in the various sections.

II. TRANSCRIPTION

A. THE PROCESS OF BEGINNING AND ENDING TRANSCRIPTION

For a gene to be expressed, its DNA must first be transcribed into RNA. Depending on the nature of the gene, this RNA may be directly functional, or may be converted to an active form by subsequent processing steps. The RNA must become associated with host ribosomes so that it then directs the synthesis of a protein. In principle, the selection of genes to be expressed at any time might be accomplished by controls that determine which genes are transcribed in their entirety into RNA or by controls that affect the subsequent fate of RNA transcripts. For example, rapid degradation of particular messenger RNAs could decrease the amount of protein specified by the mRNA. In higher cells, most message RNAs have a 7-methyl guanosine triphosphate ("cap") attached to their 5′-termini (Fig. 2). Failure of formation of the cap structure of the 5′-end of the RNA (34) or failure of addition of polyadenylic acid to the 3′-end (35) could also decrease the rate of expression of a given gene.

As a general rule, regulation of biochemical patterns occurs at an early step in the reaction scheme. In prokaryotes, this is generally true

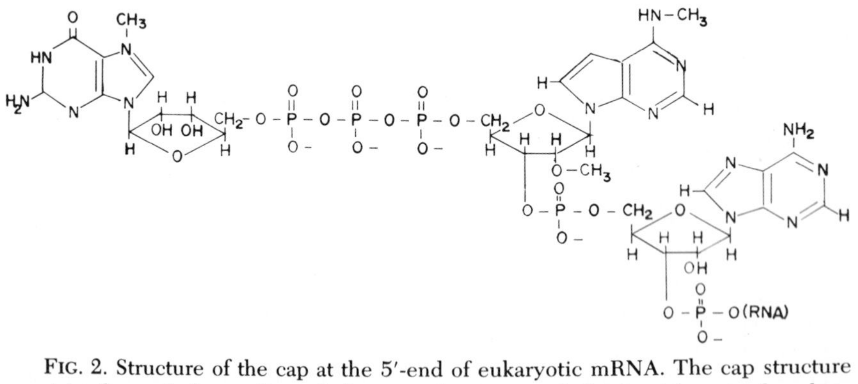

FIG. 2. Structure of the cap at the 5′-end of eukaryotic mRNA. The cap structure consists of a methylguanylic acid shown at the extreme left joined by a triphosphate bridge to the 5′-hydroxyl of the 5′-terminal residue of the RNA chain. The first residue of the RNA chain is often a 6-methyladenosine and the sugar of the first residue has a 2′–O-methyl group.

for gene expression. In a large number of cases, the primary determinant of whether a gene is expressed is whether or not the DNA corresponding to that gene is transcribed, rather than by any regulation of the subsequent utilization of the RNA. This implies that there must be specific regions where transcription is initiated. Also, since cellular genomes are made of a single or of a relatively small number of very large DNA molecules, there must be specific sites at which transcription is terminated. Knowledge of the precise structure of genes and biochemical mechanisms for initiation and termination of transcription is fundamental for understanding how gene expression is regulated. In addition to the extent that expression of individual genes or operons may vary in a discordant fashion, there must be selective factors that modify the relative efficiency of initiation or termination of RNA polymerase on specific operons. These include proteins (*repressors*) that inhibit transcription of specific genes (29), proteins that stimulate initiation of transcription of specific coordinately expressed genes (*operons*) (36), or groups of operons (37), factors that modify RNA polymerase and its specificity (38–41), and factors that prevent (42,43) or augment (44,45) termination of transcription (42,43).

Understanding the regulation of transcription in bacterial systems has progressed much more rapidly than the understanding of analogous processes in higher cells. Genetic analyses of certain prokaryotic gene systems, such as the lactose operon (46), provided early strong evidence for the concept of mRNA and the existence of genes whose primary function was to regulate the production of RNAs for specific operons. Shortly after these initial discoveries, the existence of separate segments of bacterial DNA known as *promotors* was proposed (47). The promotors themselves need not be transcribed but their presence is necessary for the initiation of transcription (48,49). The promotor sequences were proposed to be different from the sequences (*operators*) at which elements such as repressors act. These perceptions have been fully confirmed by subsequent molecular analysis.

Bacterial RNA polymerases were isolated as soluble enzymes and purified a number of years before the extensive purification of animal polymerases was achieved (50). Bacterial polymerases can be prepared conveniently in much larger amounts than animal polymerases, and most importantly their *in vitro* properties resemble those exhibited *in vivo*. They actively transcribe a variety of DNA templates, prefer double-stranded to single-stranded DNA, and require only physiological ingredients for activity.

There are at least three categories of nuclear RNA polymerases in eukaryotes (27,30). These are RNA polymerase I that transcribes ribosomal DNA, RNA polymerase II that synthesizes the bulk of heterogeneous nuclear RNA (hnRNA) and mRNAs, and RNA polymerase III that synthesizes tRNAs, 5 S RNA, and some virus-coded small RNAs. Each purified polymerase contains at least eight different polypeptides and exhibits microheterogeneity of unknown biological significance.

Animal polymerases, even when extensively purified, exhibit a number of awkward features. Some of them, when partially purified, are more active with single- than double-stranded DNA and require manganese. This ion is presumably not present in the cell in amounts corresponding to those required for the *in vitro* reactions and can decrease the selectivity of other copying reactions and enzymatic recognition of nucleic acids (51–53). In these cases, it is not clear to what extent the *in vitro* reaction parallels reactions that occur in the cell. Finally, as discussed below, it has been known for some years that bacterial RNA polymerase can initiate and terminate transcription precisely on purified DNA in the absence of accessory factors (54). Specific initiation of transcription has only been demonstrated at the sequence level *in vitro* in a very few animal cell systems, all dependent on RNA polymerase III (55–64). One favorite system for study of polymerase III action *in vitro* is the low molecular weight 5 S RNA, which is associated with ribosomes. A second useful system is provided by cells infected with human adenoviruses. Such cells synthesize remarkably large amounts of a low molecular weight virus-specified RNA called VA-RNA I (65). The function of this RNA is not known, but it is well suited for analysis of transcription *in vitro*, since the RNA is made in cell-free systems even with disrupted nuclei (66). Synthesis reinitiates repeatedly *in vitro* as judged by the incorporation of label from $[\beta\text{-}^{32}P]$- or $[\gamma\text{-}^{32}P]GTP$ (57,56). The transcription of VA-RNA has been studied in some detail, and it is clear that transcription is mediated by RNA polymerase III and that exogenously added polymerase can synthesize VA-RNA from chromatin preparations of adenovirus-infected cells. Until very recently, precise initiation of transcription had only been demonstrated when chromatin was used as template, but not when purified RNA polymerase, with or without possible accessory factors, was added to naked DNA. One of the results of this difference in the ease of manipulation of the prokaryotic and eukaryotic systems is that there is now extensive sequence and mutational information for a variety of promotors and terminators for bacterial RNA polymerase but that the sequence of initiation and termination signals has only been determined for RNA polymerase III.

The subsequent sections will, first of all, discuss the initiation of transcription of bacterial RNA polymerases and then present the more limited amount of information available about analogous events in animal systems. Following this, the sequences at promotor and transcription termination sites, repressor binding sites, and cyclic AMP binding protein recognition sites will be reviewed. In general, the discussion will deal with *E. coli* RNA polymerase, although where knowledge is available it appears that RNA polymerases from other bacteria have very similar properties.

B. THE INITIATION REACTION

Escherichia coli RNA polymerase *holoenzyme* consists of two copies of an α-subunit (MW approximately 40,000), one copy each of a β (MW 100,000), and β' (MW 100,000), and one copy of a sigma (MW 80,000) subunit (67). There is also an additional peptide, omega, found in variable amounts in RNA polymerase preparations that has not yet been shown to play a role in the function of the enzyme. The sigma subunit of RNA polymerase may be removed to generate a *core* enzyme. Core enzymes are still capable of RNA synthesis but lack the initiation specificity of the holoenzyme; sigma acts to decrease the aggregation of RNA polymerase at low ionic strength (68), to destabilize the binding of RNA polymerase to nonspecific sites on the DNA (69), to stabilize polymerase promotor complexes (70), and to maintain the rate of pyrophosphate exchange (71). It is possible but not conclusively demonstrated that there are specific sequences at promotors that are recognized by sigma.

Once transcription is initiated, sigma is released (72–74), so that the only polypeptide subunits involved in the elongation of transcripts are α-, β-, and β'-subunits. A current model for the initiation of transcription proposes that the holoenzyme first recognizes a promotor in double-stranded DNA and binds to it to form a relatively unstable *closed complex*. In a second step, an *open complex* is formed in a reaction that is believed to involve local melting of the DNA helix. The open complexes are often rather stable (estimated association constant equals 10^{14}) so that protected DNA segments can be isolated after DNase treatment. The complexes are only very slowly dissociated by agents, such as heparin and rifampicin, that rapidly inactivate free RNA polymerase. On the other hand, not all promotors form stable complexes (75).

The initial interaction of DNA and RNA polymerase and subsequent formation of an open complex can be distinguished, since DNA melting is a highly cooperative reaction that is markedly favored by warm temperatures, low salt concentration, or the presence of glycerol. At temperatures below 15°C, one may permit the first stage of reaction and yet inhibit the opening of the DNA helix and the formation of the open complex. Direct evidence for the existence of the closed complex was obtained by electron microscopy of *E. coli* RNA polymerase bound to restriction endonuclease fragments of T7 DNA. Polymerase was seen to bind specifically to promotor regions on the DNA even at temperatures below those where stable initiation complexes could be formed (76).

Promotors may differ in several independent parameters of their strength, such as the "off" rate for bound RNA polymerase, the sensitivity to salt inhibition, and the temperature at which conversion from a closed to an open complex will occur (77,78). Different promotors can compete for limiting amount of RNA polymerase with greatly varying efficiency. This behavior is not simply related to any single one of these factors. The stability of complexes of DNA and RNA polymerase may be greater with supercoiled than with relaxed DNA (79). As an example of the difficulty in assigning intrinsic affinities to promotors, initiation at the tyrosine tRNA promotor was very sensitive to salt inhibition, and the promotor polymerase complex was readily dissociated when the promotor was part of a restriction endonuclease fragment 600 base pairs long. However, transcription was less sensitive to inhibition when the template was the intact DNA of a phage that had incorporated the original restriction fragment.

Promotors are specific for particular polymerases. For example, the RNA polymerase encoded by phage T7 will recognize T7 promotors but not those of the related phage T3, and neither T7 nor T3 polymerases will recognize host cell promotors (80). RNA polymerase from uninfected *Bacillus subtilis* will recognize only early promotors of the phage SP01. When the polymerase is modified by replacement of sigma factor by a phage-coded protein (gp28), the polymerase will preferentially transcribe the "middle" genes of the phage (81,82). There is even a suggestion that bacterial polymerases of different species may not have exactly the same specificity for promotors, since T4 DNA is a less good template for *B. subtilis* polymerase than for *E. coli* polymerase (83).

The DNA helix is thought to be locally melted in the open initiation complex and during transcription. The number of base pairs of DNA helix opened per complex has been estimated as follows (84): circular double helical bacteriophage fd or λ DNA molecules were nicked so that they were relaxed, and one strand of the double helix could rotate freely about the other strand without topological constraints. Polymerase was bound to the double helical molecules, and the cut ends of the DNA were rejoined with DNA ligase. Protein molecules were then removed. If the DNA was melted by RNA polymerase, the ligation would leave DNA molecules that were underwound with respect to the relaxed deproteinized DNA molecule. As one strand of the DNA helix is wound about another to form a Watson–Crick structure in the absence of RNA polymerase, the topological constraints in the DNA molecules force the formation of negative supercoils. The amount of the supercoiling of the DNA was measured by ethidium bromide binding. The unwinding of the DNA required the presence of sigma factor and showed a marked temperature dependency. These figures led to estimates that about 7 bp were melted per RNA polymerase molecule at 50 mM KCl (84) and about 4 bp at 200 mM KCl. At 50 mM KCl, the number of complexes present exceeded the number of promotors. The amount of DNA unwinding was still proportional to the number of bound polymerase molecules, so that local melting occurred at sites that could not initiate transcription. The quantitative estimates are somewhat difficult, however, since it is not clear that all RNA polymerase molecules found on the DNA in the preparation were actively transcribing, and distortion of the double helix near the polymerase could alter the amount of net unwinding of the DNA.

Whatever the exact figure for the number of melted base pairs in a DNA–RNA polymerase complex, there is not sufficient collapse of the molecule to produce shortening of DNA fragments to an extent that is measurable by electron microscopy. This experiment has been performed by Schleif (85), using DNA fragments from the arabinose operon. In these experiments, one could further verify that the RNA polymerase molecule was transcribing the DNA, since the nascent RNA molecules were visible. However, the DNA helices might have been unwound if DNA were held in an extended state by the RNA polymerase molecule. Reformation of the DNA double helix could not be the motive force for release of nascent RNA or progression of RNA polymerase, since negatively supercoiled DNA is a better template than relaxed DNA, although the free energy of the supercoiled DNA would be progressively decreased by formation of an RNA–DNA hybrid up to several hundred base pairs long.

C. THE TERMINATION REACTION

Termination of transcription has been analyzed less extensively than initiation, although one might suspect that it could be a simpler process. It is clear that steps in termination must include reformation of the DNA double helix and release of the RNA molecule. Kinetic studies suggest that an arrest of the progression of RNA polymerase may occur without release either of free RNA polymerase or of RNA from the template (86–89). The termination of transcription and release of RNA molecules may occur spontaneously, at appropriate sequences, either *in vitro* or *in vivo*, or may require the presence of at least one additional factor.

In an important study, Roberts found that there was an additional protein in *E. coli* that inhibited transcription of λ bacteriophage DNA (44). This protein, the *rho factor*, promoted termination of transcription at specific sites. These sites corresponded to boundary points between early and delayed early message RNAs transcribed from λ phage DNA by bacterial RNA polymerase, and Roberts suggested that the regulation of transcription of the delayed early RNA by the bacteriophage N gene product was mediated by overcoming the effect of the rho factor. Figure 3 summarizes the features of λ that are referred to in the following discussion. More recent evidence suggests the N function may act as an *antiterminator* at a termination site in λ DNA that is active in the absence of rho. Therefore, N may be a general antagonist of termination, rather than an antagonist of rho (17).

Rho factor could be a multifunctional protein. Its gene maps in the *E. coli* chromosome near or within the structural gene for the β-subunit of membrane associated ATPase. Temperature-sensitive or -suppressible mutants of rho have been isolated (90,91), and these show pleiotropic effects including an inability to metabolize succinate, similar to those seen in *E. coli unc* mutants. These latter genes are known to code for the subunit of ATPase. Rho itself has an ATPase activity that is stimulated by RNA (92). This stimulation is seen principally with single-stranded polynucleotides containing cytidylic acid residues. Polyuridylic acid containing one residue in 20 of cytidylic acid will stimulate the ATPase, but polyuridylic acid itself does not stimulate the ATPase (93). The ATPase activity of rho appears necessary for termination of transcription *in vitro*. Thus, β,γ-imido or -methylene analogues of nucleoside triphosphates block rho-dependent termination of transcription on several substrates, even though they are incorporated into the growing chain. Termination can be restored by adding the normal nucleoside triphosphate to the reaction mixture even after RNA synthesis has been initiated (94). This interaction with RNA has led to the suggestion that rho may bind to nascent RNA chains and thus promote termination when appropriate sequences are transcribed. Furthermore, it was suggested that ribosomes which are translating nascent mRNA prevent the formation of a ternary polymerase–RNA–rho complex, and therefore prevent

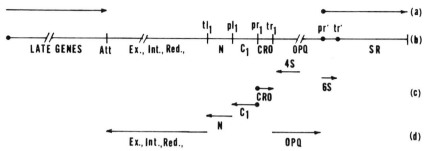

FIG. 3. Schematic representation of certain phage λ transcriptional units. The phage λ genome is drawn in linear representation with the genome ends corresponding to those found in vegetative phage. (a) Top line represents phage mRNA transcribed from left to right as indicated by arrows. (b) Att refers to λ site at which the circular phage DNA is cleaved during insertion into the bacterial genome. t_L and P_L, respectively, are the rho-dependent termination sites for leftward transcribed early message and the promotor for transcription of this early message; P_{RI} and t_{RI} refer to the promotor and rho-dependent termination site, respectively, for transcription of the rightward early message. The leftward early message includes the coding region for the gene N responsible for turn-on of delayed early message. The rightward early message includes the codons for the cro gene, a second repressor-like molecule that acts on rightward and leftward early transcription. $P_{R'}$ and $t_{R'}$ refer to the promotor and rho-independent terminator, respectively, for λ 6 S RNA that comprises a constitutively transcribed initial leader segment for the late gene region. "N" shows the location of the coding region of the N gene, whose function is to overcome termination at t_{LI} and t_{RI}, permitting expression of delayed early genes. "Cro" is the location of coding region for the cro protein. Q indicates the location of coding region for the Q gene product whose function is to permit expression of late genes. S and R refer to the two phage genes involved in cell lysis. These are the 5'-terminal genes of the late RNA transcript. The remainder of late genes are located in the left half of the genome and are presently thought to be transcribed from P_{RI} and $P_{R'}$ transcripts that extend across the circularized form of the genome up to regions close to the attachment site. The solid line represents the linear DNA itself. (c) The solid lines with arrowheads represent the early transcripts formed prior to action of λ N or Q product. These include, in addition to the 6 S RNA, a small leftward transcribed 4 S RNA, the transcripts on theP_{LI} and P_{RI} promotors and some leftward transcription of the C_1 major repressor gene, and a rightward transcript including the cro gene, an additional repressor of early transcription. The 5'-terminal sites from which the transcripts of the C_1 genes are formed during establishment of lysogeny is currently a matter for investigation. (d) The solid lines with arrowheads on the bottom row represent the additional leftward and rightward mRNAs transcribed after N gene has overcome termination at t_{LI} and t_{RI} but before Q gene action has become evident. These include the red system involved in λ recombination, the int and ex genes involved in integration and excision of λ, and the O and P genes involved in initiation of λ DNA replication.

termination of transcription (95). Nonsense mutations within a gene would prematurely release ribosomes, allowing subsequent rho-dependent termination sequences to function. These would terminate RNA synthesis so that downstream genes in the same operon would not be transcribed, accounting for polarity effects on the transcriptional level (17). Consistent with this model, there is a class of mutations that overcome polarity and which lie within the structural gene for rho (96,97). Presumably the phenotype of this mutation is a consequence of the failure to terminate transcription at certain sites in the untranslated RNA bearing nonsense mutations.

In addition, rho may function *in vivo* at termination sites that are effective with purified RNA polymerase *in vitro*. Cells with certain mutations in rho factor are partially derepressed for expression of the tryptophan operon (98,99). This suggests that rho may play a role in the termination of the transcription of the leader sequence of tryptophan operon *in vivo*, although termination occurs at this site *in vitro*, even in the absence of rho. If rho and the β-subunit of ATPase are on the same polypeptide, it may be less likely that rho confers base sequence specificity to transcription termination from DNA templates. A mutant RNA polymerase has been isolated that is able to terminate transcription *in vivo* or *in vitro* at sites that normally are rho dependent (100). This indicates that RNA polymerase itself has recognition capacity for all termination sites and further raises a question whether specific sequence recognition by rho occurs during the termination of transcription.

Additional protein factors that may be involved in termination of transcription have been partly characterized (101,102). There may be other mechanisms of termination. For example, RNA polymerase unable to initiate RNA synthesis may still bind at potential promotors and cause termination of transcription initiated at upstream promotors (103).

D. MODEL SYSTEMS FOR *IN VITRO* TRANSCRIPTION OF SPECIFIC GENES

1. *Bacteriophage Lambda*

Early studies demonstrated that bacterial RNA polymerase would selectively transcribe certain regions of the DNA of either bacteriophage λ or bacteriophage T4 *in vitro* (104,105). The regions transcribed corresponded to those regions of the phage DNA which are transcribed intracellularly by bacterial RNA polymerase before additional phage-coded regulatory factors are produced. From the earliest studies onward, λ has been a particularly favorable system for analyses of events involved in transcriptional control. The expression of the λ genome has been studied genetically (106,107), and many individual transcription units have been identified (Fig. 3). Early biochemical studies showed that if purified λ phage DNA was transcribed with bacterial RNA polymerase in the absence of detectable amounts of other factors, transcription products included large amounts of discrete low molecular weight RNAs (54,108,109), the most abundant of which was a 6 S RNA whose template lies between the genes for delayed early and late proteins of λ (110–112). The promotor for 6 S RNA is the most active promotor for λ DNA transcription *in vitro,* and the 6 S RNA termination site is sufficient for efficient RNA release *in vitro* in the absence of added factors. The sequences preceding the 6 S RNA probably represent a constitutive promotor responsible for much λ late gene expression. Switching on of late gene expression is accomplished by a delayed early gene (Q) product that may function to overcome termination at the end of 6 S RNA, allowing transcripts that begin with 6 S RNA sequence to be extended across the late region on circular DNA templates (112–114).

Within the early region of λ there is a strong promotor coupled to a termination signal that is partly rho independent, and leads to formation of 4 S RNA (109) both *in vivo* and *in vitro.* Prevention of termination of this RNA has been suggested to result in read-through to the *cI* repressor gene (115). There are two other strong promotors on λ, one for the leftward (P_L) and one for the rightward (P_R) early gene expression. Termination signals at the end of these transcripts (T_{L1} and T_{R1}, respectively) are strictly rho dependent (39), although antitermination by the product of the λ *N* gene leads to transcription of λ delayed early genes. A promotor adjacent to the repressor gene initiates leftward transcription of the gene only in the presence of low levels of repressor (116). Although other promotors or RNA polymerase binding sites exist on λ DNA, they have not been extensively characterized functionally or structurally.

Direct analysis of the transcription products found from the λ promotors gave the first demonstration that initiation of transcription *in vitro* occurred with great precision at particular nucleotides within the DNA sequence, and that low molecular weight RNAs also terminated predominantly at a single nucleotide within the DNA. These sites of initiation and termination appear to be identical *in vivo* and *in vitro* (110,113,114), so that the only component necessary for biologically meaningful initiation and termination of transcription from these portions of phage DNA were the DNA itself and the RNA polymerase holoenzyme.

2. lac *and* gal *mRNA Synthesis*

The transcription of the *E. coli* lactose and galactose operons has also been studied intensively. Both *lac* and *gal* repressors have been purified and shown to be active *in vitro*, and operator constitutive mutants are available for both operons. In both cases, transcription *in vivo* and *in vitro* is strongly dependent on the presence of cyclic AMP and cyclic AMP binding protein (37), although there are promotor mutants in the *lac* operon that relieve the requirement for cAMP for efficient transcription. In the *gal* operon there are adjacent transcription initiation sites, one of which requires cyclic AMP and cyclic AMP-binding protein *in vitro*, while the other functions in the absence of cyclic AMP (117). The latter could cause a low-level constitutive expression of *gal* genes but there is presently no direct evidence for this. In this and other cases, expression of a single operon may occur with multiple promotors with different efficiency and subject to different regulation.

3. *The Tryptophan Operon (Fig. 4)*

The tryptophan *(trp)* operons of *E. coli* and *Salmonella typhimurium* have also been the subject of extensive elegant studies (18). *In vitro* transcription of the operon by RNA polymerase holoenzyme produces a leader sequence of about 190 nucleotides that corresponds to the 5'-end of *trp* operon transcripts found *in vivo* (119). *In vivo*, transcription is also terminated at this site in cells with an adequate supply of tryptophan. Under conditions of tryptophan starvation, termination no longer occurs and transcription proceeds through the structural gene. There is also a specific *trp* repressor and operator site.

4. *Other Model Systems*

In addition to the above cases, specific initiation by *E. coli* RNA polymerase has been demonstrated on a wide variety of templates, including other bacterial operons, such as arabinose and biotin; tyrosine-tRNA genes; bacteriophage T7; the single-stranded DNA phages ϕX174, G4, and fd; several phages of *B. subtilis;* and eukaryotic DNA viruses. With bacterial and phage DNA, many initiation sites found *in vitro* correspond to sites used in intact cells. With eukaryotic viruses and bacterial RNA polymerase this is not the case.

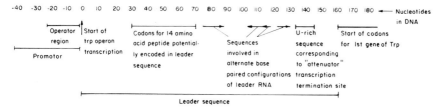

FIG. 4. Schematic representation of initial sequence of tryptophan operon. The numbers in the upper line correspond to base pairs preceding (−) or following the site at which transcription of the *trp* operon initiates (residue 0). Horizontal lines with vertical boundaries span the sequence that is thought to be involved in the indicated functions. The operator region is a site within which operator constitutive mutations have been located and presumptively the region that contains sufficient sequence information for repressor binding. There is an initiator codon followed by 14 sense codons within the leader sequence from 30 to 70 nucleotides downstream from the site of initiation of *trp* operon transcription. Beyond this there are pairs of rotationally symmetric sequences that could correspond to base-paired stem and loop structures within the mRNA. At the end of this region is a sequence in the DNA which would be transcribed to eight consecutive uridylic acids and corresponds to the site at which transcription terminates *in vivo* and *in vitro*. Beyond this region is a presumptive ribosome binding site and the initiator codon for the first functional gene of the trp operon.

E. PROMOTOR SEQUENCES

1. *Promotors for Prokaryotic RNA Polymerase*

Curiously, the first nucleotide sequence presented for a promotor was that from the fortuitous, strong initiation signal for *E. coli* RNA polymerase transcription of the DNA of the small animal virus, simian virus 40 (SV40) (120,121). This was followed very shortly by sequences obtained for early gene promotors on λ (122–124), phage T7 (125,126), phage fd (127,128), *lac* (129), *gal* (130), and tyrosine-tRNA promotors (131,132). By now there are more than 20 bacterial and bacteriophage promotor sequences published and a larger number are available in various laboratories.

Figure 5 compares the sequence of a number of *E. coli* RNA polymerase promotors grouped into those that function *in vitro* in the absence of accessory factors, such as cyclic AMP binding protein and those that require such factors.

The first promotor sequences to be examined revealed many of the features that appear to be similar in most or all promotors, but despite intensive investigation, the full spectrum of necessary and sufficient requirements for promotor sequences is not yet defined. One of the first features to be noted was that more than 30 bp of DNA preceding

SV 40	AATGC	AATTG	TTGTT	AACTT	GTTTA	TTGCA	GCTTA	TAATG	GTTAC	AAATA
λpr	ACCGT	GCGTG	TTGAC	TATTT	TACCT	CTGGC	GGTGA	TAATG	GTTGC	ATGTA
λpl	TCTGG	CGGTG	TTGAC	ATAAA	TACCA	CTGGC	GGTGA	TACTG	AGCAC	ATCAG
λpr'	CGGCA	GATAT	TGACT	TATTG	AATAA	AATTG	GGTAA	ATTTG	ACTCA	ACGAT
λpl'	CTGCC	GAAGT	TGAGT	ATTTT	TGCTG	TATTT	GTCAT	AATGA	CTCCT	GTTGA
fd gene X	ATGTT	TTTGA	TGCAA	TTCGC	TTTGC	TTCTG	ACTAT	AATAG	ACAGG	GTAAA
φx gene A	ACCGT	CAGGA	TTGAC	ACCCT	CCCAA	TTGTA	TGTTT	TCATG	CCTCC	AAATC
φx gene P	TCTCT	TGTTG	ACATT	TTAAA	AGAGC	GTGGA	TTACT	ATCTG	AGTCC	GATGC
λ Cl7	ATGC	ATTTA	TTTGC	ATACA	TTCAA	TCAAT	TGTTA	TAATT	GTTAT	CTAAG
G4 gene A		TGCTT	GACTA	ATACT	CAATG	ACCAC	TCTAA	TATGC	CTCCC	ATCAG
G4 gene B		TAGCT	TGCAA	AACAC	GTGGC	CTTAT	GGTTA	CTCTA	TGCCC	ATCGC
φx gene B	TTAAA	TAGCT	TGCAA	AATAC	GTGGC	CTTAT	GGTTA	CAGTA	TGCCC	ATCGC
Tyr RNA	AACGT	AACAC	TTTAC	AGCGG	CGCGT	CATTT	GATAT	GATGC	GCCCC	GCTTC
T7 A2				A	AGTAA	CATGC	AGTAA	GATAC	AAATC	GCTAG
T7 A3					GTA	AACAC	GGTAC	GATGT	ACCAC	ATGAA
loc UV5	ACCCC	AGGCT	TTACA	CTTTA	TGCTT	CCGGC	TCGTA	TAATG	TGTGG	AATTG
loc wt	ACCCC	AGGCT	TTACA	CTTTA	TGCTT	CCGGC	TCGTA	TGTTG	TGTGG	AATTG
gal	TTCCA	TGTCA	CACTT	TTCGC	ATCTT	TGTTA	TGCTA	TGGTT	ATTTC	ATACC
Tryp E coli	AATGA	GCTGT	TGACA	ATTAA	TCATC	GAACT	AGTTA	ACTAG	TACGC	AAGTT
ara C	ACCGT	GATTA	TAGAC	ACTTT	TGTTA	CGCGT	TTTTG	TCATG	GCTTT	GGTCC
ara BAD	GGATC	CTACC	TGAGG	CTTTT	ATCGC	AACTC	TCTAC	GTTTC	TCCAT	AC AC
λ Cl	CATCG	AATGG	CGCAA	AACCT	TTCGC	GGTAT	GGCAT	GATAG	CGCCC	GGAAG

FIG. 5. Tabulation of some known promotor sequences. The gene or source of the sequence is indicated in the left-hand column. Oligonucleotides are grouped in sets of five. The DNA sequence shown corresponds to the sequence of the RNA transcript. The underlined oligonucleotide in the first position of the last block of five is in general the oligonucleotide at which RNA transcription initiates.

the site where transcription initiated were necessary to retain faithful initiation of transcription. This has been shown in a number of experiments. Purified RNA polymerase was bound to DNA to protect the promotor sequences. The unprotected DNA was then removed by digestion with pancreatic DNase. The protected DNA fragments were about 50 bp long and contained 23–26 bases of DNA preceding the transcription initiation site (116,124–128). When the DNA fragment was deproteinized and added to fresh RNA polymerase, initiation of transcription did not occur, possibly because either or both of the RNA polymerase and the DNA undergo conformational changes between first contact and formation of the stable initiation complex. When λ DNA was bound to RNA polymerase and digested with λ exonuclease and S1 nuclease, a protected fragment 65 bp long was generated. This fragment contained more than 30 bases upstream from the RNA initiation site, and was able to bind polymerase and direct transcription (116).

In another approach, a restriction endonuclease fragment that contained 30 bases upstream of the strong *E. coli* RNA polymerase start on SV40 DNA had lost promotor activity for *E. coli* RNA polymerase (121). SV40 DNA fragments containing 60 bases upstream from the initiation site retained full promotor activity. A deletion mutant of SV40 that lacked 6–9 nucleotides 30 bases upstream from the strong initiation site still retained full promotor activity (133). Similar conclusions were drawn from mutations affecting the λ promotor. Lambda P_{L1} sequences contain a restriction cleavage site for the enzyme HindII, located 36 nucleotides preceding the site of initiation of transcription. Mutant promotors whose sequence was the same except that the HindII site had been removed by a single base change were much weaker *in vivo* and *in vitro* (116) than the "wild-" type promotor. Cleavage at this HindII site also inactivated the wild-type promotor (134). Although the P_R promotor of λ lacks a HindII site, mutations changing a base 31 or 33 nucleotides upstream from the initiation site damaged the promotor.

Although many of the promotors subsequently sequenced have not had a HindII site at this region, it appears that there is partial similarity between HindII recognition sites and an *E. coli* RNA polymerase recognition sequence in this area. This accounts for the frequent presence of a HindII site in promotors. The spectrum of sequences occurring in this region of promotors can be seen in Fig. 5. The sequence GTTG often occurs approximately 35 nucleotides preceding the polymerase start. In some promotors this is part of a larger similar sequence resembling TGTTGA(CAT). In some strong promotors such as SV40 and λ $P_{R'}$, there are tandem imperfect repeats of this or very similar sequences.

Further inspection of promotor sequences shows that a substantial subset has more extensive variation in the upstream region. The exact distance in base pairs between this "upstream site" and the transcription initiation site is somewhat variable.

There is an additional heptanucleotide sequence, ACACTTT, that is found in several promotors beginning about 38 to 30 nucleotides upstream for the RNA start. A segment of this general form occurs in *gal, lac,* and both arabinose promotors and the *Tyr*-tRNA promotor. Possibly this may be an alternative to TGTTGAC as the upstream promotor recognition site.

Takanami and his colleagues (135) raised the question whether specific sequences are needed more than 15 bp upstream from the initiation site or whether all that is needed is a sufficient length of DNA. These authors found that a promotor in fd phage DNA had an *H. hemolyticus* restriction endonuclease cleavage site 15 nucleotides upstream from the transcription initiation site. They ligated the *H. hemolyticus* restriction fragment containing the transcription initiation site to each of three DNA fragments of known sequence. In two cases they obtained effective promotors even though the sequences preceding the DNA cleavage site were different from those in the original promotor. However, inspection of the sequences about the newly formed promotors shows that sequences TTGTTGTC or CTTGAT were introduced about 30 bases before the initiation site, reconstituting sequences resembling those found in natural promotors.

Most promotors show at least partial similarity in nucleotide sequence from 5 to 12 bases preceding the initiation site. This region has been called the "Pribnow box" (126), and in canonical form contains the nucleotide sequence TATAATG in the DNA strand whose sequence would correspond to that of the RNA transcript. No single base in this region is invariant in all promotors, and in several promotors as many as three of the six nucleotides may differ from the prototype, although deviation is more frequent at some bases than at others. Base one is T in 16 of 20 and base seven is a purine in 16 of 20. Functional studies indicate these sequences are important for the initiation of transcription. The *lac UV5* promotor has the sequence TATAATG [J. Gralla, quoted Gilbert (18)]. A base change to TATATTG produces weaker promotor while the wild-type promotor with the sequence TATGTTG can only function in the presence of cAMP and cAMP binding protein. A tandem duplication of 9 bp near the rho-dependent termination site for early rightward mRNA on λ DNA creates a new promotor—the *C17* promotor (136). This duplication produces a "Pribnow box" sequence TATAATT, four to eight bases before the new transcription initiation site.

The most obvious sequence requirement for initiation of transcription is that the first base of the transcript is generally a purine. This can occasionally be bypassed *in vitro* if the concentration of purines or the ratio of purines to pyrimidines is low (132). The only potential physiological reactions in which polymerase has been seen to start with a pyrimidine are the initiation of one of the transcripts of the DNA of vaccinia virus formed by the endogenous viral polymerase (137) and possibly initiation from the *C17* promotor of λ discussed above. Under nonphysiological circumstances, such as high CTP and low GTP concentration, transcription of the tyrosine-tRNA promotor can produce RNA with CTP at the 5'-terminus. It has been suggested that absence of a pyrimidine in the appropriate position of a DNA sequence may prevent initiation, even when other structural features of a promotor are present (138).

The site of initiation of transcription is not absolutely specific, even with physiological concentrations of substrates and reaction conditions. For example, initiation of transcription of the cyclic AMP-dependent *gal* operon transcribes most often before pppAAU, but sometimes begins with pppGAAU, beginning transcription one base upstream from the forward initiation site.

Sequences beyond the initiation site differ from one another and appear not to be important for the initiation of transcription. No mutations affecting promotor efficiency have been found within the transcribed sequences. Deletions in both *gal* and *trp* that remove most of the sequences beyond the initiation site do not obviously alter the strength of the promotor. Bennet and Yanofsky (139) have isolated a deletion mutant that removes the DNA from position −5 downstream through the normal *trp* initiation site without drastically reducing *trp* operon expression.

Gilbert and his colleagues (unpublished results) have studied the contact sites between *E. coli* RNA polymerase and DNA by chemical modification of promotor DNA protected by RNA polymerase. These studies have provided evidence that contact points exist both between RNA polymerase and the Pribnow box, and between RNA polymerase and nucleotides 30 to 35 bp preceding the site of initiation of transcription. These contacts exist with firmly bound RNA polymerase, and the presently available results do not necessarily provide us with direct information about the contacts that take place during the initial stages of interaction of RNA polymerase with the promotor.

In addition to these specific sequences, one might imagine that if *E. coli* RNA polymerase initiation required melting of DNA helices, such initiation would be favored by strategically located AT-rich regions. Wells suggests that AT-rich region might destabilize adjacent sequences of double helical DNA ("telestability") (140), and calculations suggest this might occur over a range of 100 or more base pairs (141). AT-rich regions occur in or near some very strong promotors, such as that in SV40 DNA and that for the λ 6 S RNA and RNA polymerase-binding sites on λ DNA tend to lie in more readily denatured regions of the genome (142).

More than one nearby promotor may initiate transcription of the same segment of DNA, so that initiation at specific bases rather than transcription of specific regions needs to be measured to assess individual promotors. There is also a suggestion that at promotors for bacterial ribosomal RNA, several polymerases may interact in a preinitiation complex to make the promotor unusually efficient in competition with other promotors (143,144).

One difficulty in recognizing a universal promotor sequence is that promotors of a wide range of strengths exist, and sequences in natural promotors could even play a role in limiting the efficiency of transcription initiation *in vivo*. Most studies of promotion have been essentially qualitative, rather than measurements under a range of physical conditions of the relative rates or affinity for initial binding of *E. coli* RNA polymerase or measurements of rate constants for converting closed complex to open forms and for the actual formation of the first phosphodiester bond at single promotors. Perhaps when these reaction components are dissected out and compared with the sequences in the promotors, more quantitative predictions may be made about what properties of a sequence specify an efficient promotor for *E. coli* RNA polymerase.

2. *Promotors for Eukaryotic Polymerases*

The only sequences available at the moment that are known to be promotors or precise transcription initiation sites for *in vivo* RNA synthesis in eukaryotes are those for the polymerase III promotors that specify initiation of transcription of ribosomal 5 S RNA from *Xenopus laevis* and yeast (145–147), yeast tRNAs (148,149), and the adenovirus 2 VA-RNA I (150). Yeast RNA polymerase III differs in details of its properties from that of higher cells, although it may be functional at animal cell promotors (151). RNA polymerase III from animal cells is known to prefer transcription of GC-rich synthetic templates, and it is therefore not surprising to find that there are GC-rich sequences in both the *Xenopus* 5 S and the VA-RNA promotor. It is curious, but probably coincidental, that there is again a *Hind*II site about 30 nucleotides preceding the position where RNA polymerase III initiated transcription of the VA-RNA. Both *X. laevis* 5 S RNA and VA-RNA I promotors show runs of four adenylic acids, followed by guanylic acid; however, their positions relative to the sites of initiation of transcription are somewhat different. Also, there are two sites for initiation of VA-RNA separated by approximately three nucleotides (152,153). Either initiation by polymerase III is less precise than with bacterial polymerases, or there may be two types of initiation complex for VA-RNA.

F. SEQUENCES AT THE TERMINATION OF TRANSCRIPTION
(FIG. 6)

As discussed above, in the absence of rho factor or other additional factors, RNA polymerase transcription termination has been demonstrated unequivocally in a few systems *in vitro* (154). These include determination of precise 3′-terminal sequences at the ends of discrete transcripts obtained with DNA from lambdoid phages (108,109), small

DNA phages (155,156), and the *trp* operon (157), and also formation of discrete RNA species (as judged by sedimentation analysis or gel electrophoresis) with DNA from the bacteriophages T7 and T3 (158–160).

Sequences beyond the site of *in vitro* factor-independent termination of transcription have been published for λ 6 S (111,114) and 4 S RNA (161–163) and for the *trp* operon leader sequence (164,165) (Fig. 3).

The 3′-end sequences of many RNA molecules isolated from intact cells are known. However, the isolation of an RNA molecule with a specific 3′-end is, of course, no proof that this RNA was formed by termination of transcription rather than posttranscriptional processing. Processing may also occur preferentially at sites that resemble transcription termination sequences. For example, small early mRNA chains from T7 infected *E. coli* are known to be produced by RNase cleavage, yet may end in a pyrimidine and U-rich sequence. As discussed above, in the presence of *Q* gene product, transcription may proceed through the 3′-end of λ 6 S RNA into the late genes. In the presence of RNase III, it appears that the elongated RNA may be cleaved very near the site where transcription terminates in the absence of *Q* (114), further illustrating the possibilities for confusion between termination and processing.

Two features of termination sequences were emphasized in the early studies. (1) The terminal sequence of the RNA includes a run of uridylic acids; (2) the termination sites are preceded by a sequence of partial symmetry in the RNA, such that the RNA may form a hairpin loop and stem extending to within about five bases of the termination site. A third feature of the termination sequence has been emphasized more recently. Most of the known termination signals in DNA consist of an AT-rich sequence preceded by a GC-rich sequence. *In vitro* at reduced temperatures RNA polymerase pauses at certain sites, some of which lie about 10 bp beyond a GC-rich sequence (18). Mutations that change a GC to an AT base pairing abolish this pause. However, this phenomenon was seen under nonphysiological conditions, and there is no reason to believe polymerase pauses at these sites *in vivo*. Nevertheless, the sequences 9 and more bases upstream from strong termination sites are often relatively GC-rich.

In the case of the λ 6 S and 4 S RNAs, the terminal bases are a sequence of six uridylic acids, followed by a single purine. In the tryptophan operon leader sequence transcription terminates at the end of a run of seven or eight uridylic acids [six in the closely related *Salmonella typhimurium trp* operon (165)].

Two deletion mutants of the *trp* operon show effects consistent

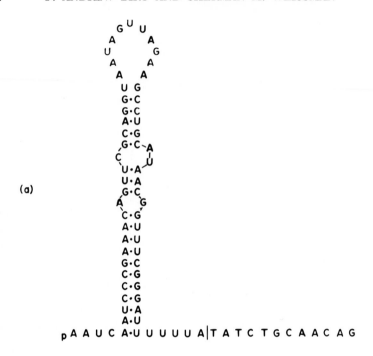

pAAUCA·UUUUUA|TATCTGCAACAG

pAUACCCA·UUUUUUUUGAACAAAATTAGAG

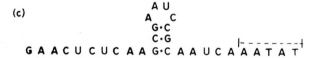

GAACUCUCAAG·CAAUCAAATAT

FIG. 6. Nucleotide sequences at the sites of certain transcription termination signals for bacterial RNA polymerase. (a) Nucleotide sequence at the site of transcription termination for the λ 6 S RNA. U is written for uridylic acid in that portion of the RNA transcribed and T is indicated in the positions corresponding to DNA beyond the transcrip-

with the assignment of a termination function to the sequences present in the 3'-ends of the transcripts (166). One deletion removed the DNA sequence in the tryptophan leader 12 bases and further downstream from the termination site, yet termination occurred normally *in vivo* and *in vitro* on this template. A second deletion removed the deoxyadenylic acids corresponding to the last four of the eight uridylic acids of the *in vitro* transcript. This deletion abolished termination *in vitro* even in the presence of added rho factor, although *in vivo* leader RNA was found with the 3'-terminal sequences $CUUUU_{OH}$ and $CUUUUG_{OH}$, even in bacteria with defective rho function. Either these are the products of posttranscriptional processing, or there are additional unidentified factors that promote termination *in vivo*.

The probable strong termination signal for transcription for ϕX174 DNA also would produce a product with six uridylic acids. This is the only site at which six consecutive deoxyadenylic acids are found on the template strand in the entire ϕX174 DNA (167). Preceding the six deoxyadenylic acids, there is a rotationally symmetric sequence in the DNA. However, the analogous site in G4 has only four deoxyadenylates followed by a deoxyguanylate (156). In ϕX174 the palindromic sequence is preceded by the gene *A* promotor, so that the gene *A* mRNA would start within the palindrome. Apparently this mRNA is oblivious to the termination signal that blocks the elongation of RNA initiated further upstream. This is further circumstantial evidence for the importance of a base-paired segment in nascent RNA as part of a transcription termination signal. A run of eight deoxyadenylic acids also occurs at the putative terminator for bacteriophage fd mRNA (140).

Transcripts that terminate in uridylic acid-rich sequences that may well represent sites for transcription termination rather than processing have been found *in vivo* (168–170). For example, 5 S RNA lies at or near the 3'-end of some ribosomal RNA primary transcripts in bacteria (57). The precursor to *B. subtilis* ribosomal 5 S RNA occurs in two forms. One form has been analyzed and is a molecule with additional sequences at both the 5'- and 3'-ends of the mature 5 S RNA

---

tion termination site. Vertical dash indicates termination and horizontal dashed line indicates region in which termination occurs. (b) Transcription termination site for *E. coli tryp* leader sequence. Symbols are as described in (a). (c) Rho-dependent transcription termination site for the tyrosine-tRNA gene. Symbols are as in (a). The triplet of bases in the short stem is the maximum perfect Watson–Crick base pairing found in the sequence immediately preceding the termination site and may not have any physical reality. However, there are possibilities for more extensive base pairing involving separated segments of sequence upstream from the termination site.

(58). The 3'-end sequence of the precursor bears remarkable resemblances to the *in vitro* RNA polymerase termination sites, including two successive runs of six uridylic acids separated by a single purine.

Finally, early studies demonstrated that deoxyadenylic acid is a poorer template for RNA polymerase than other homopolymers, and that the 3'-ends of total transcript prepared with *E. coli* RNA polymerase tended to be uridylic acid-rich (171). At the time these experiments were performed, it was suggested that RNA polymerase transcribed adenylic acid more slowly than other bases.

Base pairing or extensive secondary structure in the RNA transcript between the 3'-terminal and the subterminal sequences might energetically favor release of the RNA molecules, because in addition to reforming the DNA double helix, one would form compact RNA structures lower in free energy than the single-stranded RNA. It has also been suggested that RNA polymerase may be directly responsive to secondary structure in the nascent chain, or that termination may be the consequence of a particular pattern of strong and weak base-paired segments in the DNA (172). Oligoguanylic acid polymers self-associate strongly, and one possibility is that the interactions within guanylic acid-rich sequences may supplement or supplant the role of Watson–Crick base pairing in subterminal RNA sequences.

There are many RNA transcripts that contain internal base-paired segments at regions where transcription does not terminate. In some cases, such as ribosomal RNA, transfer RNA, prokaryotic 5 S RNA, or the unprocessed bacteriophage T7 mRNA, these may be quite long, so clearly this base pairing is not a sufficient determinant of transcription termination. Some very GC-rich regions are found internally in RNA, such as ribosomal RNA and transfer RNAs. Finally, runs of six or more uridylic acids occur internally in bacterial rRNA (173), and in a transcript of the λ *cro* gene (174). RNA containing internal regions of six or more uridylic acid is transcribed from SV40 DNA *in vitro* with bacterial RNA polymerase (175). Therefore, no one feature of rho-independent termination sequences is alone sufficient to effect termination.

Heterogeneity at the 3'-ends of transcription may be the result of any of several processes, including termination at one of two more adjacent bases, exonucleolytic action at 3'-termini, and posttranscriptional addition of nucleotide residues to the RNA. For example, extra adenylic acids may be added at the end of transcripts formed *in vitro* (82) or to RNA molecules formed by transcription termination or by precursor cleavage *in vivo* (176). As previously mentioned, the rho factor may play a supplemental role in termination of transcription even at sites where termination can occur in the absence of added rho.

In the presence of rho, some 6 S and 4 S RNA transcribed from λ DNA has an extra uridylic acid added at the 3'-end. In these two products, the uridylic acid might have been transcribed from DNA since the sequence in the DNA overlapping the RNA termini is ATA. As already indicated, termination in the *trp* leader may occur at one of two or more adjacent U residues.

In the cases where the sequences in DNA beyond transcription termination sites are known, such as the *trp* operon and the λ 6 S and 4 S RNA, these sequences do not display the striking features exhibited by the terminal portions of the transcripts and also differ from one another so that no clear common features are apparent. This suggests that, in at least some cases, the information for the termination of transcription is provided by the terminal part of the primary transcript itself and that the sequences beyond the site of termination may not contribute to this process. However, there are some similarities between sequences overlapping and extending beyond the 3'-end of the *trp* leader and those at the 3'-end of a possible termination site for Q gene mRNA as well as certain rho-dependent termination sites discussed below, so that the issue cannot be considered closed.

The sequences of the DNA and RNA at three rho-dependent transcription termination sites have been determined (136,177,178), and they differ in several respects from the sequences of rho-independent termination sites. The termination site for transcription of λ"rightward" early RNA (136) has been extensively studied. Although this rho-dependent sequence includes a uridylic acid-rich run UUUAUUUU, transcription does not terminate at the end of this run but terminates about 20 bases further. The sequences in the RNA flanking the sequence UUUAUUU site contain two complementary octanucleotides separated by six bases. Mutants that either enhance or diminish the efficiency of termination *in vivo* at this site, respectively, increase or decrease the strength of the base pairing in the stem of the suggested hairpin loop in the RNA. Transcription termination is not precise but occurs at each of the positions between 7 and 11 nucleotides after the base of the stem, producing RNA with the terminal sequence(s) CAAU(CAA).

The sequence about the rho-dependent transcription termination site at the 3'-end of the *trp* operon has been analyzed. Shortly beyond the termination codon for the last gene of the *trp* operon, the transcript contains a pair of complementary octanucleotides separated by six bases. The sequences beyond the self-complementary region are uridylic acid-rich, including the sequence UUUUAACUUUCUUU. Transcription probably ends within this sequence (T. Platt, personal communication).

A third rho-dependent termination site occurs at the end of the transcript of the precursor to *E. coli* tryosine-tRNA. The 3'-end of the Tyr-tRNA lies at the beginning of the first of 3.14 tandem repeats of a 178 bp DNA sequence. Termination occurs within the second repeat. The sequence of the RNA near the termination site has at best a complementary sequence consisting of three GC pairs separated by four bases although there is an inverted repeat of ten bases in upstream sequence (179); and there are no runs of more than four deoxyadenylic acids in the template DNA. However, a heptanucleotide sequence in the DNA template at the termination site (CAATCAA) is identical to a heptanucleotide at the λ rightward early RNA (*tr*) termination site. The sequence in the first tandem repeat differs slightly, including the change of CAATCAA to CAATTAA, although it is not known whether this change is responsible for the lack of termination in the first repeat. These unanticipated results raise the possibility that there could be more than one type of recognition sequence or base-pairing pattern for rho-dependent termination.

In addition to rho, there is some evidence for other transcription termination effects *in vivo* in *E. coli*. As discussed above, the *trp* attenuator with a deletion of the last four AT pairs of the termination site is read through in the presence of rho *in vitro*, but at least partly terminates transcription *in vivo*. Data at the level of sequence analysis are lacking, but transcription termination at the end of T3 early genes cannot be demonstrated *in vitro*, even though it occurs *in vivo* both for wild-type *E. coli* and in *E. coli* with rho mutants (180).

G. TERMINATION OF TRANSCRIPTION IN EUKARYOTES

The only transcripts for which termination sites are established for higher cells are those low molecular weight RNA molecules known to be transcribed by RNA polymerase III (Fig. 7). Sequences are now available for the DNA encoding and beyond the 3'-end of 5 S RNA from *Xenopus laevis* (181), and *Saccharomyces cerevisiae* (146,147), for yeast tyrosine- (148) and phenylalanine-tRNA (149) genes, and for the sequences embedding the 3'-ends of VA-RNA (182). Even in these cases, the establishment of the sites of termination of primary transcripts are not entirely unambiguous. For example, larger RNA molecules containing 5 S RNA have been detected in cells in culture (183) or in heat-shocked *Drosophila* (184) cells, and these RNAs presumably have extra sequences at their 3'-ends.

The normal VA-RNA terminus has the sequence(s) CUCCUU-(UUU) that is vaguely reminiscent of the U$_6$ sequence at the end of

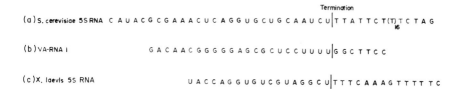

FIG. 7. Representative transcription termination sites for RNA polymerase III. (a) Presumptive site of termination of transcription of *Saccharomyces cerevisiae* 5 S RNA. (b) Site of first transcription termination for VA-RNA I from adenovirus 2-infected cells. (c) Site of transcription termination for *Xenopus laevis* 5 S RNA. The actual sites of transcription termination are aligned with one another and are indicated by a vertical dash in sequence. The transcribed sequence preceding the termination site is drawn with uridylic acids (U) to correspond to RNA and the sequence beyond the termination site is drawn with thymidylic acid (T) corresponding to thymidylic acid in the DNA.

rho-independent bacterial termination sites. This sequence is preceded by a very guanylic acid-rich region. There is a short palindrome in the subterminal sequence of VA-RNA. In the DNA extending about 30 to 40 nucleotides beyond the 3′-end of the mature VA-RNA, there is a very GC-rich sequence, followed by six deoxyadenylic acids, followed by a guanylic acid. *In vitro* and perhaps *in vivo* part of the primary transcript extends beyond the major VA terminus and up to at least the beginning of the run of uridylic acids complementary to the six deoxyadenylic acid residues (185). Therefore, there appears to be a second termination site for transcription of VA-RNA, at least in the *in vitro* system, in the region containing the sequence UUUUUUG. In this respect and in the GC-rich content of the preceding sequences, this termination signal closely resembles the rho-independent bacterial termination signals. On the other hand, the possibility for base pairing within terminal sequences in this region is less prominent here than in the bacterial systems discussed above. Perhaps the tendency of guanylic acid-rich sequences to self-associate could provide a functional alternative to joining between complementary bases.

The sequences predicted from the DNA for extended transcripts of three cloned yeast tyrosine-tRNA genes are AUUUUUUUG in two cases and in one case AUUUUUUUGUUUUUUA (148). The latter is a set of 13 nucleotides identical to the sequence at the 3′-end of the *B. subtilis* 5 S RNA precursor (169). Similarly located runs of six thymidylic acid occur in some of the cloned yeast phenylalanine-tRNA genes (134). The sequences at the 3′-end of the *X. laevis* 5 S RNA are less remarkable, containing only four successive uridylic acids, but there is a *Drosophila* 5 S precursor with extra sequences at the 3′-end terminating in six uridylic acids (186). The yeast 5 S RNA gene is remarkable in a respect different from that seen for VA-RNA and the tRNA genes. The sequences immediately beyond the terminus of the 5 S RNA contain a run of 18 successive deoxyadenylic acids on the strand from which the 5 S RNA is transcribed, so that there could be a run of 18 uridylic acids in the extended transcript. Sequences immediately preceding the uridylic acid run are not remarkably rich in guanylic and cytidylic acid residues.

In all these small RNA molecules transcribed by polymerase III, the mature RNA shows extensive base pairing between its 5'- and 3'-terminal sequences. One possibility is that this pairing also occurs during RNA synthesis and substitutes for the subterminal stem and loop structures found in the prokaryotic transcripts.

In the case of RNA polymerase II transcripts, there is no direct proof that any 3'-end of a cytoplasmic RNA represents or lies near a transcription termination site. However, the sequences of the SV40 genome are known in their entirety, and the 3'-ends of the cytoplasmic early and late messenger RNAs have been located on the genome (187–189). Immediately beyond the 3'-end of cytoplasmic late messenger RNA, elongated transcripts would have a relatively uridylic acid-rich sequence, followed by a sequence that has 13 out of 17 of its bases guanylic acids. Following this would be a sequence of six uridylic acids, followed by an adenylic acid. This sequence has similarities to the bacterial termination signals and particularly the strong termination signal beyond the 3'-end of the VA-RNA. At one location near the 3'-end of early SV40 messenger RNA there is a run of seven uridylic acids in the RNA. The 3'-end of early RNA is difficult to locate as precisely as the 3'-end of late, since so little of the RNA is available. It is not certain whether this run of uridylic acids occurs at the end of the RNA or is included within the 3'-terminal sequence of RNA. Several runs of six uridylic acids occur internally in SV40 mRNA, and it is clear that in themselves they are insufficient to cause termination of transcription. Also longer transcripts can be found in the nuclei that span the 3'-terminal uridylic acid-rich sequences so that termination may not be obligatory.

The hexanucleotide sequence AAUAAA, is found between about 12 and 20 bases preceding the polyadenylic acids of SV40, adenovirus, and a number of host cell mRNAs (187,190–196). The regular occurrence of such a sequence is statistically most unlikely. This sequence can also occur internally in mRNA (189,197), so that it alone need not specify either transcription termination or sites of cleavage and polyadenylation of mRNA. Part, but not all, of the polyadenylic acid in mRNA is associated with specific proteins. Specific sequences at the 3'-end of mRNA could, therefore, serve as partial signals in transcription termination, polyadenylation of mRNA, or for the association of poly(A) and 3'-terminal segments of mRNA with specific binding proteins.

In summary, available evidence suggests that there may be definite homologies between strong termination signals for eukaryotic polymerases (especially polymerase III) and those for bacterial polymerases. There is a suggestion that base pairing in the subterminal RNA sequences may be less essential in the eukaryotic than in the prokaryotic signals and that guanylic acid-rich regions may sometimes be more prominent than in most of the bacterial systems.

III. REGULATION OF TRANSCRIPTION BY SEQUENCE-SPECIFIC DNA BINDING PROTEINS

One of the most advanced areas of study of the molecular basis for regulation of gene expression is the analysis of the interaction of repressor molecules, such as those for the lactose and arabinose operon, and λ early genes, with specific operator sites in DNA (29). These are favorable systems biologically because they exemplify the extraordinary specificity required of regulatory mechanisms in the cell. They are also favorable chemically, since these repressors contain only a single type of polypeptide chain and the binding is a thermodynamically and kinetically reversible process. Certain general features and problems have emerged concerning the interaction of the repressors with DNA. First, the binding constants of the repressors with specific DNA sites are very high ($K = 10^{13}\ M^{-1}$). This binding constant is determined by the ratio of two rate constants: the rate at which repressor molecules attach to the fragments of DNA containing the operator ($k_a = 7 \times 10^9\ M^{-1}\ \text{sec}^{-1}$) and the rate at which repressor molecule is released ($k_{\text{off}} = 6 \times 10^{-4}\ \text{sec}^{-1}$, corresponding to a half-life of 20 minutes at low ionic strength). Inducers may bind either to free repressor or to repressor bound to DNA. The off rate is increased, but the association rate is not substantially altered by binding of inducer. The overall effect is that binding of inducer decreases the binding constant for the operator by about 10^3.

The rate of attachment of repressor molecules in solution to DNA could, in principle, be limited by the frequency of direct contact between the operator site and repressor molecules. If this process were limited by free diffusion rates of DNA and of repressor in solution, the maximum estimated k_a has been estimated as up to two orders of magnitude less than that observed. To achieve the observed equilibrium constant for the binding of *lac* repressor and *lac* operator, the estimated diffusion limited k_{on} would imply a k_{off} rate 1 to 2 logs higher than the measured rate. Therefore, there may be additional factors accelerating the "on" rate above that expected if the repressor molecules all existed freely as randomly dispersed soluble molecules in the solution and were subject only to diffusion. At the same time, repressor has a substantial affinity for nonspecific sites on DNA that might compete with specific binding to the operator.

Several proposals have been made to account for rapid association of repressor and operator and the similar problems concerning promotor recognition by RNA polymerase. First, it has been suggested that if long DNA molecules have behavior that approaches that of random coils, then the concentration of nucleotides within the domain defined by the radius of gyration of a single DNA molecule would generally be much higher than in the solution as a whole (198). Therefore, repressor molecules attached to any point on the DNA molecule are in the presence of much higher concentrations of the operator than would be estimated from total solution content, and the probability of intramolecular domain transfer would be high compared with the probability of transfer of repressor directly from solution to an operator. An alternative suggestion is that the repressor actually slides along the DNA molecule from nonspecific binding sites to specific operator sites so that diffusion occurs in only one dimension. This suggestion has been criticized, and it appears less likely. Both of the above possibilities fail to account for the fact that synthetic operator DNAs of chain length less than 30 show binding constants for repressor nearly as high as those of large DNA fragments containing the repressor (199,200). An alternative that has been suggested is that electrostatic attraction between repressor and operator accelerates the rate of the interaction (201).

Another important aspect of the physiology of repressor control of gene expression is that the repressor molecule has a high affinity for nonspecific DNA sequences. At pH 7.4 and 37°C the general affinity of *lac* repressor for DNA is about 3×10^{-9} of that for the specific DNA sequence. When one considers the number of repressor molecules in each bacterial cell, and the content of DNA in the cell, only 4% of the repressor in the cell is not DNA bound, even in the induced state. In the absence of binding to nonspecific DNA, all the cellular operator DNA would be complexed with repressor, even in the presence of inducer (202,203). There appears to be a large degree of overlap between the repressor site involved in binding nonspecific and operator DNA. Competition between nonspecific binding and specific operator binding is probably an important factor in determining the level of repression achieved for any given operon, and mutations that increase the affinity of repressor for both nonspecific and operator DNA may lead to constitutive production of *gal* genes, presumably because the repressor is tied up intracellularly by the nonspecific interactions with DNA.

The general affinity of repressors for DNA makes it more difficult to use repressor to isolate operator DNA *in vitro*. The problem, as has been discussed extensively by Alberts (204), becomes more serious in animal cells, where the ratio of nonspecific to specific DNA is almost 10^3 higher. It is not certain whether repressor molecules could be designed so that they have a very much higher ratio of affinity for specific to nonspecific DNA by way of lowering the binding constants for nonspecific sequences. Increasing the binding constants for specific sequences would produce remarkably long off times. Solutions for this problem have been considered that would permit repressor control of gene expression in cells with complex genomes. For example, there could be some sort of effective sequestration of the bulk of DNA in any differentiated cell. Alternatively, there could be active processes for removing the repressor from nonspecific DNA molecules, cooperative binding of proteins to control sites, and/or a substantially larger number of repressor molecules per cell than in bacteria.

A number of lines of evidence summarized elsewhere indicate that the best characterized bacterial repressors interact with double-stranded DNA without denaturing the DNA. For example, measurements of DNA unwinding by *lac* repressor show an unwinding angle of only about 90° (205). Several lines of experimental evidence indicate that repressor does not obligatorily enter into the major groove when it binds nonspecifically to DNA. One can modify the dimensions of the major groove by several chemical reactions, such as methylation or alkylation of the N-7 amino group of guanine, and even introduce substituted mercurated nucleotides (206). In the latter case, the substituent fills the major groove and may project as much as 10 Å above it without interfering with repressor interaction.

The possible contacts and forces involved in repressor (and other protein) interaction with DNA sequences have been extensively discussed. The nonspecific interaction of DNA with *lac* repressor can be accounted for energetically by the entropy gained on release of about eleven counterions from the DNA for each molecule of repressor bound. The additional free energy of specific interaction of repressors with operator DNA appears not to be a result of ion release, and operator repressor binding has actually been estimated to involve less ion release (about eight counterions) than in the nonspecific reaction (207–209).

One category of models for base-specific regulation has been proposed in which peptide chains conform to DNA helical parameters and fit into the minor groove of double-helical DNA (210–213). In these proposals there are two amino acids in each peptide chain for each DNA base pair (Fig. 8). Two antiparallel peptide chains held together by hydrogen bonds between amide nitrogen and carbonyl groups of alternate amino acid residues form a "β" ribbon that could lie within the minor groove of the double helix. In one model hydrogen bonding is suggested between amide groups of every other amino acid and 3'-oxygens of the polynucleotide backbone. In an alternative ingenious development of this type of model (213), it is proposed that in one configuration of the peptide chain alternate peptide linkages are base paired to the thymidine or cytosine carbonyl group or the ring N^3 nitrogen of adenine. In an alternative conformation of the peptide chain, peptide carboxyl groups could base pair to the two amino groups of guanine in the minor groove. The later base pairs might be effectively competed against by some amino acid side chains, such as

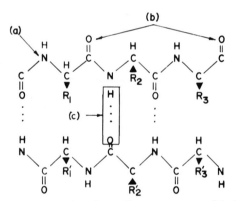

FIG. 8. Schematic representation of peptide arrangement of β ribbon conformation of two antiparallel peptide chains arranged in antiparallel β ribbon conformation. Three dots indicate the hydrogen bond between carbonyl oxygen in one peptide chain and hydrogen attached to amide nitrogen in the other peptide chain. R_1, R_2, R_3, $R_{1'}$, $R_{2'}$, and $R_{3'}$ indicate side chains of amino acids. (a) Indicates the site at which Kim and his colleagues proposed hydrogen bonding can occur between nitrogen residues and sugar hydroxyls of the nucleic acid (210). (b) Indicates carbonyl residues which could form hydrogen bonds with heterocyclic ring hydrogens of bases AT in DNA according to Gursky and his colleagues. (c) Indicates hydrogen bonds which could be broken so that the peptide chain might re-form hydrogen bonds with the guanine basis of DNA according to Gursky et al. (213).

that of serine, so that a "code" could be written in which certain amino acids would favor interaction with AT pairs and others permit interaction with guanylic acid.

An alternative type of model for sequence-specific protein–DNA interactions would be that each subunit of the recognition protein sees only one face of the helix instead of being wound around the DNA. Only bases exposed in one projection of the minor groove would be accessible for interaction with one protein, but more than one subunit could interact with differing faces of the same segment of helical nucleic acid (214).

Any model involving base discrimination by hydrogen bonding interactions between amino acids and the portions of bases exposed in the minor groove is subject to certain restrictions. For example, potential hydrogen-bond-forming atoms in the minor groove are located in very similar positions regardless of whether the base on a given strand is an A or a T. Some further discrimination could be obtained by forming two hydrogen bonds, such as is possible with arginine, asparagine, or glutamine side chains, but distinction between adenine and thymine would require additional interactions. These and other considerations have been lucidly presented by Rich and his colleagues (212, 215).

Some of the proposed models may prove to have substantial validity and insight. However, as with tRNA structure and recognition by aminoacyl-tRNA synthetase, a satisfactory understanding of DNA–repressor interaction will await successfull crystallographic analysis of the components and the protein–nucleic acid complex.

In the *lac* repressor, and perhaps the λ repressor, a relatively short amino acid sequence (less than 60 residues of the *lac* repressor) at the amino terminal end of the protein is sufficient for DNA binding (216, 217). Even a relatively small peptide encoded in the λ *cro* gene (218, 219) can function as a site-specific repressor. These results are consistent with the nucleic acid studies referred to below, indicating that repressors interact with relatively short segments of DNA.

A. OPERATOR SEQUENCES (FIG. 9)

One of the very striking features about the sequences of the *lac* operator (220) that has been found in the sequence of *trp* operator (221, 222), the probable *gal* operator (223), and partly in the λ operators (116) is that they have imperfect twofold axes of symmetry (Fig. 9). Thirteen out of fourteen bases on the one arm of the *lac* repressor are complementary to bases in the other half of the repressor sequence.

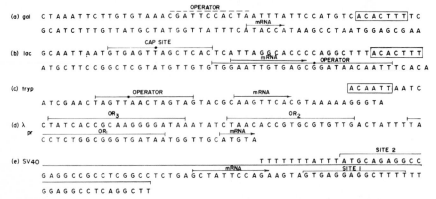

FIG. 9. Repressor binding sites in operator regions of certain prokaryotic operons compared with T antigen binding site of SV40 DNA. (a) Galactose operon of *E. coli*. (b) Lactose operon of *E. coli*. (c) Tryptophan operon of *E. coli*. (d) λ rightward promotor and operator binding sites. (e) SV40 T antigen binding sites. Sequences drawn in each case represent the plus strand of DNA. The word "operator" with dashed lines underneath it signifies the region that is protected by repressor and/or contains mutations that affect the operator site. Short vertical line followed by long arrow indicates the nucleotide with which mRNA transcription begins. Enclosed boxes show similarities of upstream sequences in promotors. "Cap site" and lines demarcate the sequences thought to be involved in binding of cyclic AMP binding protein. In λ OR3, OR2, and OR1 indicate the sites which bind λ repressor, respectively, at the highest intermediate and lower concentrations of repressor. Site 1 and site 2 in SV40 indicate the sites that bind, respectively, higher concentrations and lower concentrations of T antigen. The location of the mRNA start in SV40 is approximately estimated by reverse transcriptase copying of early mRNA and sequence analysis of the resulting DNA product.

Initially this was thought to reflect a symmetry in the interaction between the dimer molecule and the DNA. However, a number of lines of evidence indicate that this is not strictly correct. For example, the effect of *lac* operator mutations on strength of repressor binding does not correlate with their effects on symmetry, and in one case a mutation increasing symmetry decreased the binding affinity (18). Studies of protection of guanylic acid or adenylic acid from chemical methylations on the 7-amino group and the 2-amino group, respectively, show that the bases at which methylation is blocked or enhanced in the repressor operator complex are not symmetrically distributed around the dyad axis of the operator DNA [W. Gilbert, quoted in Müller-Hill *et al.* (217)]. Finally, short synthetic oligonucleotides (20 bp) that lack the furthest bases on each side of the symmetry bind the wild-type *lac* repressor with an affinity similar to that of the intact operator. Statistically, the extra symmetry observed beyond that known to interact with the operator is significant, but at the moment there are little or no published data indicating a role for these sequences in the interaction with wild-type repressor *in vivo* or *in vitro*. The λ operators do not show as prominent rotational symmetry as *lac* or *trp* operators, but translational symmetries also occur in the λ operon. Rightward transcription from the λ *Pr* promotor is regulated by repressor binding proximal to the site where transcription is initiated. There are three tandem repeats of a repressor binding sequence with affinities such that a small concentration of repressor will first bind to the operator nearest the transcription initiation site for the rightward transcription. At higher concentrations, two and then all three sites may be occupied by repressors. The extreme complexity shown in the region is indicated both by the fact that the operator and all promotor sequences for rightward transcription of *cro* message and leftward transcription of λ repressor message interpenetrate and that initiation of transcription of the repressor message may begin at a region adjacent to the operator binding site most distal from the rightward *cro* message transcription initiation site (Fig. 3). Repressor binding in this region can, therefore, both inhibit rightward transcription and either stimulate at low concentrations or inhibit at higher concentration the leftward transcription from the same region of DNA. In this case the common feature for operator recognition sequences appears to be a decanucleotide with some sequence variation.

Methylation experiments performed with the λ repressor and operator showed that there was no symmetrical pattern of base protection. Unlike the *lac* operator, methylation of adenylic acid in the λ sequence was not affected. Since the N-7 of guanine is in the major groove, while the N-3 of adenine is in the minor groove in this experiment, the only detectable contact of repressor with base, occurred in the major groove (224).

B. CYCLIC AMP BINDING PROTEIN

The expression of a number of bacterial operons including *lac* and *gal* is also regulated by intracellular levels of cyclic AMP. A binding protein for cyclic AMP is also a specific DNA binding protein, and in the presence of cAMP it can activate certain promotors (37). Interestingly, these promotors can also be activated *in vitro* by high concentrations of glycerol (225). In the *lac* control region, cAMP binding protein recognizes a sequence in which 12 of 14 nucleotides are related by a dyad axis about 61 base pairs upstream from the RNA initiation site (226). The cap recognition site for *gal* may also involve a partially symmetric sequence upstream from the initiation site. For example, 8 of 10 base chains centered about 70 bases upstream from the *gal* mRNA start are the same as bases at the corresponding site in *lac*. There is also a partly symmetric sequence in *gal* about 60 bases upstream from the RNA start, but mutations in this site are O⁻, so that this may be the *gal* operator. If so, then *gal* repressor may act to block cAMP binding protein interaction with DNA. Too few examples are available at the moment to identify the range of sequences that permit specific binding of this protein to DNA.

C. LAMBDA N PROTEIN

As discussed above, the N protein of λ is a positive gene effector that acts on both λ major rightward and leftward transcription by preventing transcription termination. Site-specific binding of N to DNA has not been observed yet. However, the two sites potentially recognized by N have been sequenced (227,228). They are very similar to each other and both are rotationally symmetric sequences. Mutations are known at one of the sites that block N action.

D. THE SV40 "A" PROTEIN (FIG. 9)

It may be that the more complex eukaryotes have sacrificed the ability for very rapid response or selective alteration of rates of protein synthesis in at least some gene systems in return for the greater range of genetic information available at their command. However, there is strongly suggestive evidence in one system that protein molecules may modulate transcription of specific nuclear genes in a manner similar to that in which bacterial repressor works. One of the major products of the SV40 early genes is the "A" protein. This protein binds selectively to SV40 DNA near the origin of DNA replication (229–232). Although it is difficult to purify useful amounts of the A protein, cells infected with certain adenovirus–SV40 hybrid viruses produce abundant amounts of a closely related protein. At low concentration, that protein protects a single segment of DNA within the template for the 5'-terminal sequence of SV40 early mRNA just upstream from the initiation codon for the A protein (233). Viruses that are temperature sensitive because of mutation in the A protein are available. *In vivo* cells infected with the mutant virus and exposed to nonpermissive temperatures produce a 50-fold increase in the amount of early mRNA synthesized compared with that of cells infected with wild-type virus (231). The above results are consistent with a model in which A protein regulates the production of its own mRNA by binding to DNA to inhibit transcription in an arrangement formally similar to that of *lac* repressor and mRNA initiation site. The ever-present spectre of regulated termination of the leader sequence cannot be rigorously ruled out until *in vitro* transcription systems responsive to A protein are developed.

IV. DNA SYNTHESIS

A. DEFINITION AND LOCATION OF REPLICATION ORIGINS

DNA replication initiates at a very limited number of specific sites on the genome of DNA viruses and bacteriophages, bacterial DNA, and mitochondrial DNA. This is also assumed to be the case for the DNA of eukaryotes, although there are more initiation sites in the genomes of higher cells than in prokaryotic systems. Initiation sites may be specific in their requirements for initiation proteins as well as in their requirements for nucleotide sequences. The "conventional" DNA polymerases are essentially unable to initiate synthesis of new strands of DNA. A key step in the initiation of DNA replication is the provision of an appropriate primer molecule that has a 3′-hydroxyl to which the initial nucleotides that are transcribed from the parental DNA strain may be added. Such primers for initiation may themselves consist either of DNA, RNA, or even of single nucleotides.

Several ways have been suggested for the generation of DNA primers. A linear DNA molecule may have an inverted repeat at its 3′-terminus such that the molecule can fold back on itself on its own 3′-end to act as primer. This may be the mode of initiation of parvovirus DNA replication (234–237). Alternatively, primers may be generated by appropriate proteins that nick one strand of a DNA molecule to generate a 3′-hydroxyl group (238,239). RNA primers can apparently be generated either by RNA polymerase acting at appropriate sites in the DNA (240) or by another more specialized type of nucleoside triphosphate polymerase (241–244), the prototype of which is the DNA G protein of *E. coli*. This enzyme may act both to form the primer for initiation of DNA synthesis, as on some small DNA phages, and also to form the primer RNA for the "Okazaki" fragments which are synthesized during the elongation reaction. To do this, multiple initiation sites for DNA G synthesis must occur on the "lagging" DNA strand. These sites could be created by other proteins (specifically the DNA B protein in *E. coli*) moving along the displaced parental DNA strand in the 5′ to 3′ direction (243).

Another mode of initiation of replication has been appreciated relatively recently. There is suggestive evidence that in the replication of small RNA viruses the initial primer may be provided by an RNA nucleotide covalently attached to a protein (245).

A protein (the "Bellet protein") is covalently linked to the 5'-end of each strand of adenovirus DNA (245). The nucleotide to which the protein is coupled is not represented by a complementary base on the other DNA strand (246). Replication begins at the extreme ends of the DNA molecule, and a currently favored model is that newly synthesized Bellet protein covalently attaches to a nucleotide and associates with the preexisting Bellet protein molecule on the 5'-end of the parental strand (Fig. 10). The nucleotide covalently attached to the new protein then acts as a primer to which the growing DNA strand is added. This ensures that the daughter strand of DNA also has a protein at its 5'-end. The other strand of parental DNA is displaced by the replication process. However, adenovirus DNA has a terminal inverted repeat such that the two ends of each single strand of DNA can base pair, generating a double-stranded segment of DNA at which replication of the displaced strand can initiate.

The prolongation of replicating forks in DNA probably involves a relatively limited number of standard mechanisms. For example, many of the simpler E. coli DNA bacteriophages probably use host enzymes for the elongation of growing DNA strands. Similarly, the small intranuclear viruses in animal cells also probably use host enzymes for elongation. During elongation, replication must reinitiate at multiple sites on at least one of the two parental DNA strands (the lagging strand), since daughter strands can only be elongated in the 5'- to 3'-direction. If there is any sequence specificity for such reinitiation on bacterial DNA, it may be for short sequences and has not yet been defined. Phage T7 codes for a protein, the gene 4 product which, like DNA G protein, generates the primer for the Okazaki fragment synthesis on the "lagging strand." However, the gene 4 protein may have sequence specificity because it preferentially forms primers with the initial sequence ACCA (247).

The sequences involved in initiation of DNA replication have been located with various degrees of resolution in several systems. Early studies demonstrated that certain genes of E. coli and B. subtilis replicated before others, providing an approximate location for the replication origin. These studies also showed, at least partially, bidirectional replication of the bacterial chromosome from a single origin (25). More precise localizations have been possible in small viral or bacteriophage genomes. Pulse labeling studies can locate the origin in cases when the labeled DNA precursor can be introduced into the cell for a period of time that is short relative to the time for an en-

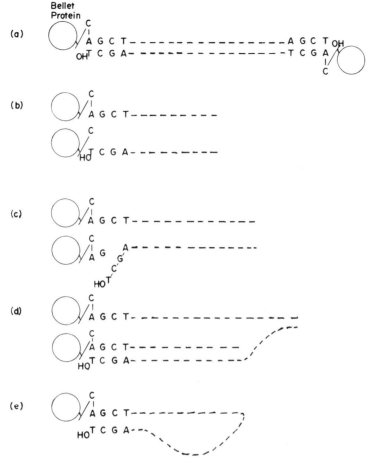

FIG. 10. Scheme for replication of adenovirus DNA. (a) The Bellet protein indicated as a large open circle is attached perhaps by a phosphate residue to a single deoxynucleotide here arbitrarily chosen as deoxycytidylic acid. This nucleotide is joined to the 5′-end nucleotide of each strand of adenovirus DNA, here indicated as a deoxyadenylic acid. The two ends of each strand of adenovirus DNA are complementary to one another. (b) In the postulated first step in adenovirus DNA replication, the second molecule of Bellet protein with an attached mononucleotide associates with the Bellet protein attached to the parental DNA strand. (c) Nucleotides are added to the mononucleotide attached to the incoming molecule Bellet protein. The added nucleotides are complementary to the first nucleotides of the DNA strand whose 3′-hydroxyl is adjacent to the incoming Bellet protein. (d) With the mononucleotide attached, the Bellet protein acts as primer. The replication of the lower DNA strand continues gradually, displacing the upper parental DNA strand. (e) After complete displacement of one of the parental DNA strands it then folds over to form a large single-stranded DNA loop with a panhandle at the ends formed by complementary 5′-terminal nucleotide sequences. The Bellet protein remains attached to the 5′-end of the single-stranded DNA. The structure at the left end of the DNA molecule in (e) is formally similar to that at the left end of the DNA molecule in (a) and presumably can initiate another round of DNA synthesis by the mechanism shown in (b) through (d).

tire replication cycle. If fully replicated molecules are isolated after a pulse label, the most radioactive portions of the DNA will be those synthesized in the last phase of replication. For example, in the first application of this method, Nathans and his colleagues located the origin of replication of SV40 virus to a unique site on the genome and demonstrated that replication proceeded bidirectionally (248).

Resolution to within 50 to 100 bp can be accomplished by electron microscopy. If replication begins internally in a double-stranded DNA molecule, replication "loops" can be seen, and it is possible to measure the size of these structures. Their lengths reflect the progression of growing forms away from a unique origin. The loops can be oriented relative to fixed positions on the genome by partial denaturation mapping, by cleavage of the replicating forms with restriction endonucleases, or by other means. In this way it is possible to extrapolate back to the position at which the smallest replication loops are located. As an example of this approach, one may cite the work of Salzman and his colleagues (249), who cleaved replicating SV40 circles at a unique site with *Eco*RI restriction endonuclease and measured the distance from the center of the loop to the ends of the linear molecules. All loops centered around a single position located at 0.67 fractional genome length from the end of the DNA. These studies were performed in parallel with the work of Nathans, and the results were in complete agreement on localization and pattern of replication of SV40 molecules. The supercoiled DNA of some bacterial plasmids may exist intracellularly as a nucleoprotein complex. When the isolated complex is treated with ionic detergents, one DNA strand is nicked at a unique site. This nick permits relaxation of the supercoiling, hence the term "relaxation complex" (250). It has commonly been assumed that the nick may be related to DNA replication, although the situation may be more complicated. For example, in the plasmid *Col*E1, the origin of vegetative replication is about 200 bases away from the nick (251). Possibly the relaxation site itself may serve as an alternate origin of replication such as might be used for chromosome mobilization during transfer of DNA from one bacterial cell to another. In bacteriophage ϕX174, a phage gene product, the A protein, nicks the DNA at a specific position that has been shown to lie between bases 4305 (C) and 4306 (A) on the ϕX174 sequence (238). This generates the free 3'-hydroxyl that acts as primer for initiation of replication. In other single-stranded DNA bacteriophages, complexes of host proteins are involved in the initiation step of DNA replication by synthesizing an RNA primer whose sequence may be analyzed. In cases such as fd DNA, a unique binding site on the DNA chromosome for an initiator protein and a unique sequence for the "primer" RNA both serve to pinpoint the origin of replication.

In another approach, the origin of DNA replication may be located by isolating a series of deletion mutants of a genome that has a unique origin for replication. These mutants may require helper function for replication by proteins provided by the helper viruses if, and only if, the defective molecules contain the cis functions essential for replication. In an extension of the earlier studies cited above, both Nathans' (252) group, Berg (253), and Shenk (254) have managed to locate the origin of replication of SV40 virus to a sequence of less than 100 nucleotides.

As a variation of this approach, it is possible to propagate mutants with defective replication origins either because the mutations occur in integrated viruses whose replication is driven by the host origins of replication (231) or because they are built in as second origins of hybrid genomes that are capable of initiating replication from a different site (254). In this way, small deletion mutations within the origin of replication of λ bacteriophage (255) and SV40 virus have been propagated, so that it was possible to locate them precisely and sequence them.

B. THE SEQUENCE OF ORIGINS OF REPLICATION (FIG. 11)

From the abundance of mechanisms for initiation of DNA replication briefly alluded to above, one might expect a great diversity of sequences at replication origins. In addition, replication may be closely coupled to the control of the genes near the origin of replication, and this could also contribute to the diversity of origin sequences. This expectation has been partially disappointed. One of the first molecules in which the precise position of the origin of replication was located is ϕX174 (238). As discussed above, replication may initiate by site-specific nicking of the DNA by the A protein. The nick lies within the A protein gene, itself, so that the initiation protein inactivates its own coding DNA. The sequence about the site of the nick is relatively featureless (167). The replication of phage fd is initiated by a different means, using host RNA polymerase and DNA binding protein. The origin of replication of fd (256) and the related phage f1 (257) includes three nearby regions of imperfect palindromes in the single-stranded DNA. Phage G4 is a single-stranded DNA phage that initiates replication by another mechanism. In this case, complementary strand synthesis requires DNA G and DNA binding proteins and is initiated with an RNA primer. The sequence of the primer has been deter-

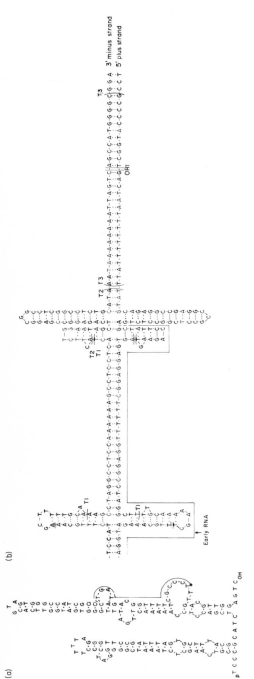

FIG. 11. DNA sequence at the origin of replication. (a) DNA sequence of single-stranded bacteriophage fd showing the suggested hairpin loop formations in the stems. The solid line over a portion of the right-hand stem indicates the spaces from which the RNA primer for initiation of DNA synthesis originated. The solid circle at one end of the line corresponds to the 5'-end of the RNA primer. (b) Nucleotide sequence of the origin of replication of SV40 DNA. Solid line under sequence indicates the approximate extent of early mRNA. Sequences T3, T2, and T1 enclosed by parentheses indicate the sequences protected, respectively, by high, intermediate, and low concentrations of adenovirus SV40 hybrid protein related to T antigen. Hybrid DNA is folded out in a cruciform structure to indicate the occurrence of potential symmetries in the DNA rather than to imply that this physical structure actually exists in the DNA molecule. Plus strand refers to the strand whose sequence corresponds to that of SV40 early mRNA.

mined and is complementary to a region of this DNA that can be drawn with stable stem and loop structures (258) (Fig. 11). Replication of the plasmid *Col*E1 has been demonstrated *in vitro* and also involves the synthesis of a primer RNA. Sequences about the initiation site have been determined and include some limited palindromes (251,259). The sequence about the relaxation nick is also known and includes regions of partial symmetry (260). The λ origin sequence has also been analyzed and contains a long (261) thymidylic acid-rich stretch whose 3'-end is adjacent to a sequence with extensive rotational symmetry.

In animal cells the sequences about the origin of replication of certain of the small DNA viruses, specifically SV40 (262), polyoma (263, 264), and BK (265) virus, have been determined. These viruses are known to be organized on very similar principles and are probably evolutionarily related to one another. The origins of replication of these viruses lie near the 5'-ends of early and late mRNA, and also exhibit several prominent structural features. SV40 was the first to be analyzed and has been the most intensively characterized. There are three prominent palindromes at or near the origin of DNA replication, one of which is very GC-rich, and extends over 27 bp with perfect symmetry so that single strands could form a very stable hairpin loop. This sequence is adjacent to the 5'-end of a long A-rich sequence, including eight successive deoxyadenylic acid residues. The template for the 5'-end of early SV40 RNA lies at one end of the long palindromic sequence on the opposite side from the A-rich sequence (Fig. 1). Late RNA is transcribed in the opposite direction from early RNAs, and the 5'-end of part of this RNA lies within the AT-rich region. Thus there is extensive overlap of sequences involved in mRNA production and in the initiation of DNA replication, but their functional interactions remain to be elucidated. The palindrome furthest from the AT-rich region can be deleted to produce a mutant virus that still replicates in the presence of helper virus. On the other hand, deletion within the longest palindrome completely inactivates the origin of replication. In the BK virus origin there is a similar pattern with nine deoxyadenylic acids adjacent to a palindromic sequence. This palindrome is even longer than the 27 base palindromic sequence of SV40, although it is not perfect. Similarly, in polyomavirus, despite a general lack of nucleic acid homology to SV40, there is an analogous palindrome adjacent to an AT-rich region.

The only other DNA origin that functions in the nucleus of animal cells and whose sequences are available at present is the sequence of the 5'-termini of adenovirus 2 (266) and 5 (267) DNA. As mentioned above, the replication of viruses probably begins by way of a protein with a covalently attached primer, and it is not clear that this origin of replication functions in a way similar to that of the papovaviruses. However, there are certain limited homologies in the sequences near the origin of replication of adenovirus and SV40, including some short blocks of GC-rich sequence. One end of the adenovirus DNA is a terminal inverted repetition of the sequence of the other end, and one could formally consider these as the two halves of a long palindrome that has been cleaved at the center of its loop. However, the recognition of legitimate homologies between origins of replication of the two classes of viruses will depend on functional analysis of the process of replication and determination of whether similar sequences are really functioning in similar ways or recognized by the same protein.

Thus, rotationally symmetric sequences occur in many of the replication origins that have been analyzed. These could reflect the symmetry in initiation of bidirectional replication on double-stranded DNA. For example, Kornberg and associates have shown that DNA B gene product of *E. coli* is a protein that can be bound to single-stranded phage ϕX174 DNA and in turn create a "promotor" at which DNA G protein synthesizes primer RNA. Kornberg suggests that similar mechanisms may be involved in double-stranded DNA synthesis (242). A sequence at the origin could permit a protein such as DNA B to bind to each strand of the DNA, and the first event in initiation of DNA replication would be identical on the two strands. Alternatively, identical sequences on both strands of DNA could provide attachment, for each strand to the same structures, such as cell membranes and centrioles, and permit chromosome segregation. In single-stranded DNA phages palindromic sequences have unique base-paired regions that may contribute to the selection of specific sites for synthesis of an RNA primer.

V. CODON SELECTION IN mRNA

The preceding sections have dealt largely with sequences that are found infrequently in nucleic acids, and which interact in a highly selective manner with specific proteins to modulate the extent of the replication or expression of genetic material. Other more pervasive properties of gene structure have also become apparent, although their role in controlling gene expression is unknown. The results of recent analyses of the primary structure of extensive sections of small DNA viruses and mRNA show that there are additional pressures on the coding regions of genes beyond that imposed by the amino acid sequence of the protein for which they code. Of the 20 amino acids for which specific coding triplets exist, two are coded for by unique base triplets (AUG for methionine and UGG for tryptophan). The remaining 18 amino acids can be coded for by any one of two, three, four, or in the cases of serine, leucine, and arginine, one of six triplets. Inspection of the sequences available for the RNA bacteriophage MS2 (9), for ϕX174 phage (167), for SV40 (189,197), and for mRNAs of rabbit (268) and human hemoglobin (269), sea urchin histones (270,271), growth hormone (194), insulin (195), and human chorionic somatomammotrophin (272) show that there are some strong preferences both for and against the utilization of some codons (Fig. 12). Certain of these preferences may be related to general requirements for primary structure of DNA. For example, the dinucleotide CG occurs very rarely in SV40 cytoplasm mRNA and is also known to occur quite infrequently in bulk animal cell DNA (273). In many eukaryotic mRNAs the dinucleotide CG seems to be biased against regardless of whether the C corresponds to the first, second, or third position of the codon, although this bias may be less than that which occurs in the bulk DNA or in the hnRNA (274). Curiously, CG doublets occur in multiple locations in human α-globin mRNA (275) and do not appear to be at all selected against in the presumably untranslated sequences of SV40 near the origin of replication.

In some cases, the nucleotide selection may be related to some evolutionary pressure expressed at the level of translation and common to a variety of codons. For example, in both SV40 and ϕX174 DNA, there is a considerable preference for uridylic acids in the third position of codons, so that almost 40% of all codons in ϕX174 and a similar percentage in SV40 end with uridylic acid. This bias is not seen in the cellular mRNA. A more selective and even more striking bias is seen in SV40 message, where there is a 20-fold preference for UU versus UC in positions 2 and 3 of codons, even though in each case codons with UU or UC would code for the same amino acid. Since the same bias does not occur as prominently for U in the first position and C in the second, or for U in the third or C in the first position, this would appear to be a bias exerted at the level of translation rather than at the level of primary structure of message or of DNA.

Marked specific selection may also exist for and against certain

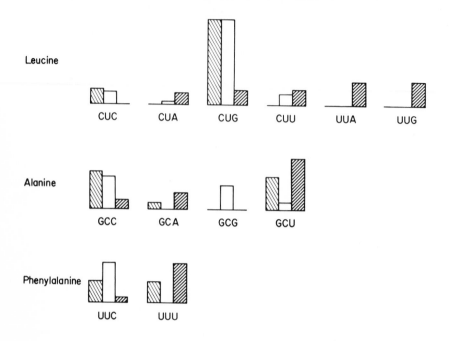

FIG. 12. Graph representing relative utilization of synonym codons for three amino acids in three mRNAs functioning in animal cells. The leftmost of the three bars (diagonally slashed) corresponds to the relative use of the different synonym codons in hemoglobin β-chain mRNA; the center bar (empty) corresponds to the relative utilization of the same synonym codons in α-globin mRNA; and the right vertical bar (diagonally slashed) corresponds to the relative utilization of various codons in SV40 mRNA. The first line shows the relative use of the six leucine codons in the three mRNAs and emphasizes the relative abundance of CUG leucine codons in both globin mRNAs and the absence of the UU purine codons. In contrast, the SV40 mRNA shows relatively abundant use of UU purine codons and no preferred use of CUG codons. The second line corresponds to the use of each of four synonym codons for alanine. Of note is the relative lack of GCC and abundance of GCU codons in SV40 mRNA as compared with α-globin mRNA and to a lesser extent with β-globin mRNA. Also of note is the use of GCG alanine codons in α-globin but not in β-globin or SV40 mRNA and the preference of the α-globin mRNA for GCC over GCU codons. The third line represents relative use of each of the two synonym codons for phenylalanine. This line contrasts the apparent indifference of the β-globin mRNA for C and U terminated codons, the preference of α-globin mRNA for C terminated codons, and the preference of the SV40 mRNA for codons with U in the third position.

codons for a particular amino acid. Arginine codons AGA and AGG (and to a somewhat lesser extent CGA and CGG) are uncommon in ϕX174 and MS2. The codon AUA for isoleucine is entirely missing in the coat protein gene for MS2, although it occurs in other genes of MS2. There is also a marked deficiency of the codon AUA in ϕX174, so that it occurs at $^1/_6$ the level of AUC and $^1/_{22}$ the level of AUU. The arginyl- and isoleucyl-tRNAs responsive to AGG and AUA, respectively, are present at low levels in bacterial cells (W. Fiers, personal communication; 276). A mutant of MS2 was found that had replaced methionine with isoleucine in the coat protein. Although this could have arisen by a single base transition (AUG–AUA), a transversion actually occurred so that the isoleucine codon was AUU rather than AUA, providing further circumstantial evidence that AUA codons hinder the expression of the coat protein gene (277). In the SV40 virus there are 22 occurrences of AUA, compared with 44 occurrences of AUU and no occurrences of the AUC codon for isoleucine. In contrast, there are a number of AUC codons in those cellular mRNAs (even though the globins lack isoleucine) but no occurrences of AUA. In the cellular mRNAs there may be a general bias against the UA in position 2 and 3 of codons, since GUA is not used and the UUA is only used once. In normal hemoglobin mRNAs there is no CAA for glutamine and there is a very marked bias in favor of CUG for leucine, as compared with the other five leucine codons. The CUG leucine codon is also preferred in the other cellular mRNAs, but CUG occurs at less than random frequency among the leucine codons of SV40 mRNA. The fibroin mRNA from the silk gland of the moth *Bombyx mori* shows further specialization in codon usage. Within analytic error, only GCU is used for alanine, UCA for serine, and principally GGU or GGA for glycine (278). These comparisons obtained with fibroin, hemoglobin and other host cell mRNAs, and SV40 mRNAs indicate that different messages functioning in eukaryotic cells may have a range of substantially different biases in preferential codon utilization for individual amino acids.

The relative abundance of tRNA species responsive to each of a number of codons has been determined for reticulocytes (279–282) and the silk gland (279,283–285). In both cases there appears to be a good correlation between abundance of codons in the major mRNA of the specialized tissue and the abundance of the tRNA responsive to that particular codon. For example, in reticulocytes there is very little CAA-responsive glutaminyl-tRNA (286). There is even a suggestion that the tRNA composition in a single type of specialized cells may change when the occurrence of a particular codon changes. For example, it has been reported that there is two- to threefold more isoleucyl-tRNA in sheep reticulocytes synthesizing forms of hemoglobin that contain isoleucine than in sheep reticulocytes synthesizing hemoglobins that lack isoleucine (287).

On first consideration, the simplest mechanism for causing the abundance of a tRNA to be proportional to the frequency of a codon in mRNA would be if tRNAs were somehow stabilized during their association with message in the ribosome, so that differential degradation would account for selective retention of heavily used tRNAs. This simple possibility is not true, at least for the silk gland. The rates of degradation of minor and major species of silk gland tRNAs were similar to one another, and the rates of synthesis of major tRNA species after refeeding were at least tenfold higher than the rates of synthesis of minor species (279). Altman (personal communication, 288) has examined tRNA precursors in the silk gland of *Bombyx mori* and finds that the relative abundance of the precursors for the principal species of tRNA are at least qualitatively in proportion to the relative amounts of the tRNA in this highly specialized gland. This implies that selection must occur during synthesis or stabilization of the precursor itself rather than of the mature tRNA. As far as is known, the precursor does not engage in any direct reaction with the mRNA. The alternative is some feedback or correlative mechanism for tRNA and messenger synthesis, such as autoregulation of the transcription of specific tRNAs by the level of free (not ribosome associated) tRNA. This clearly requires more intensive investigation.

Several explanations have been advanced for the nonrandom selection of codons in different mRNAs. In the case of the small RNA bacteriophages, where the differential use of synonym condons was first observed in detail, an initial suspicion that there was extensive secondary structure in the RNA was borne out by the complete sequence analysis of MS2, electron microscopy of RNA phages, and other studies. An early suggestion was that the nucleotides in third position of codons were chosen in such a way as to maximize the stability of base pairing within the sequences of the phage RNA. This may be partly so, but it is not at all clear that this explanation would account for the selective bias against codons in the coat protein gene as compared with other genes of MS2. Also, third positions of codons are not preferentially found in the base-paired regions of proposed secondary structures for MS2 RNA. Base pairing appears to be less extensive in the mRNAs transcribed from the DNA phages and viruses, and also appears to be a less adequate explanation for some of the biases described above.

Pieczinek initially suggested a very intriguing hypothesis that codon recognition by tRNA may involve interaction of more than three bases in mRNA with tRNA molecules. For example, the nature of the base 3'-terminals to the third base in a codon of mRNA might affect the ease with which a given tRNA is bound to that codon. An interesting speculation incorporating this idea is that in early evolution each tRNA interacted with five adjacent bases on mRNA, including the triplet codon and in alternative conformations the two bases preceding or following the codon (289). Attempts to relate bases in mRNA position plus or minus 1 and 2 with particular codons in SV40 show relatively few correlations, and some of these are a consequence of preference for certain amino acid sequences such as lysyllysine. To explain the differences in codon utilization in ϕX174, as compared to the RNA phages or in the different mRNAs in eukaryotic cells, on the basis of tRNA–mRNA interactions beyond triplet codon–anticodon pairing, one would have to assume that there are differences between the prokaryotic and eukaryotic tRNAs responsive to the same codon. Furthermore, to account for differences among animal mRNAs, one would have to presume that there is something very unlikely, such as differences in the exact structure of optimal ribosome message–tRNA interactions or different forms of tRNA responsive to the same codon in differentiated cells.

Another simple hypothesis has been that codon selection occurs in such a way as to obtain an optimally stable codon–anticodon interaction determined principally by the triplet interactions. This proposal is consistent with a considerable amount of the codon bias in some of the prokaryotic systems where, for example, there is a preference for CCU over CCC, since CCC would interact with three guanosines in tRNA and provide unusually strong base pairing compared with most triplet interaction. The preference for UUC over UUU is much less strong. On the other hand, the same rule does not seem to be operative in SV40 mRNA or in globin mRNA, and if this is a factor in codon selection it must be only one of the determinants. In general, a bias for U over C in the third position of codons could be related to a preference for G:U structure in the wobble position. There are no tRNAs so far described with unmodified adenylic acid in the first position of the anticodon so that there is no tRNA which would react only with the codons of the form NN'U and not NN'C. Consequently, preference for this hypothetical tRNA does not seem to be an attractive explanation for the U over C bias in the third position in codons of the viral message.

Another possibility is that codon selection might be biased so as to have an abundance of one residue at every third position in the mRNA in order to favor mRNA interaction with RNA binding proteins. In tobacco mosaic virus (TMV) each successive set of three bases forms a three-sided cage enclosing an α-helical segment of the structural protein, so that every third base is in the same position relative to a protein subunit (290). The selection of pairs for the G third residue of RNA has been proposed for the TMV RNA sequence used in the nucleation reaction (291) initiating encapsidation of the RNA. In this regard, in addition to interacting with cytoplasmic ribosomes, it appears that eukaryotic mRNA also enters intracellular ribonucleoprotein complexes (292).

Another recurrent hypothesis is that there are one or more codons whose rates of recognition by tRNAs are sufficiently slow that they retard or modulate the rate of translation of specific mRNAs. The simplest form of this hypothesis is that the rate of reaction is limited by the abundance of tRNA responsive to some codons. To investigate this possibility, the fraction of various aminoacyl-tRNAs actually bound to reticulocyte ribosomes has been measured. Leucyl-tRNA was unusual in that a relatively high fraction (65%) was bound to ribosomes (287).This number could have been even higher for CUG-responsive tRNA because isoaccepting species of leucyl-tRNA were not separately analyzed and CUG is the overwhelmingly abundant codon for leucine in both α- and β-chains. Kinetic studies did not reveal pause points during the synthesis of the globin chains, but globin has a high leucine content and those studies might not have been sensitive enough to detect multiple slow points in chain assembly. Nevertheless, the hypothesis is rather conjectural at the moment and without direct experimental support. If the modulation hypothesis were true, then one might expect that mRNAs whose protein is needed, such as the coat protein of the RNA bacteriophage MS2, would have a different bias in codon utilization than mRNAs for proteins that a cell might make only in limited amounts. This could be true for MS2, but does not account for the difference in codon utilization between the mRNAs for the major structural proteins of SV40, on the one hand, and the hemoglobin β- and α-chains, on the other, since all of these are proteins that are made in large amounts.

Further tests of this hypothesis can come from analysis of mutant hemoglobins. In most cases, the amino acid substitutions in hemoglobin can be accounted for by mutations that produce single base changes, although until the mutant messages themselves are analyzed, it must be proved that no compensatory second base change has occurred. Certain of these mutations would be expected to produce hemoglobin messages with codons that are not used in the normal message. There is one group of mutations where compensatory base changes are very unlikely. These are single base changes that alter the termination codon for the α- or β-globin of human hemoglobin. Several such mutations are known (293). In each case, a longer globin peptide chain is synthesized with amino acid sequence which can be predicted from the known nucleotide sequence of the 3'-terminal untranslated regions of the globin message. Translation of this 3'-terminal region of the β-chain uses several relatively uncommon codons (294,295), including a CAA for glutamine. It seems most unlikely that all of these condons would have been mutationally changed to more favorable forms to correspond with the more abundant codons present in the region coding for the normal hemoglobin β-chains. These hemoglobins may be synthesized at rates considerably more than half that of the normal β-chain, despite the occurrence of the unfavored codons (295). These results suggest that if modulation of rates of translation by codon utilization does occur at all within the differentiated cell, such modulation must either be limited to a few of the codons or must represent a quantitatively small effect. It would be interesting to examine reticulocytes that are synthesizing elongated hemoglobin to see if the CAA-responsive tRNA were present more abundantly there than in normal reticulocytes.

There are some hints that different tRNA responses to the same codons may play different physiological roles. For example, in the posterior silk gland of B. mori, there are two abundant tRNA[Ala] responsive to GCU, GCC, and GCA, differing by a C to ψ change in the anticodon arm. During fibroin synthesis, the species with ψ is more abundant, whereas the C containing component is the main component of carcass tissue (283). Also, there is a relative block to the translation of globin mRNA by extracts from interferon-treated cells. It has been reported that this block can be corrected by a minor component of CUG-responsive tRNA[Leu] but not by the major component (296). Structural data are not available to judge whether the two tRNAs might be encoded by different genes.

VI. A BRIEF OVERVIEW OF OTHER SEQUENCES THAT PLAY ROLES IN GENE EXPRESSION

A. RIBOSOME BINDING SITES

The sequences involved in the initiation of translation have been the subject of intensive study by methods of sequence analysis, biochemical analysis of the translation process in vitro, and genetic analysis. This subject has been authoritatively reviewed by Steitz (16) and will be discussed only briefly here. The ribosome binding site on mRNA was first investigated by forming initiation complexes with radioactive phage mRNA and unlabeled bacterial ribosomes. After digestion with ribonucleases, fragments of about 30–40 nucleotides including the translation initiation codons could be isolated (297–299). The significance of the sequences of the protected fragments was partly clarified by the suggestion of Shine and Dalgarno (300) that a purine-rich sequence in the mRNA preceding the translation start could base pair with the 3'-end of the 16 S RNA of the small ribosomal subunit.

Several lines of evidence support the Shine–Dalgarno hypothesis, and the predicted base-paired complex between ribosomal and mRNA has been evaluated. However, there is no quantitative correlation between the extent of base pairing possible between the sequence complementary to 16 S RNA, on the one hand, and the efficiency of initiation of translation, on the other. Additional factors that may play a role include the ability of the translation initiation site to interact with ribosomal proteins, such as the protein S1 (301), which is itself an RNA melting protein, and the secondary structure of the mRNA that may tie up initiation codons or translation initiation sequences (302). Recent studies suggest that the nucleotide beyond the AUG initiation condon may also be important in determining the efficiency with which initiation complexes are formed. The coat protein initiation site of the RNA phage has the sequence AUG, mutants with AUGA were several times as efficient as the wild type in forming initiation complex, and a mutant with AUAA was one-third as efficient as wild type, even though the fragment lacked initiation codon (303).

In spite of the limitations that make it difficult to predict quantitatively the relative efficiency of different translation initiation sites, their features are well enough understood so that they can be of considerable use in predicting which potential initiation codons within a nucleotide sequence might be used to initiate protein synthesis in prokaryotic systems. The requirement for the initiation of translation in eukaryotic systems is different. Capped structures at the 5'-ends of mRNAs play an important role. For example, certain prokaryotic messages may be translated with great efficiency in animal cells if they are enzymatically capped *in vitro* (304). Complementarity between the extreme 3'-end of 18 S ribosomal RNA and the ribosome binding site has so far been a less prominent feature of animal cell mRNAs than of bacterial mRNAs, and there is a possibility that complementarity between subterminal 18 S RNA sequences and the ribosome binding site may also be important (305).

Most prokaryotic and all eukaryotic mRNAs examined to date have had various lengths of 5'-untranslated sequences. The single exception is the mRNA which produces repressor for maintenance of λ lysogeny (116). Here the 5'-triphosphate is attached to the initiation codon for translation of one form of repressor, where presumably this constitutes a very inefficient ribosome binding and therefore limits the rate of repressor production so that a small amount of repressor destruction could activate lysogenic λ. In prokaryotes some of the long 5'-RNA leader sequences may include regulatory regions, such as the attenuator previously discussed and the leader sequences for the λ late mRNA. In eukaryotic cellular and viral mRNAs 5'-untranslated sequences as short as ten or eleven nucleotides have been found (306). These are presumably sufficient to provide any necessary aditional ribosomal interaction beyond that of a capped structure and the initiation codon itself. Longer 5'-terminal sequences of the order of 40 to 50 bases are found in some cellular mRNAs such as globin mRNAs (307–311) and viral RNA (312–314). The longest reported leader sequence of efficiently translated mRNA is that of the VP1 structural protein of SV40 that contains 238 nucleotides preceding the first initiation codon for the structural protein (315). The internal sequences in the SV40 leader could themselves code for a short peptide and also could permit extensive internal base pairing within the leader. The functions of these leaders, other than potentially providing signals for capping or initiation of translation, is entirely unknown, although it is tempting to speculate that in some cases they may influence the stability of the message.

B. UNTRANSLATED 3'-REGIONS OF mRNAs

The 3'-terminal untranslated sequences in animal messages are often longer than those of the 5'-end. The occurrence of a subterminal hexanucleotide AUAAA has been discussed previously, in relation to transcription termination. However, there are specific proteins associated with poly(A) in the subterminal sequences of animal cell messages, and possibly this common sequence may furnish a binding site for such proteins. In addition, Brawerman and his colleagues have suggested that subterminal sequences in the individual mRNAs may affect the stability of the poly(A) and presumably therefore the stability and level of expression of individual messages. The subterminal regions, which may extend to more than 200 nucleotides in length, not only vary in sequence from message to message but also show considerable variation between mRNAs, for the analogous protein from different species, as between the β-globin mRNAs of rabbit and man. It is known that they are not necessary for translation, since Efstratiadis and his colleagues (personal communication) have shown that a synthetic β-globin mRNA lacking the 3′-terminal sequences can be translated efficiently.

C. REPEATED SEQUENCES

Moderately and highly reiterated DNA sequences have been known for a long time to exist in eukaryotic DNA. Some of these so-called satellite sequences consist of many exact or approximate copies of quite simple sequences, arranged in a pattern of tandem reiteration (316,317). Some highly reiterated sequences show longer range repeats (318–323) which often may be recognized by a subunit structure based on about 180 bp (324). The primary structure of many of these sequences has been investigated, and they are quite variable from species to species. The DNA bearing these sequences is sometimes concentrated at particular regions within the chromosomes as in the heterochromatin (325), but their detailed function is as yet entirely unknown (326). Curiously, tandem duplicated sequences have also been found in small papoviruses (327), and these sequences again are more variable between the related viruses or even within strains of a single virus than are other portions of the viral genome. Until very recently no tandem repeated sequences were known in wild-type bacteriophages except for multiple copies of DNA encoding or ribosomal and tRNA genes. However, recently Landy and his colleagues have investigated the DNA sequence about E. coli tyrosine-tRNA gene and found that it contained a threefold tandem repeat of 170-odd bases, the first copy of which overlaps the 3′-end of the tyrosine-tRNA gene itself. Presumably more examples of this type of sequence will become available in the future, and the powerful genetic methods of analysis of bacterial genes may aid in evaluating their functions.

D. NUCLEIC ACID REARRANGEMENTS

In prokaryotes it is well established that there are genetic elements which have the propensity for recombination with one or even with a great many different sites in chromsomal DNA. Recombinational events may result in mutations or in promotor or gene transfer. A very few sites have been described in which local DNA rearrangement may occur in the normal regulation of specific genes (328–331). In a number of cases, such as the amplification of ribosomal genes in

Xenopus (332,333) or dihydrofolate reductase gene (334) in a mouse cell line and for the production of specific immunoglobulins (335, 336), DNA recombination may modify gene expression in animal cells.

Within the last 2 to 3 years there has been a remarkable development in eukaryotic molecular genetics with the realization that in viral and in host cell genes mRNA may not represent a faithful transcript of contiguous stretches of DNA (337–346) and may be assembled from transcripts derived from noncontiguous DNA regions. Functionally, this can result in a number of events. mRNAs can be monocistronic but, at least in viral messages, there may be potential for translating more than one protein from the same mRNA. In this case only the protein coded for by the region nearest the 5'-end of the message will be made. One way in which the downstream sequences can be made available for translation is by construction of an mRNA that retains at least one of the 5'-terminal untranslated sequences but deleted the initial coding sequences for the primary protein to bring the downstream initiation codon nearer the 5'-end of message and allow it to form initiation complexes. In both the viral and the host cell systems, the actual coding sequence for a single polypeptide chain may be reconstituted at the RNA level by bringing together in phase two or possibly more sets of codons transcribed from noncontiguous DNA sequences. This mechanism can enable a single region of DNA to code for a peptide domain that becomes a part of two or more polypeptides. The deletions from mRNA may also occur in such a fashion as to remove 3'-terminal untranslated sequences. Splicing events occur in the formation of at least some yeast tRNAs with the posttranscriptional removal short of internal sequences from the precursor molecule (347). A similar process may occur in some ribosomal RNA and viral and globin mRNAs. The sequences about the sites at which such RNA splicing occurs are known only for yeast tRNAs and SV40 mRNAs and in part for a mouse β-globin mRNA. There are no self-evident strong homologies in primary structure, even between the various tRNAs, or the different splicing sites in SV40 mRNA, nor is there an obvious extensive and perfect base pairing possibility about either donor or acceptor ends of such spliced RNA, but rather complex patterns of base pairing have been suggested between segments of donor and acceptor RNA regions (Fig. 13). This raises a possibility that there may be considerable subtlety and complexity in the mechanisms that achieve these splices, and at the moment this constitutes one of the most exciting areas of molecular genetics.

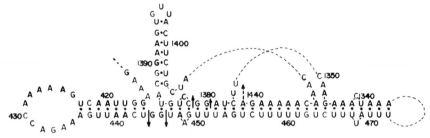

FIG. 13. Hypothetical base pair scheme for sequences about the leader and about the main body of the SV40 VP1 (16 S) mRNA. This mRNA is formed by joining the leader sequence of 201–203 nucleotides to the 5′-end of a continuous transcript derived from DNA 960 bp away from that encoding the leader sequence. The continuous transcript includes 36 nucleotides preceding the initiation codon for the major structural protein and all the codons of the major structural protein and the 3′-untranslated portion of the mRNA. As in all splicing reactions observed to date there is ambiguity in the precise position of the splice because the 3′-end of the leader (nucleotids 443, 444, 445) overlaps by one or two purines with the 5′-end of the acceptor segment of RNA (nucleotides 1379–1381) The final 16 S mRNA is formed by joining the sequences demarcated by the two downward pointing vertical arrows to the 5′-end of the sequences beginning between the two upward pointing arrows near residue 1380. Dashed lines represent mRNA loops of various sizes connecting different segments to the base-paired region to form this base pairing. It is important that looping and pairing segments are formed in the correct sequence to avoid topological entanglement. It should be noted that this base-pairing scheme is quite hypothetical, although most of it is consistent with the known ability of various deletion mutants of SV40 to make the splice necessary to form 16 S mRNA.

VII. RAPID NUCLEIC ACID SEQUENCING METHODS

At present the knowledge of nucleic acid structure has undergone a series of revolutions, so that it is difficult to recall now that less than 15 years ago there was no knowledge of the primary structue of nucleic acids and that investigation of even a relatively small molecule, such as a tRNA, was a major scientific achievement. The last 3 years have seen particularly remarkable developments in nucleic acid sequencing methodology. In the authors' opinion the heart of the major part of these developments has been the discovery and application of highly specific restriction endonucleases. These enzymes permitted the preparation of relatively large amounts of discrete well-defined sections of nucleic acid that could in principle be analyzed by a number of routes. A second powerful tool has been the use of radioactive labeling for sequencing studies, rather than the dependence on optical and strictly chemical methods for analysis. The recent major development has been the appreciation of the usefulness of electrophoresis in acrylamide gels for separating nucleic acid fragments of the same sequence but which differ in length by a single base residue and extend in chain length to over 200 bases. Several very powerful methods have been developed for sequencing DNA by the use of such gels. In one method, that of Sanger and Coulson (4), a DNA primer which may be either a restriction fragment or a synthetic oligonucleotide is annealed to a complementary strand of DNA whose sequence is to be determined. The primer is then extended, first with a few radioactive nucleotides, so that one can then follow the subsequent fate of DNA molecules attached to primer. The second step of the extension is accomplished by either a proteolytic fragment of *E. coli* polymerase that lacks 5'-exonucleolytic activity or by T4 phage DNA polymerase in separate reactions, each of which produces a series of fragments which terminate at a specific base. When the *E. coli* polymerase is used, each of the four reaction mixes includes only three of the four deoxynucleotides and the enzyme will stop when the next residue to be added is missing from the incubation mix. When the T4 polymerase is used, each of the four reaction mixes includes only one of the four deoxynucleotides. The polymerase will degrade the extended chain by virtue of its 3'-exonucleolytic activity, but will stop when it reaches the base residue that is included in the reaction mix and catalyze an exchange reaction between that base and the free triphosphate. By randomizing the amount which the primer was elongated in the first step of the reaction, one obtains a random series of elongated radioactive DNA molecules which consist of the primer extended for various distances along the 3'-end and terminated at sites that correspond to specific residues in the sequence. For example, if the second step extension has been performed with a mixture of A, G, and C, then the product will consist of a series of fragments extending from the 5'-end of the primer up to the last residue before each deoxythymidylic acid in the template strand. Fractionation of these four sets of fragments in a gel enables one to read large parts of the DNA sequence in a rather straightforward way.

A second noteworthy approach has been the application of chemical cleavage methods for DNA fragments labeled at either the 5'- or 3'-terminus. This was developed by Maxam and Gilbert (3), so that for practical purposes, one could obtain specific cleavages of DNA strands either at guanylic acid alone or at adenylic acids and, to a lesser extent, at guanylic acids, at both cytidylic and thymidylic acid, or preferentially at cytidylic acids. Cleavage at the purines was achieved by a methylation followed by a depurination, either under alkaline conditions for deoxyguanylic acid. Cleavage at pyrimidines was accomplished by hydrazinolysis followed by piperidinolysis of the modified DNA. Remarkable specificity could be achieved for cleavage of the chain at deoxycytidylic acid residues if the initial reactions were conducted at high salt concentrations. This method is convenient and offers an advantage over the Sanger and Coulson method in that it displays runs of homopolymers as well-separated bands which the Sanger method does not, and also is readily applied to double-stranded restriction endonuclease fragments of DNA.

Most recently, yet more convenient methods have been developed by Sanger (5) and his colleagues, in which the extension reaction may be conducted in a single step, elongating a DNA fragment on a template, using a mixture of ordinary deoxynucleoside triphosphates and a chain terminating analogue such as a dideoxynucleoside triphosphate or the arabinosyl nucleoside triphosphates. Whenever the dideoxynucleotide is incorporated into the growing DNA chain, the chain terminates. Therefore, when these reactions are performed separately with one of each of the four dideoxynucleoside triphosphates and run side by side in a gel, one may see a display of radioactive bands corresponding to each position of a given dideoxynucleotide in the DNA. Very thin gels are run at high voltage and with narrow slots. In this way, sequences of two hundred or more nucleotides can be displayed in a few hours.

In principle, of course, the same gel separation could be applied to RNA in situations where it is possible to obtain discrete RNA molecules with label on the 5′-terminus. RNA chemical cleavages have proved more difficult partly because alkaline hydrolysis conditions cannot be used because of owing to the alkaline lability of the 3′-phosphodiester bond. Three groups have recently described approaches for RNA sequencing by use of different nucleases and fractionation of partial degradation products (348–350). Limited digestion with T_1 ribonucleases under conditions favoring destabilization of RNA secondary structure generates a series of fragments corresponding to each position guanylic acid in RNA. Digestion with U_2 ribonuclease under appropriate conditions is reported to provide quite specific cleavages at adenylic acids. The distinctions between cytidylic acids and uridylic acids have proved somewhat more difficult, although there are enzymes that show partial promise in this regard, and the application of chemical modification procedures has not yet been described. The application of these procedures will be successful to the extent that it is possible to prepare discrete RNA fragments with label at their 5′- or 3′-termini in the absence of RNA-specific restriction endonucleases. At this time, the method is particularly suitable for analysis of small

RNAs such as tRNAs and 5 S RNA, which can be sequenced in a single gel.

Either the original "plus-and-minus" method of Sanger and Coulson or the dideoxynucleotide modification can be applied to RNA if a specific DNA (either synthetic oligonucleotides or restriction endonuclease fragments) primer is available and the extension is performed by reverse transcriptase rather than DNA polymerase.

The successful use of the specific DNA cleavage methods has been principally limited to methylation and hydrazinolysis reaction, but in principle any method which cleaved at specific bases or combination of bases could be used to provide sequence information, and laboratories are searching for cleavage methods which are as rapid and simple to perform as the dideoxynucleotide extension technique.

REFERENCES

1. Sinsheimer, R. L., *Annu. Rev. Biochem.* **46**, 415–438 (1977).
2. Zabeau, M., and Roberts, R.J., This volume, Chapter I.
3. Maxam, A. M., and Gilbert, W., *Proc. Natl. Acad. Sci. U.S.A.* **74**, 560–564 (1977).
4. Sanger, F., and Coulson, A. R., *J. Mol. Biol.* **94**, 441–448 (1975).
5. Sanger, F., Nicklen, S., and Coulson, A. R., *Proc. Natl. Acad. Sci. U.S.A.* **74**, 5463–5467 (1977).
6. Barnes, W. M., *J. Mol. Biol.* **119**, 83–99 (1978).
7. Woese, C. R., "The Genetic Code." Harper, New York, 1967.
8. Steege, D. A., *Proc. Natl. Acad. Sci. U.S.A.* **74**, 4163–4167 (1977).
9. Fiers, W., Contreras, R., Duerinck, F., Haegeman, G., Iserentant, D., Merregaert, J., MinJou, W., Molemans, F., Raeymakers, A., Van den Berghe, A., Volckaert, G., and Ysebaert, M., *Nature (London)* **260**, 500–507 (1976).
10. Ghosh, H. P., Söll, D., and Khorana, H. G., *J. Mol. Biol.* **25**, 275–298 (1967).
11. Fluck, M. M., Salser, W., and Epstein, R. H., *Mol. Gen. Genet.* **151**, 137–149 (1977).
12. Salser, W., *Mol. Gen. Genet.* **1**, 125–130 (1969).
13. Salser, W., Fluck, W. M., and Epstein, R. H., *Cold Spring Harbor Symp. Quant. Biol.* **34**, 513–520 (1969).
14. Feinstein, S. I., and Altman, S., *Genetics* **88**, 201–219 (1978).
15. Champe, S. P., and Benzer, S., *Proc. Natl. Acad. Sci. U.S.A.* **48**, 532–546 (1962).
16. Steitz, J. A., in "Biological Regulation and Control" (R. Goldberger, ed.), pp. 349–399. Plenum, New York, 1978.
17. Adhya, S., and Gottesman, M., *Annu. Rev. Biochem.* **47**, 967–996 (1978).
18. Gilbert, W., in "RNA Polymerase" (R. Losick and M. Chamberlin, eds.), pp. 193–205. Cold Spring Harbor Press, Cold Spring Harbor, New York, 1976.
19. Roberts, J., in "RNA Polymerase" (R. Losick and M. Chamberlin, eds.), pp. 247–272. Cold Spring Harbor Press, Cold Spring Harbor, New York, 1976.
20. Chamberlin, M. J., *Annu. Rev. Biochem.* **43**, 721–775 (1974).
21. Chamberlin, M. J., in "RNA Polymerase" (R. Lodish and M. Chamberlin, eds.), pp. 17–68. Cold Spring Harbor Press, Cold Spring Harbor, New York, 1976.

22. Doi, R. H., *Bacteriol. Rev.* **41**, 568–594 (1977).
23. Wu, R., Jay, E., and Roychoudary, R., *Methods Cancer Res.* **12**, 87–176 (1976).
24. Smith, J. D., *Prog. Nucleic Acid Res. Mol. Biol.* **16**, 25–73 (1976).
25. Kornberg, A., *in* "DNA Synthesis" (W. H. Freeman), San Francisco, California, 1974.
26. Chargaff, E., *Prog. Nucleic Acid Res. Mol. Biol.* **16**, 1–24 (1976).
27. Roeder, R. G., *in* "RNA Polymerase" (R. Losick and M. Chamberlin, eds.), pp. 285–330. Cold Spring Harbor Press, Cold Spring Harbor, New York, 1976.
28. Biswas, B. B., Ganguly, A., and Das, A., *Prog. Nucleic Acid Res. Mol. Biol.* **15**, 145–184 (1975).
29. Bourgeois, S., and Pfahl, M., *Adv. Protein Chem.* **30**, 1–99 (1976).
30. Chambon, P., *Annu. Rev. Biochem.* **44**, 613–638 (1975).
31. Lodish, H. F., *Annu. Rev. Biochem.* **45**, 39–72 (1976).
32. Perry, R. P., *Annu. Rev. Biochem.* **45**, 605–629 (1976).
33. Darnell, J. E., *Prog. Nucleic Acid Res. Mol. Biol.* **19**, 493–511 (1976).
34. Shatkin, A. J., *Cell* **9**, 645–653 (1976).
35. Brawerman, G., *Prog. Nucleic Acid Res. Mol. Biol.* **17**, 117–148 (1976).
36. Pastan, I., and Adhya, S., *Bacteriol. Rev.* **40**, 527–551 (1976).
37. Englesberg, E., Squires, C., and Meronk, F., Jr., *Proc. Natl. Acad. Sci. U.S.A.* **62**, 1100–1107 (1969).
38. Geiduschek, E. P., Armelin, M. C. S., Petrusek, R., Beard, C., Duffy, J. J., and Johnson, G. G., *J. Mol. Biol.* **117**, 825–842 (1977).
39. Losick, R., and Pero, J., *in* "RNA Polymerase" (R. Losick and M. Chamberlin, eds.), pp. 227–246. Cold Spring Harbor Press, Cold Spring Harbor, New York, 1976.
40. Rabussay, D., and Geiduschek, E. P., *Proc. Natl. Acad. Sci. U.S.A.* **74**, 5305–5309 (1977).
41. Whiteley, H. R., Spiegelman, G. B., Lawrie, J. M., and Hiatt, W. R., *in* "RNA Polymerase" (R. Losick and M. Chamberlin, eds.), pp. 587–600. Cold Spring Harbor Press, Cold Spring Harbor, New York, 1976.
42. Adhya, S., Gottesman, M., and de Crombrugghe, B., *Proc. Natl. Acad. Sci. U.S.A.* **71**, 2534–2538 (1974).
43. Epp, C., and Pearson, M. L., *in* "RNA Polymerase" (R. Lodish and M. Chamberlin, eds.), pp. 667–691. Cold Spring Harbor Press, Cold Spring Harbor, New York, 1976.
44. Roberts, J. W., *Nature (London)* **224**, 1168–1174 (1969).
45. Lee, F., and Yanofsky, C., *Proc. Natl. Acad. Sci. U.S.A.* **74**, 4365–4369 (1977).
46. Beckwith, J. R., and Zipser, D., eds., "Lactose Operon." Cold Spring Harbor Press, Cold Spring Harbor, New York, 1972.
47. Jacob, F., Ullman, A., and Monod, J., *C. R. Hebd. Seances Acad. Sci.* **258**, 3125–3128 (1964).
48. Beckwith, J., *Science* **156**, 597–604 (1967).
49. Epstein, W., and Beckwith, R. J., *Annu. Rev. Biochem.* **37**, 411–436 (1968).
50. Burgess, R. R., *in* "RNA Polymerase" (R. Losick and M. Chamberlin, eds.), pp. 1169–1201. Cold Spring Harbor Press, Cold Spring Harbor, New York, 1976.
51. Hsu, M.-T., and Berg, P., *Biochemistry* **17**, 131–138 (1977).
52. Paddock, G. V., Heindell, H. C., and Salser, W., *Proc. Natl. Acad. Sci. U.S.A.* **71**, 5017–5021 (1974).
53. Melgar, E., and Goldthwait, D. A., *J. Biol. Chem.* **243**, 4409–4416 (1968).
54. Larsen, C. J., Lebowitz, P., Weissman, S. M., and DuBuy, B., *Cold Spring Harbor Symp. Quant. Biol.* **35**, 35–46 (1970).

55. Ohe, K., and Weissman, S. M., *J. Biol. Chem.* **246**, 6991–7009 (1971).
56. Söderlund, H., Petterson, U., Vennström, B., Philipson, L., and Matthews, M. B., *Cell* **7**, 585–593 (1976).
57. Price, R., and Penman, S., *J. Mol. Biol.* **70**, 435–450 (1972).
58. Weinman, R., Raskas, H. J., and Roeder, R. G., *Proc. Natl. Acad. Sci. U.S.A.* **71**, 3426–3430 (1974).
59. Jaehning, J. A., and Roeder, R. G., *J. Biol. Chem.* **252**, 8753–8761 (1977).
60. Parker, C. S., and Roeder, R. G., *Proc. Natl. Acad. Sci. U.S.A.* **74**, 44–48 (1977).
61. Sklar, V. E. F., and Roeder, R. B., *Cell* **10**, 405–414 (1977).
62. Harris, B., and Roeder, R., submitted for publication.
63. Van Henlen, H., and Retel, J., *Eur. J. Biochem.* **79**, 579–588 (1977).
64. Yamamoto, M., and Siefart, K. H., *Biochemistry* **17**, 457–461 (1978).
65. Reich, P., Rose, J., Forget, B. G., and Weissman, S. M., *J. Mol. Biol.* **17**, 428–439 (1966).
66. Ohe, K., Weissman, S. M., and Cooke, R., *J. Biol. Chem.* **244**, 5320–5332 (1969).
67. Berg, D., and Chamberlin, M., *Biochemistry* **9**, 5055–5064 (1970).
68. Mutler, K., *Mol. Gen. Genet.* **111**, 273 (1971).
69. Hinkle, D. C., and Chamberlin, M. J., *J. Mol. Biol.* **70**, 187–195 (1972).
70. Hinkle, D. C., and Chamberlin, M. J., *J. Mol. Biol.* **70**, 157–185 (1972).
71. Krakow, J. S., and Fronk, E., *J. Biol. Chem.* **244**, 5988–5993 (1969).
72. Krakow, J. S., and van der Helm, K., *Cold Spring Harbor Symp. Quant. Biol.* **35**, 73–83 (1970).
73. Gerand, S. T., Johnson, S. C., and Brezia, J. B., *Biochemistry* **11**, 989–997 (1972).
74. Wu, C. W., Yarbrough, L. R., Hatlet, Z., and Wu, F. Y. U., *Proc. Natl. Acad. Sci. U.S.A.* **72**, 3019–3023 (1975).
75. Kleppe, R., and Khorana, H. G., *J. Biol. Chem.* **275**, 6149–6156 (1972).
76. Williams, R. C., and Chamberlin, M. J., *Proc. Natl. Acad. Sci. U.S.A.* **74**, 3740–3744 (1977).
77. Dausse, J. T., Sentenac, A., and Fromageot, P., *Eur. J. Biochem.* **31**, 394–404 (1972).
78. Stahl, S. J., and Chamberlin, M. J., *J. Mol. Biol.* **112**, 577–601 (1977).
79. Warner, C. K., and Schaller, H., *FEBS Lett.* **74**, 215–219 (1977).
80. Bautz, E. K., in "RNA Polymerase" (R. Losick and M. Chamberlin, eds.), pp. 273–284. Cold Spring Harbor Press, Cold Spring Harbor, New York, 1976.
81. Duffy, J. J., and Geiduschek, E. P., *Nature (London)* **270**, 28–32 (1977).
82. Tjian, R., and Pero, J., *Nature (London)* **262**, 753–757 (1976).
83. Pene, J., and Barrow-Carraway, J., *J. Bacteriol.* **111**, 15–23 (1972).
84. Wang, J. C., Jacobsen, J. H., and Saucier, J. M., *Nucleic Acids Res.* **4**, 1225–1241 (1977).
85. Hirsh, J., and Schleif, R., *Cell* **11**, 545–550 (1977).
86. Howard, B. H., de Crombrugghe, B., and Rosenberg, M., *Nucleic Acids Res.* **4**, 827–842 (1977).
87. Darlix, J.-L., and Horaist, M., *Nature (London)* **256**, 288–292 (1975).
88. Darlix, J.-L., and Fromageot, P., *Biochimie* **54**, 47–54 (1972).
89. Darlix, J.-L., *Biochimie* **56**, 693–701 (1974).
90. Das, A., Court, D., and Adhya, S., *Proc. Natl. Acad. Sci. U.S.A.* **73**, 1959–1963 (1976).
91. Inoko, H., Shigesada, K., and Imai, M., *Proc. Natl. Acad. Sci. U.S.A.* **74**, 1162–1166 (1977).
92. Lowery-Goldhammer, C., and Richardson, J. P., *Proc. Natl. Acad. Sci. U.S.A.* **71**, 2003–2007 (1974).

93. Lowery, C., and Richardson, J. P., *J. Biol. Chem.* **252**, 1381–1385 (1977).
94. Galluppi, G., Lowery, C., and Richardson, J. P., *in* "RNA Polymerase" (R. Losick and M. Chamberlin, eds.), pp. 657–665. Cold Spring Harbor Press, Cold Spring Harbor, New York, 1976.
95. Adhya, S., Gottesman, M., and de Crombrugghe, B., *Proc. Natl. Acad. Sci. U.S.A.* **71**, 2534–2538 (1974).
96. Richardson, J. P. R., Grimley, C., and Lowery, C., *Proc. Natl. Acad. Sci. U.S.A.* **72**, 1725–1728.
97. Ratner, D., *in* "RNA Polymerase" (R. Losick and M. Chamberlin, eds.), pp. 645–655. Cold Spring Harbor Press, Cold Spring Harbor, New York, 1976.
98. Korn, L. J., and Yanofsky, C., *J. Mol. Biol.* **106**, 231–241 (1976).
99. Yanofsky, C., *J. Mol. Biol.* **113**, 663–677 (1977).
100. Guarente, L. P., and Beckwith, J., *Proc. Natl. Acad. Sci. U.S.A.* **75**, 294–297 (1978).
101. Schafer, R., and Zillig, W. K., *Eur. J. Biochem.* **33**, 201–206 (1973).
102. Yang, H.-L., and Zubay, G., *Biochem. Biophys. Res. Commun.* **56**, 725–731 (1974).
103. Stahl, S. J., and Chamberlin, M. J., *J. Mol. Biol.* **112**, 577–601 (1977).
104. Khesin, R. B., and Shemiakin, M. F., *Biokhimiya* **27**, 761–779 (1962).
105. Khesin, R. B., Shemiakin, M. F., Gorlenko, Zh. M., Bogdonova, S. L., and Afanas'eva, T. P., *Biokhimiya* **27**, 1092–1105 (1962).
106. Herskowitz, T., *Annu. Rev. Genet.* **7**, 289–324 (1973).
107. Hershey, A. D., ed., "Bacteriophage Lambda." Cold Spring Harbor Press, Cold Spring Harbor, New York, 1971.
108. Lebowitz, P., Weissman, S. M., and Radding, C. M., *J. Biol. Chem.* **246**, 5120–5139 (1971).
109. Dahlberg, J. E., and Blattner, F. R., *Fed. Proc., Fed. Am. Soc. Exp. Biol.* **32**, 664 (1973).
110. Blattner, F. R., and Dahlberg, J. E., *Nature (London), New Biol.* **237**, 227–232 (1972).
111. Sklar, J., Yot, P., and Weissman, S. M., *Proc. Natl. Acad. Sci. U.S.A.* **72**, 1817–1821 (1975).
112. Roberts, J. W., *Proc. Natl. Acad. Sci. U.S.A.* **72**, 3300–3304 (1975).
113. Sklar, J., Ph.D. Thesis, Yale University, New Haven, Connecticut (1977).
114. Sklar, J., and Weissman, S. M., in preparation.
115. Honigman, A., Hu, S.-L., Chase, R., and Szybalski, W., *Nature (London)* **262**, 112–116 (1976).
116. Ptashne, M., Backman, K., Humayun, M. Z., Jeffrey, A., Maurer, R., Mayer, B., and Sauer, R. T., *Science* **194**, 156–161 (1976).
117. Musso, R. E., DiLauro, R., Adhya, S., and de Crombrugghe, B., *Cell* **12**, 847–854 (1977).
118. Bertrand, K., Korn, L., Lee, F., Platt, T., Squires, C. L., Squires, C., and Yanofsky, C., *Science* **189**, 22–26 (1975).
119. Bertrand, K., Korn, L. U., Lee, F., and Yanofsky, C., *J. Mol. Biol.* **117**, 227–247 (1977).
120. Dhar, R., Weissman, S. M., Zain, B. S., Pan, J., and Lewis, A. M., Jr., *Nucleic Acids Res.* **1**, 595–613 (1974).
121. Zain, B. S., Weissman, S. M., Dhar, R., and Pan, J., *Nucleic Acids Res.* **1**, 577–594 (1974).
122. Maniatis, T., Ptashne, M., Barrell, B. G., and Donelson, J., *Nature (London)* **250**, 394–397 (1974).
123. Maniatis, T., Jeffrey, A., and Kleid, D. G., *Proc. Natl. Acad. Sci. U.S.A.* **72**, 1184–1188 (1975).

124. Walz, A., and Pirotta, V., *Nature (London)* **254**, 118–121 (1975).
125. Pribnow, D., *Proc. Natl. Acad. Sci. U.S.A.* **72**, 784–788 (1975).
126. Pribnow, D., *J. Mol. Biol.* **99**, 419–443 (1975).
127. Schaller, H., Gray, C., and Hermann, K., *Proc. Natl. Acad. Sci. U.S.A.* **72**, 737–741 (1975).
128. Sugimoto, K., Okamoto, T., Sugisaki, H., and Takanami, M., *Nature (London)* **253**, 410–414 (1975).
129. Dickson, R. C., Abelson, J., Barnes, W. M., and Reznikoff, W. S., *Science* **187**, 27–35 (1975).
130. Sklar, J., Weissman, S. M., Musso, R. E., DiLauro, R., and de Crombrugghe, B., *J. Biol. Chem.* **252**, 3538–3547 (1977).
131. Sekiya, T., and Khorana, H. G., *Proc. Natl. Acad. Sci. U.S.A.* **71**, 2978–2982 (1974).
132. Kupper, H., Contreras, R., Khorana, H. G., and Landy, A., *in* "RNA Polymerase" (R. Losick and M. Chamberlin, eds.), pp. 473–484. Cold Spring Harbor Press, Cold Spring Harbor, New York, 1976.
133. Dhar, R., Shenk, T., and Weissman, S. M., in preparation.
134. Allet, B., Roberts, R. J., Gesteland, R. F., and Solem, R., *Nature (London)* **249**, 217–221 (1974).
135. Okamoto, T., Sugimoto, K., Sugisaki, H., and Takanami, M., *Nucleic Acids Res.* **4**, 2213–2222 (1977).
136. Rosenberg, M., Court, D., Wulff, D. L., Shimatake, H., and Brady, C., *Nature (London)* **272**, 414–423 (1978).
137. Kates, J., and Beeson, J., *J. Mol. Biol.* **50**, 1–18 (1970).
138. Mueller, K., Oebbecke, C., and Forster, G., *Cell* **10**, 121–130 (1977).
139. Bennet, G. N., and Yanofsky, C., in press.
140. Burd, J. F., Wartell, R. M., Dodgson, J. B., and Wells, R. D., *J. Biol. Chem.* **250**, 5109–5113 (1975).
141. Wartell, R. M., *Nucleic Acids Res.* **4**, 2779–2797 (1977).
142. Jones, B. B., Chan, H., Rothstein, S., Wells, R. D., and Reznikoff, M., *Proc. Natl. Acad. Sci. U.S.A.* **74**, 4914–4918 (1977).
143. Venetianer, P., Sumize, J., and Volkaert, A., *Control Ribosome Synth. Proc. Alfred Benzon Symp., 9th, 1976* pp. 252–267 (1976).
144. Mueller, K., Oebbecke, C., and Forster, G., *Cell* **10**, 121–130 (1977).
145. Federoff, N., *Cell* **13**, 701–717 (1978).
146. Valenzuela, P., Ball, G. I., Masiarz, F. R., DeGennaro, L. J., and Rutter, W. J., *Nature (London)* **267**, 641–643 (1977).
147. Maxam, A. M., Tizard, R., Skryabin, K. G., and Gilbert, W., *Nature (London)* **267**, 643–645 (1977).
148. Goodman, H. M., Olson, M. V., and Hall, B. D., *Proc. Natl. Acad. Sci. U.S.A.* **74**, 5453–5457 (1977).
149. Valenzuela, P., Venegas, A., Weinberg, F., Bishop, R., and Rutter, W. J., *Proc. Natl. Acad. Sci. U.S.A.* **75**, 190–194 (1978).
150. Pan, J., Celma, M. L., and Weissman, S. M., *J. Biol. Chem.* **252**, 9047–9054 (1977).
151. Parker, J. S., Jaehning, J. A., and Roeder, R., *Cold Spring Harbor Symp. Quant. Biol.* **42**, 577–587 (1977).
152. Celma, M. L., Pan, J., and Weissman, S. M., *J. Biol. Chem.* **252**, 9043–9046 (1977).
153. Vennström, B., Pettersson, U., and Philipson, L., *Nucleic Acids Res.* **5**, 195–204 (1978).
154. Larsen, C. J., Lebowitz, P., Weissman, S. M., and DuBuy, D., *Cold Spring Harbor Symp. Quant. Biol.* **35**, 35–46 (1970).

155. Sugimoto, H., Sagesaki, H., Okamoto, T., and Takanami, M., *J. Mol. Biol.* **111**, 487–500 (1977).
156. Quoted in Godson, G. N., Barrell, B. G., Staden, R., and Feddes, J. C., *Nature (London)* **276**, 236–247 (1978).
157. Lee, F., Squires, C. L., Squires, C., and Yanofsky, C., *J. Mol. Biol.* **103**, 383–393 (1976).
158. Chakraborty, P. R., Salvo, R. A., Majumder, H. K., and Maitra, U., *J. Biol. Chem.* **252**, 6485–6493 (1977).
159. Maitra, U., Lockwood, A. H., Dubnoff, J. S., and Guha, A., *Cold Spring Harbor Symp. Quant. Biol.* **35**, 143–156 (1970).
160. Millette, R. L., Trotter, C. D., Herrlich, P., and Schweiger, M., *Cold Spring Harbor Symp. Quant. Biol.* **35**, 135–142 (1970).
161. Rosenberg, M., de Crombrugghe, B., and Musso, R., *Proc. Natl. Acad. Sci. U.S.A.* **73**, 717–721 (1976).
162. Kleid, D., Humayun, Z., Jeffrey, A., and Ptashne, M., *Proc. Natl. Acad. Sci. U.S.A.* **73**, 293–297 (1976).
163. Kössel, H., Scherer, G., and Hobom, G., *Nature (London)* **265**, 117–121 (1977).
164. Squires, C., Lee, F., Bertrand, K., Squires, C. L., Bronson, M. J., and Yanofsky, C., *J. Mol. Biol.* **103**, 351–381 (1976).
165. Bertrand, K., and Yanofsky, C., *J. Mol. Biol.* **103**, 339–349 (1976).
166. Bertrand, K., Korn, L. J., Lee, P., and Yanofsky, C., *J. Mol. Biol.* **117**, 227–247 (1977).
167. Sanger, F., Air, G. M., Barrell, B. G., Brown, N. L., Coulson, A. R., Fiddes, J. C., Hutchison, C. A., III, Slocombe, P. M., and Smith, M., *Nature (London)* **265**, 687–698.
168. Ikemura, T., and Dahlberg, J. E., *J. Biol. Chem.* **248**, 5024–5032 (1973).
169. Sogin, M., Pace, N., Rosenberg, M., and Weissman, S. M., *J. Biol. Chem.* **251**, 3480–3488 (1976).
170. Pieczenik, G., Barrell, B. G., and Gefter, M. L. *Arch. Biochem. Biophys.* **152**, 152–165 (1972).
171. Goldberg, A. R., and Hurm, F. J., *J. Biol. Chem.* **247**, 5637–5645 (1972).
172. McMahon, J. E., and Tinoco, I., Jr., *Nature (London)* **271**, 275–277 (1978).
173. Branlant, C., and Ebel, J. P., *J. Mol. Biol.* **111**, 215–256 (1977).
174. Roberts, T. M., Shimatake, H., Brady, C., and Rosenberg, M., *Nature (London)* **270**, 274–275 (1977).
175. Subramanian, K. N., Dhar, R., Weissman, S. M., and Ghosh, P. K., *J. Biol. Chem.* **252**, 340–354 (1977).
176. Kramer, R. A., Rosenberg, M., and Steitz, J. A., *J. Mol. Biol.* **89**, 767–776 (1974).
177. Egan, J., and Landy, A., *J. Biol. Chem* **253**, 3007–3022 (1978).
178. Platt, T., personal communication.
179. Kupper, H., Sekiya, T., Rosenberg, M., Egan, J., and Landy, A., *Nature (London)* (in press).
180. Kiefer, M., Neff, N., and Chamberlin, M., *J. Virol.* **22**, 548–552 (1977).
181. Brown, R. D., and Brown, D. D., *J. Mol. Biol.* **102**, 1–14 (1976).
182. Celma, M. L., Pan, J., and Weissman, S. M., *J. Biol. Chem.* **252**, 9032–9042 (1977).
183. Denis, H., and Wegnez, M., *Biochimie* **55**, 1137–1151 (1973).
184. Rubin, G. M., and Hogness, D. S., *Cell* **6**, 207–213 (1975).
185. Harris, B., and Roeder, R., personal communication.
186. Jacq, B., Jourdan, R., and Jordan, B. R. P., *J. Mol. Biol.* **117**, 785–795 (1977).

187. Dhar, R., Subramanian, K. N., Zain, B. S., Levine, A., Patch, C., and Weissman, S. M., *Colloq. Inst. Nat. Sante Rech. Med.* **47**, 25–32 (1975).
188. Thimmappaya, B., Dhar, R., Zain, B. S., and Weissman, S. M., *J. Biol. Chem.* **253**, 1613–1618 (1978).
189. Fiers, W., Contreras, R., Haegeman, G., Rogiers, R., Van de Voorde, A., Van Heuverswyn. H., Van Heereweghe, J., Volckaert, G., and Ysebaert, M. *Nature (London)* **273**, 113–120 (1978).
190. Proudfoot, N. J., and Brownlee, G. G., *Nature (London)* **252**, 359–362 (1974).
191. Milstein, C., Brownlee, G. G., Cartwright, E. M., Jarvis, J. M., and Proudfoot, N. J., *Nature (London)* **252**, 354–359 (1974).
192. Proudfoot, N. J., and Brownlee, G. G., *Nature (London)* **263**, 211–214 (1976).
193. Hamlyn, P. H., Gillam, S., Smith, M., and Milstein, C., *Nucleic Acids Res.* **4**, 1123–1134 (1977).
194. Seeburg, P. H., Shine, J., Martial, J. A., Baxter, J. D., and Goodman, H. M., *Nature (London)* **270**, 486–494 (1977).
195. Ullrich, A., Shine, J., Chirgwin, J., Pictet, R., Tischer, E., Rutter, W. J., and Goodman, H. M., *Science* **196**, 1313–1319 (1977).
196. Ziff, E., and Fraser, N., *J. Virol.* **25**, 897–906 (1978).
197. Reddy, V. B., Thimmappaya, B., Dhar, R., Subramanian, K. N., Zain, B. S., Pan, J., Ghosh, P. K., Celma, M. L., and Weissman, S. M., *Science* **200**, 494–502 (1978).
198. von Hippel, P. H., *J. Cell. Physiol.* **74**, Suppl. 1, 235–238 (1969).
199. Goeddel, D. V., Yansura, D. G., and Caruthers, M. H., *Proc. Natl. Acad. Sci. U.S.A.* **74**, 3292–3269 (1977).
200. Bahl, C. P., Wu, R., Stawinsky, J., and Narang, S. A., *Proc. Natl. Acad. Sci. U.S.A.* **74**, 966–970 (1977).
201. Richter, P. H., and Eigen, M., *Biophys. Chem.* **2**, 253–263 (1974).
202. Wang, A. C., Revzin, A., Butler, A. P., and von Hippel, P. H., *Nucleic Acids Res.* **4**, 1579–1593 (1977).
203. Kao-Huang, Y., Revzin, A., Butler, A. P., O'Conner, P., Noble, D. W., and von Hippel, P. H., *Proc. Natl. Acad. Sci. U.S.A.* **74**, 4228–4232 (1977).
204. Yamamoto, K. R., and Alberts, B., *Cell* **4**, 301–310 (1975).
205. Wang, J. C., Barkley, M. D., and Bourgeois, S., *Nature (London)* **251**, 247–249 (1974).
206. Richmond, T. J., and Steitz, T. A., *J. Mol. Biol.* **103**, 25–38 (1976).
207. Record, M. T., Jr., Lohman, T. M., and de Haseth, P., *J. Mol. Biol.* **107**, 145–158 (1976).
208. Record, M. T., Jr., de Haseth, P. L., and Lohman, T. M., *Biochemistry* **16**, 4791–4802 (1977).
209. de Haseth, P. L., Lohman, T. M., and Record, T. M., Jr., *Biochemistry* **16**, 4783–4790 (1977).
210. Church, G. M., Sussman, J. L., and Kim, S.-H., *Proc. Natl. Acad. Sci. U.S.A.* **74**, 1458–1462 (1977).
211. Carter, C. W., and Kraut, J., *Proc. Natl. Acad. Sci. U.S.A.* **71**, 283–287 (1974).
212. Seeman, N. C., Rosenberg, J. M., and Rich, A., *Proc. Natl. Acad. Sci. U.S.A.* **73**, 804–808 (1976).
213. Gursky, G. V., Tumanyan, V. G., Zasedatelev, A. S., Zhose, A. L., Grokhovski, S. L., and Gottskh, B. P., in "Nucleic Acid-Protein Recognition" (H. J. Vogel, ed.), pp. 189–217. Academic Press, New York, 1977.
214. O'Neill, M. C., *Nucleic Acids Res.* **4**, 4439–4463 (1977).
215. Rich, A., Seeman, N. S., and Rosenberg, J. M., in "Nucleic Acid-Protein Recognition" (H. J. Vogel, ed.), pp. 362–374. Academic Press, New York, 1977.

216. Jovin, T. M., Geisler, N., and Weber, K., *Nature (London)* **269**, 668–672 (1977).
217. Müller-Hill, B., Gronenbora, B., Kanin, J., Schlotmann, M., and Beyreuther, K., *in* "Nucleic Acid-Protein Recognition" (H. J. Vogel, ed.), pp. 219–236. Academic Press, New York, 1977.
218. Takeda, Y., Folkmanis, A., and Echols, H., *J. Biol. Chem.* **252**, 6177–6183 (1977).
219. Hsiang, M. W., Cole, R., Takeday, Y., and Echols, H., *Nature (London)* **270**, 275–277 (1977).
220. Gilbert, W., Maizels, N., and Maxam, A., *Cold Spring Harbor Symp. Quant. Biol.* **38**, 845–855 (1973).
221. Bennet, G. N., Schweingruber, M. E., Brown, K. D., Squires, C., and Yanofsky, C., *Proc. Natl. Acad. Sci. U.S.A.* **73**, 2351–2355 (1976).
222. Brown, K. D., Bennet, G. N., Lee, F., Schweingruber, M. E., and Yanofsky, C., *J. Mol. Biol.* (in press).
223. Musso, R., DiLauro, R., Rosenberg, M., and de Crombrugghe, B., *Proc. Natl. Acad. Sci. U.S.A.* **74**, 106–110 (1977).
224. Humayun, Z., Kleid, D., and Ptashne, M., *Nucleic Acids Res.* **4**, 1595–1608 (1977).
225. Nakanishi, S., Adhya, S., Gottesman, M., and Pastan, I., *J. Biol. Chem.* **249**, 4050–4056 (1974).
226. Dickson, R. C., Abelson, J., Johnson, P., Reznikoff, W. S., and Barnes, W. M., *J. Mol. Biol.* **111**, 65–75 (1977).
227. Rosenberg, M., Court, D., Shimatake, H., and Wulff, D. L., *Nature (London)* **272**, 414–423 (1978).
228. Dahlberg, J. E., and Blattner, F. R., *Nucleic Acids Res.* **2**, 1441–1458 (1975).
229. Griffith, J., Dieckmann, M. D., and Berg, P., *J. Virol.* **15**, 167–172 (1975).
230. Jessel, D., Landau, T., Hudson, J., Lalor, T., Tenen, D., and Livingston, D. M., *Cell* **8**, 535–545 (1976).
231. Reed, S. I., Ferguson, J., Davis, R. W., and Stark, G. R., *Proc. Natl. Acad. Sci. U.S.A.* **72**, 1605–1609 (1975).
232. Tjian, R., *Cell* **13**, 165–179 (1978).
233. Reed, S. I., Stark, G. R., and Alwine, J. C., *Proc. Natl. Acad. Sci. U.S.A.* **73**, 3083–3087 (1976).
234. Bourgignon, G. J., Tattersall, P. J., and Ward, D. C., *J. Virol.* **20**, 290–306 (1976).
235. Tattersall, P., and Ward, D. C., *Nature (London)* **263**, 106–109 (1976).
236. Straus, S. E., Sebring, E. D., and Rose, J. A., *Proc. Natl. Acad. Sci. U.S.A.* **73**, 742–746 (1976).
237. Spear, I. S., Fife, K. H., Hauswirth, W. W., Jones, C. J., and Berns, K. I., *J. Virol.* **24**, 627–634 (1977).
238. Langeveld, S. A., van Mansfeld, D. M., Baas, P. D., Jansz, H. S., Van Arkel, G. A., and Weisbeck, P. J., *Nature (London)* **271**, 417–420 (1970).
239. Henry, T. J., and Knippers, R., *Proc. Natl. Acad. Sci. U.S.A.* **71**, 1549–1553 (1974).
240. Geider, K., Beck, E., and Schaller, H., *Proc. Natl. Acad. Sci. U.S.A.* **75**, 645–649 (1978).
241. Wickner, W., Brutlag, D., Schekman, R., and Kornberg, A., *Proc. Natl. Acad. Sci. U.S.A.* **69**, 965–969 (1972).
242. McMacken, R., Ueda, K., and Kornberg, A., *Proc. Natl. Acad. Sci. U.S.A.* **74**, 4190–4194 (1977).
243. Wickner, S., *Proc. Natl. Acad. Sci. U.S.A.* **74**, 2815–2819 (1977).
244. Flanegan, J. B., and Baltimore, D., *Proc. Natl. Acad. Sci. U.S.A.* **74**, 3677–3680 (1977).

245. Rekosh, D. M. K., Russell, W. C., Bellett, A. J. D., and Robinson, A. J., *Cell* 11, 283–296 (1977).
246. Roberts, R., personal communication.
247. Scherzinger, E., Lanka, E., and Hillenbrand, G., *Nucleic Acids Res.* 4, 4151–4163 (1977).
248. Nathans, D., and Danna, K. J., *Nature (London), New Biol.* 236, 200–202 (1972).
249. Thoren, M. M., Sebring, E. D., and Salzman, N. P., *J. Virol.* 10, 462–468 (1972).
250. Clewell, D., and Helinski, D., *Proc. Natl. Acad. Sci. U.S.A.* 62, 1159–1166 (1969).
251. Tomizawa, J. I., Ohmori, H., and Bird, R. E., *Proc. Natl. Acad. Sci. U.S.A.* 74, 1865–1869 (1977).
252. Gutai, M., and Nathans, D., personal communication.
253. Shenk, T., Subramanian, K. N., and Berg, P., personal communication.
254. Shenk, T., *Cell* 13, 791–798 (1978).
255. Furth, M. E., Blattner, F. E., McLeester, C., and Dove, W. F., *Science* 198, 1046–1051 (1977).
256. Gray, C. P., Sommer, R., Polke, C., Beck, E., and Schaller, H., *Proc. Natl. Acad. Sci. U.S.A.* 75, 50–53 (1978).
257. Ravetch. J. V., Horiuchi, K., and Zinder, N. D., *Proc. Natl. Acad. Sci. U.S.A.* 74, 4219–4222 (1977).
258. Godson, G. N., Barrell, B. G., Staden, R., and Fiddes, J. C., *Nature (London)* 276, 236–247 (1978).
259. Bastia, D., *Nucleic Acids Res.* 4, 3123–3142 (1977).
260. Bastia, D., *J. Mol. Biol.* 124, 601–639 (1978).
261. Denniston-Thompson, K., Moore, D. D., Kruger, K. E., Furth, M. E., and Blattner, F. R., *Science* 198, 1051–1056 (1977).
262. Subramanian, K. N., Dhar, R., and Weissman, S. M., *J. Biol. Chem.* 252, 355–367 (1977).
263. Soeda, E., Miura, K.-I., Nakaso, A., and Kimura, G., *FEBS Lett.* 79, 383–389 (1977).
264. Friedmann, T., LaPorte, P., and Esty, A., *J. Biol. Chem.* 253, 6561–6567 (1978).
265. Dhar, R., Lai, C.-J., and Khoury, G., *Cell* 13, 345–358 (1978).
266. Arrand, J., and Richards, R., *J. Mol. Biol.* (in press).
267. Steenbergh, P. H., Maat, J., Van Ormondt, H., and Sussenbach, J. S., *Nucleic Acids Res.* 4, 4371–4390 (1977).
268. Efstratiadis, A., Kafatos, F. P., and Maniatis, T., *Cell* 10, 271–305 (1977).
269. Marotta, C. A., Wilson, J. T., Forget, B. G., and Weissman, S. M., *J. Biol. Chem.* 252, 5040–5053 (III).
270. Birnstiel, M. L., Schaffner, W., and Smith, H. O., *Nature (London)* 266, 603–607 (1977).
271. Kedes, L., personal communication.
272. Shine, J., Seeburg, P. H., Martial, J. A., Baxter, J. D., and Goodman, H. M., *Nature (London)* 270, 494–499 (1977).
273. Russel, G. J., Walker, P. M. B., Elton, R. A., and Subak-Sharpe, J. H., *J. Mol. Biol.* 180, 1–23 (1976).
274. Fraser, N. U., Burdon, R., and Elton, R. A., *Nucleic Acids Res.* 2, 2131–2146 (1975).
275. Wilson, J. T., unpublished results.
276. Haroda, F., and Nishimura, S., *Biochemistry* 13, 300–307 (1974).
277. MinJou, W., Van Montagu, M., and Fiers, W., *Biochem. Biophys. Res. Commun.* 73, 1083–1093 (1976).
278. Suzuki, Y., and Brown, D. D., *J. Mol. Biol.* 63, 409–429 (1972).

244   P. ANDREW BIRO AND SHERMAN M. WEISSMAN

279. Fournier, A., Chavancy, G., and Garel, J.-P., *Biochem. Biophys. Res. Commun.* **72**, 1187–1194 (1976).
280. Garel, J.-P., *J. Theor. Biol.* **43**, 211–225 (1974).
281. Smith, D. W. E., *Science* **190**, 529–535 (1975).
282. Hatfield, D., and Matthews, R. C., unpublished data.
283. Meza, L., Araya, A., Leon, G., Krauskopf, M., Siddiqui, M. A. Q., and Garel, J.-P., *FEBS Lett.* **77**, 255–260 (1977).
284. Zuniga, M. C., and Steitz, J. A., *Nucleic Acids Res.* **4**, 4175–4196 (1977).
285. Sprague, K. U., Hagenbuchle, O., and Zuniga, M. C., *Cell* **11**, 561–570 (1977).
286. Hilse, K., and Rudloff, E., *FEBS Lett.* **60**, 380–383 (1975).
287. Litt, M., and Kabat, D., *J. Biol. Chem.* **247**, 6659–6664 (1972).
288. Garber, R. L., Siddiqui, M. A. Q., and Altman, S., *Proc. Natl. Acad. Sci. U.S.A.* **75**, 635–639 (1978).
289. Crick, F. C., Brenner, S., and Pieczenik, G., *Symp. Origins Life* (in press).
290. Stubbs, G., Warren, S., and Holmes, K., *Nature (London)* **267**, 216–221 (1977).
291. Jonard, G., Richards, K. E., Guilley, H., and Hirth, L., *Cell* **11**, 483–494 (1977).
292. Karn, J., Vidalli, G., Boffa, L. C., and Allfrey, V., *J. Biol. Chem.* **252**, 7307–7322 (1977).
293. Bunn, H., Forget, B. G., and Ranney, H. M., "Human Hemoglobinopathies." Saunders, Philadelphia, Pennsylvania, 1977.
294. Forget, B., Marotta, C. A., Weissman, S. M., and Cohen-Solal, M., *Proc. Natl. Acad. Sci. U.S.A.* **72**, 3614–3618 (1975).
295. Wilson, J. T., de Riel, J. K., Forget, B. G., Marotta, C. A., and Weissman, S. M., *Nucleic Acids Res.* **4**, 2353–2368 (1977).
296. Zilberstein, A., Dudock, B., Berissi, H., and Revel, M. *J. Mol. Biol.* **108**, 43–54 (1976).
297. Steitz, J. A., *Nature (London)* **224**, 957–964 (1969).
298. Hindley, J., and Staples, D. H., *Nature (London)* **224**, 964–967 (1969).
299. Gupta, S. L., Chen, J., Schaefer, L., Lengyel, P., and Weissman, S. M., *Biochem. Biophys. Res. Commun.* **39**, 883–888 (1970).
300. Shine, J., and Dalgarno, L., *Nature (London)* **254**, 34–38 (1975).
301. Draper, D. E., Pratt, C. W., and von Hippel, P. H., *Proc. Natl. Acad. Sci. U.S.A.* **74**, 4786–4790 (1977).
302. MinJou, W., Haegeman, M., Ysebaert, M., and Fiers, W., *Nature (London)* **237**, 82–88 (1972).
303. Taniguchi, T., and Weissmann, C., *J. Mol. Biol.* **118**, 533–565 (1978).
304. Rosenberg, M., personal communication.
305. Steitz, J., personal communication.
306. Dasgupta, R., Shih, D. S., Saris, C., and Kaesberg, P., *Nature (London)* **256**, 624–628 (1975).
307. Chang, J. C., Temple, G. F., Poon, R., Neumann, K. H., and Kon, Y. W., *Proc. Natl. Acad. Sci. U.S.A.* **74**, 5145–5149 (1977).
308. Baralle, F., *Nature (London)* **267**, 279–281 (1977).
309. Baralle, F. E., *Cell* **10**, 549–558 (1977).
310. Lockard, R. E., and RajBhandary, U. L., *Cell* **9**, 747–760 (1976).
311. Baralle, F. E., *Cell* **12**, 1085–1095 (1977).
312. Kozak, M., and Shatkin, A., *Cell* **13**, 201–212 (1978).
313. Rose, J. K., *Proc. Natl. Acad. Sci. U.S.A.* **74**, 3672–3676 (1977).
314. Koper-Zwarthoff, E. C., Lockard, R. E., Alzner-de Weerd, B., RajBhandary, U. L., and Bol, J. T., *Proc. Natl. Acad. Sci. U.S.A.* **74**, 5504–5508 (1977).

315. Ghosh, P., Reddy, V. B., Swinscoe, J., Choudary, P. V., Lebowitz, P., and Weiss-man, S. M., *J. Biol. Chem.* **253**, 3643–3647 (1978).
316. Southern, E. M., *Nature (London)* **227**, 794–798 (1970).
317. Biro, P. A., Carr-Brown, A., Southern, E. M., and Walker, P. M. B., *J. Mol. Biol.* **94**, 71–86 (1975).
318. Southern, E. M., *J. Mol. Biol.* **94**, 51–69 (1975).
319. Igo-Kemenes, T., Golil, W., and Zachau, H. G., *Nucleic Acids Res.* **4**, 3387–3400 (1977).
320. Carlson, M., and Brutlag, D., *Cell* **11**, 371–381 (1977).
321. Roizes, G., *Nucleic Acids Res.* **3**, 2677–2696 (1977).
322. Botchan, M. R., *Nature (London)* **251**, 288–292 (1974).
323. Rosenberg, H., Singer, M., and Rosenberg, M., *Science* **200**, 394–402 (1978).
324. Maio, J. J., Brown, F. L., and Musich, P. R., *J. Mol. Biol.* **117**, 637–655 (1977).
325. Gall, J. G., and Pardue, M. L., *Proc. Natl. Acad. Sci. U.S.A.* **63**, 378–383 (1969).
326. Fry, K., and Salser, W., *Cell* **12**, 1069–1084 (1977).
327. Subramanian, K. N., Reddy, V. B., and Weissman, S. M., *Cell* **10**, 497–507 (1977).
328. Bukhari, A. I., and Ambrosia, L., *Nature (London)* **271**, 573–575 (1978).
329. Kamp, D., Kohmann, R., Zapser, D., Broder, T. R., and Chou, W. T., *Nature (London)* **271**, 575–577 (1978).
330. Faden, M., Huisman, O., and Toussaint, A., *Nature (London)* **271**, 580–582 (1978).
331. Zieg, J., Silverman, M., Hilmen, M., and Simer, P., *Science* **198**, 170–172 (1978).
332. Bird, A. P., Rochaix, J., and Bakken, A., *in* "Molecular Cytogenetics" (B. A. Hamlako, and J. Papaconstantinou, eds.), pp. 49–58. Plenum, New York, 1973.
333. Tobler, H., *Biochem. Anim. Dev.* **3**, 91–143 (1975).
334. Alt, F. W., and Schimke, R. T., *J. Biol. Chem.* **253**, 1357–1370 (1978).
335. Hozumi, N., and Tonegawa, S., *Proc. Natl. Acad. Sci. U.S.A.* **73**, 3628–3632 (1976).
336. Rabbitts, T. H., and Forster, A., *Cell* **13**, 319–327 (1978).
337. Dhar, R., Subramanian, K. N., Pan, J., and Weissman, S. M., *Proc. Natl. Acad. Sci. U.S.A.* **74**, 827–831 (1977).
338. Chou, L. T., Gelinas, R. E., Broker, T. R., and Roberts, R. J., *Cell* **12**, 1–8 (1977).
339. Berget, S. M., Moore, C., and Sharp, P. A. *Proc. Natl. Acad. Sci. U.S.A.* **74**, 3171–3175 (1977).
340. Celma, M. L., Dhar, R., Pan, J., and Weissman, S. M., *Nucleic Acids Res.* **4**, 2549–2559 (1977).
341. Breatnach, R., Mandel, J. L., and Chambon, P., *Nature (London)* **270**, 314–319 (1977).
342. Aloni, Y., Dhar, R., Laub, O., Horowitz, M., and Khoury, G., *Proc. Natl. Acad. Sci. U.S.A.* **74**, 3686–3690 (1977).
343. Hsu, M. T., and Ford, J., *Proc. Natl. Acad. Sci. U.S.A.* **74**, 4982–4985 (1977).
344. Lavi, S., *Proc. Natl. Acad. Sci. U.S.A.* **74**, 5323–5327 (1977).
345. Tilghman, S. M., Tiemeier, D. C., Polsky, F., Edgell, M. H., Seidman, J. G., Leder, A., Enquist, L. W., Norman, B., and Leder, P., *Proc. Natl. Acad. Sci. U.S.A.* **74**, 4406–4410 (1977).
346. Westphal, H., and Lai, S.-P., *J. Mol. Biol.* **116**, 525–548 (1977).
347. Knapp, G., Beckmann, J. S., Johnson, P. F., Fuhrman, S. A., and Abelson, J., *Cell* **14**, 221–236 (1978).
348. Brownlee, G. G., *Nature (London)* **269**, 833–836 (1977).
349. Donis-Keller, H., Maxam, A. M., and Gilbert, W., *Nucleic Acids Res.* **4**, 2527–2538 (1977).
350. Randerath, K., *Nucleic Acids Res.* **4**, 3444–3454 (1977).

Chapter VI

Nucleosomes: Composition and Substructure

RANDOLPH L. RILL

I. INTRODUCTION

Central to the elucidation of molecular mechanisms of operation of the eukaryotic genome are questions about modes of DNA packaging in chromatin fibrils. In a typical mammalian cell nearly 2 m of DNA is confined within a nucleus less than 10 μm (10^{-5} m) in diameter. Compaction to this degree is accomplished through the intervention of structural proteins, whose interactions with DNA and each other provide the energy required to alter the statistical configuration of the free DNA chain. Superimposed on necessities for active modification of DNA configuration are requirements for carrying out, or maintaining the potential for carrying out, genetic functions such as replica-

247

tion, transcription, and repair. Each of these processes imposes steric constraints on the mode of packaging related to requirements for recognition and manipulation of specific DNA sequences, and must involve at least transient alterations in the local nucleoprotein structure.

In accord with the functional complexity of the genome, this compaction of DNA is accomplished not through a single folding scheme, but instead through multiple levels of folding and association of the nucleoprotein fibril. Intuition leads to the expectation of correlations between fibril packaging and gene function. At a certain level, these expectations have been borne out through examinations by electron microscopy, particularly when coupled with autoradiography. Obvious examples are the gathering together of interphase chromatin into morphologically distinguishable, highly condensed metaphase chromosomes to facilitate apportioning of DNA during cell division; the association of nascent RNA transcription products with diffuse euchromatin (and periheterochromatin), rather than the dense heterochromatin of interphase cells; and the puffing of transcriptionally active regions of giant dipterian salivary gland chromosomes.

Despite distinguishable, and in certain cases marked, morphological variations in the chromosomal material with cell cycle and between species, at one level all of eukaryotic chromatin appears to be remarkably similar. This is at what may be termed the primary level of chromatin structure, where the first degree of compaction of DNA occurs on account of interactions with highly basic histones, the dominant structural proteins of chromatin. Over the past few years it has become clear that four of the five common histones (H2A, H2B, H3, and H4) form highly specific complexes, located at semiperiodic intervals along DNA, that serve as globular "cores" about which a specific length of DNA is wrapped. Each core and its associated DNA defines a subunit, now commonly called a nucleosome, of the primary chromatin fibril. (A more precise definition of the nucleosome will be given below.) The fundamental concept of fibrils of nucleosomes and the folding of these fibrils to yield higher levels of compaction are described in Chapter VII. Here the focus will be on details of the isolation, characterization, and internal structural features of nucleosomes.

References contained herein were collected with an eye toward completeness, but in the main are intended to be illustrative. A remarkable outpouring of high-quality work gives adequate witness to the intensity and excitement of this field over the past few years. Omissions signify little but inadequacies in dealing with such volume in reasonable space and time. For further details the reader is referred to several other reviews (Elgin and Weintraub, 1975; Felsenfeld, 1975, 1978; Van Holde and Isenberg, 1975; Kornberg, 1977).

II. HISTONES

A. OCCURRENCE, ISOLATION, AND CLASSIFICATION

Histones are small basic proteins that occur universally in all multicellular organisms in about equal proportion, by weight, to nuclear DNA. They also have been found in some of the simplest single-celled organisms considered to be eukaryotic, such as yeast and other fungi (Felden *et al.*, 1976; Goff, 1976; Morris, 1976; Nelson *et al.*, 1977a) and *Tetrahymena* (Gorovsky *et al.*, 1973; Gorovsky and Keevert, 1975a), but are absent from bacteria and viruses, with the important exception of certain mammalian DNA viruses (e.g., SV40, polyoma) that incorporate host histones (Roblin *et al.*, 1971; Frearson and Crawford, 1972; Lake *et al.*, 1973; see also several articles in *Cold Spring Harbor Symp. Quant. Biol.* **39**, 1974).

The term histone was first coined by Kossel in 1884 to refer to proteins extracted from nuclei with dilute hydrochloric acid. [For an interesting historical account of early studies of histones and nucleoproteins, see Luck (1964).] Modern procedures for isolating histones still frequently employ hydrochloric or sulfuric acids for extraction, and ethanol or acetone for precipitation of the respective salts (Johns, 1971; Hardison and Chalkley, 1977). A note of caution is needed here because acid extraction is notorously nonquantitative and partially selective, unless special steps are taken. This difficulty has led to considerable variation in estimates of total histone and the relative amounts of individual histones in chromatin from various sources. Recent application of DNase I (Bafus *et al.*, 1978) or cationic detergents (Shmatchenko and Varshavsky, 1978) for DNA removal may alleviate this difficulty. Concern over the harshness of acid extraction has led to development of milder extraction procedures employing high salt concentrations for the dissociation of histones; and protamine, hydroxyapatite, or gel permeation columns for the removal of DNA [e.g., see van der Westhuyzen and von Holt (1971)].

Early attempts to fractionate and characterize histones were complicated by their susceptibility to proteolysis, propensity to aggregate, and multiplicity resulting from posttranscriptional modifications (e.g., acetylation and phosphorylation), and thus led to gross overestimates of the number of distinct histone species. In fact there are only five major classes of histones, now designated according to the convention of the Ciba Foundation Symposium as H1, H2A, H2B, H3, and H4 (Bradbury, 1975). Traditionally these have been classified in terms of their relative lysine/arginine contents as lysine-rich (H1), slightly lysine-rich (H2A, H2B), and arginine-rich (H3, H4). Other histones are unique to certain specialized cells, such as H5 of nucleated erythrocytes and H6(T) of trout testes. Both of these uncommon histones are lysine-rich, appear to replace H1 during cell development, and to play a structural role similar in certain respects to H1 (see below). Some fundamental properties of histones are summarized in Table I.

These five major histone classes occur universally among differentiating eukaryotes, including primitive species such as sea urchins (Cohen *et al.*, 1975) and slime molds (Bradbury *et al.*, 1973a). Less is known about the histone content of single-celled organisms considered to be eukaryotic because, among other factors, of difficulties in isolating clean nuclei, proteolytic degradation, and large-scale contamination of chromatin by ribonucleoproteins in rapidly dividing cells. Analyses of isolated nucleosomes have shown that yeast (*Saccharomyces cerevisiae*) definitely contains all four core histones, as well as proteins comparable to H1 (Nelson *et al.*, 1977a; Rill and Nelson, 1978). All five histones have also been found in other fungi and ciliates (Felden *et al.*, 1976; Goff, 1976), but appear absent in dinoflagellates (Rizzo and Nooden, 1972). Curiously, the transcriptionally active macronucleus of the ciliate *Tetrahymena* contains all five histone classes, while the inactive micronucleus reportedly lacks the electrophoretic equivalents of H1 and H3 (Gorovsky and Keevert, 1975a).

Taken as a whole, estimates of the relative amounts of histones from a variety of sources suggest that core histones are present in approximately equimolar amounts in chromatin, and such equimolarity is demanded if nucleosomes are accepted as the sole structural arrangement of histones. In reality, literature values usually deviate from equimolarity, often grossly so. The absence of such basic information about chromatin, although unsettling, is understandable in view of past difficulties in quantitatively extracting undegraded histones, in completely resolving histones according to class by electrophoresis, and in determining accurate staining constants. Taking these difficulties into account, Olins *et al.* (1976a) and Joffe *et al.* (1977) have shown that the core histones are present in equimolar amounts in chicken erythrocytes and erythroblasts, respectively, within a precision of 20%. The molar ratio of H1 to core histones is usually taken as 0.5/1, but again significant variations are found. In chicken erythrocytes the total amount of H1 and H5 is equimolar with

TABLE I

CHARACTERIZATION OF THE HISTONES[a]

Class	Fraction	$\dfrac{\text{Lys} + \text{Arg}}{\text{Glu} + \text{Asp}}$	Lys/Arg ratio	Total residues	Molecular weight	N-terminal	C-terminal
Very lysine-rich	H1 (I, f1, KAP)	7.11	22.0	~215	~21,500	Ac-Ser	Lys
Lysine-rich	H2A (IIb1, f2a2, ALK)	2.89	1.17	129	14,004	Ac-Ser	Lys
	H2B (IIb2, f2b, KSA)	2.80	2.50	125	13,774	Pro	Lys
Arginine-rich	H3 (III, f3, ARK)	3.44	0.72	135	15,324	Ala	Ala
	H4 (IV, f2a1, GRK)	3.57	0.79	102	11,282	Ac-Ser	Gly

[a] All data for histones of calf thymus. Adapted from Elgin and Weintraub (1975).

core histones. At present it is not clear if this is a special case, or if total lysine-rich histones are generally equimolar to core histones.

B. PRIMARY STRUCTURE OF CORE HISTONES

The amino acid sequences of histones H3 and H4 (Figs. 1 and 2) have been remarkably conserved. Comparisons of the pea and calf H3 and H4 sequences show differences in only 3 and 2% of the residues, respectively (DeLange and Smith, 1975). The latter difference corresponds to an estimated mutation rate of 0.06 per 100 residues per 10^8 years, by far the lowest yet observed. [For comparison, note that the mutation rates of cytochrome c, insulin, hemoglobin, and pancreatic ribonuclease are 3, 4, 14, and 33 per 100 residues per 10^8 year, respectively (see Dayhoff, 1972).] Although all changes in H3 and H4 sequences are conservative, the replacement of Ser to Cys-96 in rabbits and higher mammals is of potential functional significance, e.g., for interactions with nonhistone proteins, but is unlikely to be related to

HISTONE H3

```
                                  10                                          20
H₂N-Ala-Arg-Thr-Lys-Gln-Thr-Ala-Arg-Lys-Ser-Thr-Gly-Gly-Lys-Ala-Pro-Arg-Lys-Gln-Leu-

                                  30                                          40
    Ala-Thr-Lys-Ala-Ala-Arg-Lys-Ser-Ala-Pro-Ala-Thr-Gly-Gly-Val-Lys-Lys-Pro-His-Arg-

                                  50                                          60
   │Tyr│Arg-Pro-Gly-Thr-Val-Ala-Leu-Arg-Glu-Ile-Arg│Arg│Tyr-Gln-Lys-Ser-Thr-Glu-Leu-
   │Phe│                                            │Lys│

                                  70                                          80
    Leu-Ile-Arg-Lys-Leu-Pro-Phe-Gln-Arg-Leu-Val-Arg-Glu-Ile-Ala-Gln-Asp-Phe-Lys-Thr-

                                  90                                          100
    Asp-Leu-Arg-Phe-Gln-Ser-Ser-Ala-Val│Met│Ala-Seu-Gln-Glu-Ala│Cys│Glu-Ala-Tyr-Leu-
                                        │Ser│                    │Ala│

                                  110                                         120
    Val-Gly-Leu-Phe-Glu-Asp-Thr-Asn-Leu│Cys│Ala-Ile-His-Ala-Lys-Arg-Val-Thr-Ile-Met-
                                        │ ?  │

                                  130
    Pro-Lys-Asp-Ile-Gln-Leu-Ala-Arg-Arg-Ile-Arg-Gly-Glu-Arg-Ala-COOH
```

FIG. 1. Amino acid sequence of calf histone H3 (DeLange *et al.*, 1972, 1973; Olson *et al.*, 1972). Complete sequence data are also available for H3 of chicken (Brandt and von Holt, 1972, 1974a,b), carp (Hooper *et al.*, 1973), shark (Brandt *et al.*, 1974b), and pea (Patthy *et al.*, 1973). Partial data are available for H3 from trout (Candido and Dixon, 1972), *Drosophila* (unpublished data, cited in Elgin and Weintraub, 1975), sea urchin (Brandt *et al.*, 1974a), mollusk (Brandt *et al.*, 1974a), and cycad (Brandt *et al.*, 1974a). All known substitutions are indicated in blocks. There are no known insertions or deletions in H3. Yeast H3 contains no cysteine (Nelson *et al.*, 1977a); the substitution is unknown.

HISTONE H4

```
                           10                                          20
Ac-Ser-Gly-Arg-Gly-Lys-Gly-Gly-Lys-Gly-Leu-Gly-Lys-Gly-Gly-Ala-Lys-Arg-His-Arg-Lys-

                           30                                          40
Val-Leu-Arg-Asp-Asn-Ile-Gln-Gly-Ile-Thr-Lys-Pro-Ala-Ile-Arg-Arg-Leu-Ala-Arg-Arg-

                           50                                          60
Gly-Gly-Val-Lys-Arg-Ile-Ser-Gly-Leu-Ile-Tyr-Glu-Glu-Thr-Arg-Gly-Val-Leu-Lys-Val
                                                                             Ile

                           70                                          80
Phe-Leu-Glu-Asn-Val-Ile-Arg-Asp-Ala-Val-Thr-Tyr-Thr-Glu-His-Ala-Lys-Arg-Lys-Thr-
                                                   Ser              Arg

                           90                                         100
Val-Thr-Ala-Met-Asp-Val-Val-Tyr-Ala-Leu-Lys-Arg-Gln-Gly-Arg-Thr-Leu-Tyr-Gly-Phe-
```

Gly-Gly-COOH

FIG. 2. Amino acid sequence of calf histone H4 (DeLange *et al.*, 1969; Ogawa *et al.*, 1969). Complete sequence data are also available for H4 of rat (Sautiere *et al.*, 1971a), pig (Sautiere *et al.*, 1971b), and pea (DeLange *et al.*, 1969). Partial data are available for H4 from bovine lymphosarcoma (Desai *et al.*, 1969), Novikoff hepatoma (Desai *et al.*, 1969), trout (Dixon *et al.*, 1975), and sea urchin (Strickland *et al.*, 1974). All known substitutions are indicated in blocks. There are no known insertions or deletions in H4.

nucleosome structure. However, nonhistone interactions also seem unlikely, since Cys-96 is inaccessible to chemical reagents (Palau and Pedros, 1972). Likewise, Cys-110 probably is not crucial for nucleosome structure, since yeast H3 lacks cysteine. As pointed out by De-Lange and Smith (1975), such evolutionary stability implies that all regions of these chains interact with other invariant portions of chromatin structure, or, in other words, H3 and H4 must determine critical structural features of nucleosomes.

The invariance of H3 and H4 is even more remarkable in view of the markedly asymmetric distributions of basic and other conformationally distinctive residues along the chain. This asymmetry is most evident if the sequences are illustrated as in Fig. 3 (adapted from Van Holde and Isenberg, 1975). In both cases the N-terminal third to half of the chains are highly basic. For example, the N-terminal 49 residues of each chain have a net +16 charge (counting histidines) and contain the majority of structure-breaking residues, glycine and proline. A lesser concentration of basic and structure-breaking residues occurs in the C-terminal 10–20 residues, most notably in H3. For obvious reasons these sequence regions are generally assumed to be the primary sites of DNA binding. There is an apparent nonrandom spacing of basic residues. Adjacent pairs of basic residues are particularly common, as are spacings of basics are four residue intervals (i.e., at

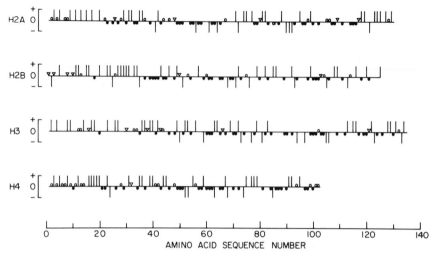

FIG. 3. The distributions of specific classes of amino acid residues along the chains of core histones. Bars projecting upward to (+) denote basic residues (Lys, Arg, His); bars projecting downward to (−) denote acidic residues (Asp, Glu). Upward projecting, open symbols denote structure-destabilizing residues Gly (○) and Pro (▽); downward projecting, closed symbols (●) denote hydrophobic, structure-promoting residues (Val, Leu, Ile, Met, Phe, Tyr).

positions $i + 1$ and $i + 3$ from residue i). The evolutionary stability of these regions argues that the steric requirements for DNA binding are surprisingly stringent, requiring preservation of this spacing, and even of the specific contact and intercontact residues (e.g., Arg/Lys replacements are not tolerated, bulkier residues cannot replace Gly, etc). Such stringency is not required of the other histones (see below). The remaining, more central chain regions contain the majority of bulky, hydrophobic groups and are generally similar in composition to typical globular proteins. Although part of the conservation of these regions is likely required for self-interactions, some is apparently due to requirements for recognition by other histones.

The general features of the H2A and H2B sequences (Fig. 3) are quite similar to those of H3 and H4. Again, each has a long highly basic N-terminal "tail" with sufficient basic residues to neutralize 7 (H2A) or 9 (H2B) DNA base pairs, a potentially globular central region, and a short, moderately (H2B) to highly (H2A) basic C-terminal region. Although the relative distributions of amino acids in all four histones are generally similar, a considerably lower degree of sequence conservation of H2A and H2B has been apparent for some time. Early electrophoretic studies showed considerable variations in mobilities of vertebrate H2A and H2B from different species (Panyim *et al.*, 1971). More recently, Cohen *et al.* (1975) have shown that there is a programmed switching of electrophoretically distinct H2A and H2B subfractions during different stages of sea urchin embryo development. A similar switching may occur in other organisms as well (Blankstein and Levy, 1976; Franklin and Zweidler, 1977). Comparisons of available sequences, though few (Figs. 4 and 5), suggest that most of the alterations in H2A and H2B occur in the basic, N-terminal, and C-terminal regions and involve deletions, insertions, and nonconservative as well as conservative substitutions. Particularly striking is the partial sequence of pea H2B derived from amino acid analyses of tryptic peptides. Compositions of 18 out of 27 peptides correlate well with the calf H2B sequence beyond residue 34, and indicate a relatively high degree of conservation. In contrast, no definitive correlations can be made between the compositions of the remaining 9, lysine-rich, peptides and the N-terminal calf sequence. Pea H2A sequence information is not available, but the amino acid compositions of pea and calf H2A indicate displacement of at least 14–15 residues between the sequences (Hayashi *et al.*, 1977).

HISTONE H2A

```
                                 10                              20
Ac-Ser-Gly-Arg-Gly-Lys[Gln]Gly[Gly]Lys[Ala-Arg]Ala-Lys-Ala-Lys[Thr]Arg-Ser-Ser-Arg-
                       Thr     Ala     Gly,Lys                 Ser

                                 30                              40
Ala-Gly-Leu-Gln-Phe-Pro-Val-Gly-Arg-Val-His-Arg[Leu]Leu-Arg-Lys-Gly-Asn-Tyr-Ala-
                                                Phe

                                 50                              60
[Glu]Arg-Val-Gly-Ala-Gly-Ala-Pro-Val-Tyr-Leu-Ala-Ala-Val-Leu-Glu[Tyr]Leu-Thr-Ala-
 Asn

                                 70                              80
Glu-Ile-Leu-Glu-Leu-Ala-Gly-Asn-Ala-Ala-Arg-Asp-Asn-Lys-Lys-Thr-Arg[Ile]Ile-Pro-

                                 90                              100
Arg-His-Leu-Gln-Leu-Ala[Ile]Arg-Asn-Asp-Glu-Glu-Leu-Asn-Lys-Leu-Leu-Gly[Lys]Val-
                        Val                                             Arg
                                                                        Gly
                                 110                             120
Thr-Ile-Ala-Gln-Gly-Gly-Val-Leu-Pro-Asn-Ile-Gln-Ala-Val-Leu-Leu-Pro-Lys-Lys-Thr-

                              129
Glu[Ser-His-His]Lys-Ala-Lys[Gly]Lys-COOH
                            Val-Ala
```

FIG. 4. Amino acid sequence of calf histone H2A (Yeoman *et al.*, 1972; Sautiere *et al.*, 1974). Complete sequence data are also available for H2A of rat (Sautiere, 1975; Laine *et al.*, 1976) and trout (Bailey and Dixon, 1973). Only amino acid composition data are available for pea H2A (Hayashi *et al.*, 1977). Deviations (substitutions and insertions) of these sequences from the calf sequence are indicated in the solid outline blocks; deletions are indicated by the dotted outline blocks.

HISTONE H2B

```
                                                          10
Calf        HN-Pro Glu Pro Ala Lys Ser Ala Pro ----Ala-Pro Lys-Lys ---- Gly Ser-Lys Lys-Ala ---
Trout       HN-Pro Glx Pro Ala Lys Ser Ala Pro ----------- Lys-Lys --- Gly Ser-Lys Lys-Ala ---
Drosophila  HN-Pro ---- Pro ---- Lys Thr Ala Gly Lys-Ala-Ala Lys-Lys Ala Gly --------- Lys-Ala Glx
Pea         (Ala,Glx,Pro,Ala-Lys)              (Lys,Lys,Pro,Lys)(Lys,Leu,Pro,Lys) (Lys,---,---,

                       20                                          30
Calf        ------ Val Thr-Lys ---- Ala Gln Lys Lys Asp Gly-Lys-Lys-Arg-Lys-Arg-Ser-Arg-Lys-Glu-
Trout       ------ Val Thr-Lys Thr-Ala Gly Lys ---- Gly Gly Lys-Lys-Arg-Lys-Arg-Ser-Arg-Lys-Glu-
Drosophila  Lys-Asx Ilu Thr-Lys Thr---- Asx Lys Lys
Pea         Lys,Asx                (Gly,Glx,Lys)   (Asx,Ile,Lys)(Lys,Arg)(Lys,Arg)(Lys,Lys)-Asx

                              40                             50
Calf        Ser Tyr-Ser-Val --- Tyr-Val-Tyr Lys-Val-Leu-Lys-Gln-Val-His-Pro-Asp Thr Gly-Ile-Ser
Trout       Ser Thr-Ala-Ile*
Pea         Ser Tyr-Thr-Val Lys Ile-Ile-Phe Lys-Val-Leu-Lys-Gln-Val-His-Pro-Asp Ile Gly-Ile-Ser

                       60                                70
Calf        Ser-Lys-Ala-Met-Gly-Ile-Met-Asn-Ser-Phe-Val-Asn-Asp-Ile-Phe-Glu Arg-Ile Ala Gly Glu
Pea         Ser-Lys-Ala-Met-Gly-Ile-Met-Asn-Ser-Phe-Val-Asn-Asp-Ile-Phe-Glu Lys-Leu Ala Ser Glu

                       80                             90
Calf        Ala-Ser-Arg-Leu-Ala His Tyr-Asn-Lys Arg Ser-Thr-Ile-Thr Ser Arg-Glu-Ile-Gln-Thr-Ala
Pea         Ala-Ser-Arg-Leu-Ala Arg Tyr-Asn-Lys Lys Ser-Thr-Ile-Thr Pro Arg-Glu-Ile-Gln-Thr-Ala

                       100                           110
Calf        Val-Arg-Leu-Leu-Leu-Pro-Gly-Glu Leu Ala-Lys-His Ala-Val Ser-Glu-Gly-Thr-Lys-Ala-Val
Pea         Val-Arg-Leu-Leu-Leu-Pro-Gly-Glu Val Ala-Lys-His Lys-Ile Ser-Glu-Ala-Thr-Lys-Ala-Val

                       120           125
Calf        Thr-Lys Tyr Thr-Ser Ser-Lys COOH
Pea         Thr-Lys Phe Thr-Ser Gly-Ala COOH
```

FIG. 5. Amino acid sequences of histone H2B from calf (Iwai *et al.*, 1972), trout (Candido and Dixon, 1972; Kootstra and Barley, 1976), *Drosophila* (Elgin, Goodfleisch, and Hood, unpublished data cited by Elgin and Weintraub, 1975), and pea (Hayashi *et al.*, 1977). Insertions and substitutions are indicated in solid outline blocks; deletions are indicated in dotted outline blocks. Peptides from the pea H2B for which only composition data are available are enclosed in parentheses. In general, these peptides from the pea N-terminal region appear to align better with the *Drosophila* than with the calf sequence. The asterisk denotes that trout and calf sequences are identical beyond residue 39 except for an Ala to Ser substitution at residue 77.

Considered alone, the highly conserved core histone sequences, especially of H3 and H4, are suggestive of stringently controlled, conformationally fixed structural elements. In fact, an enormous variety of conformers are made possible through the device of posttranscriptional modifications which include acetylation (*N*-acetyllysine), phosphorylation (phosphoserine), and methylation (mono-, di-, and trimethyllysine). Each modification introduces new steric constraints, alters the charge or basicity of the groups involved, and thus must significantly affect neighboring DNA and/or protein contacts. The subject is exceedingly complex and cannot be treated in detail here. Suffice it to say that all histones are subject to *in vivo* modifications. Significantly, the most sequence-conserved histones, H3 and H4, are the most highly modified. For example, H4 can be acetylated at up to five distinct sites, methylated at two sites, and phosphorylated at one site, yielding 240 possible variants if all combinations of modifications are allowed [for reviews, see Bradbury and The Biophysics Group

(1975); Dixon *et al.* (1975); Elgin and Weintraub (1975); Ord and Stocken (1975); and Ruiz-Carillo *et al.* (1975)].

Major advances have been made in establishing the temporal modification of specific residues in different histones. In general, histone modifications are controlled by families of enzymes, rather than a single acetylase, kinase, deacetylase, phosphatase, etc. Turnover at most sites is rapid, with the exception of methylation and N-terminus acetylation. Multiple modifications of a single histone chain are not necessarily performed in concert, but instead may occur at distinctively different cell cycle stages or differentially, in response to external stimuli such as hormones. Certain modifications occur in the cytoplasm, and may or may not be retained in the nuclear chromatin, while others occur *in situ*. Histone modifications have been implicated in a variety of functions such as processing of histone precursors, histone transport from cytoplasm to nucleus, chromatin assembly, gene activation, facilitation of transcription and replication, and premitotic chromatin condensation. Whether these modifications are causitive or facilitative of any given process remains a major question. Clearly, in spite of sequence conservation, considerable fine tuning of histone conformations is possible, and is presumably essential for chromatin assembly and function. In this regard, it is most interesting to note that all modifications characterized to date occur in the basic regions of histones chains, suggesting that this fine tuning involves mainly alterations of histone–DNA affinities and contacts, rather than of histone–histone contacts that appear essential for defining the stoichiometry and globularity of the nucleosome core (see below).

C. HISTONE CONFORMATIONS AND SELF-ASSOCIATION PROPERTIES

Unlike familiar globular proteins, histones exist in native conformations only in the very special environment of the chromatin structure defined by neighboring histones and requirements for DNA binding. Since typical spectroscopic methods cannot distinguish between histones, only estimates of average, general structural features such as α-helix or β-sheet content can be obtained by studies of whole chromatin, and even these may be complicated by an uncertain background spectrum of DNA. Questions pertaining to native histone conformation and mechanisms of chromatin assembly have been approached by examining the conformational properties of individual histones, with the expectation that these properties will reflect at least part of the histone contribution to the native structure.

Conformational properties of histones have been studied extensively, both theoretically and experimentally. The "theoretical" approach, in its simplest form, involves educated guessing of the likely conformation features of local sequence regions based on general principles deduced from comparisons with polypeptides and typical globular proteins. Such deductions are considerably aided by two-dimensional projections of the chain emphasizing various spatial relationships between side chains. The plots shown previously (Fig. 3) emphasized the markedly asymmetric distributions of basic amino acids along the chains that lead to an expectation of a bipartite character for histone conformations that persists in present models for nucleosome structure and assembly. That is, in the absence of external influences (e.g., DNA), sequences containing the majority of basic residues are expected to resemble extended random coils, due to charge repulsions, while regions rich in apolar residues have potential for helix or sheet formation. This latter potential for structure is apparent if histone sequences are represented either on helical wheels (Fig. 6) or as helical surfaces (Fig. 7).

The available experimental evidence supports this general view of histone conformations, but must be considered keeping two important aspects of histone chemistry in mind. The first is that individual histone fractions are almost invariably isolated under conditions that denature proteins. Second, all histones excepting H1 aggregate severely in the presence of modest concentrations of salt. As pointed out by Isenberg (1977), neither of these factors presents insurmountable difficulties in principle, since denatured proteins renature under suitable conditions, and the extent and rate of aggregation are minimized at low protein concentrations. In practice they render detailed comparisons of different studies difficult. Nonetheless many general aspects of histone conformations seem clear.

Acid-extracted histones at high dilutions ($\sim 10^{-5} M$ histone) in solutions of low ionic strength ($\mu \leq 10^{-3}$) retain little secondary structure. Addition of simple salts above a critical level (varying with salt and histone) generally causes two major conformational effects—rapid chain folding and α-helix formation, followed by slow aggregation with β-sheet formation—that are distinguished by comparisons of fluorescence anisotropy, circular dichroism, and light scattering properties. The observed order of rates of aggregation at these concentrations is reportedly H4 $\cong$ H3 > H2A > H2B (no aggregation observed), with the actual rates dependent strongly on temperature and salt concentration (D'Anna and Isenberg, 1972, 1974a,c; Li et al., 1972; Wickett et al., 1972; Small et al., 1973; Smerdon and Isenberg, 1973, 1974; for review, see Isenberg, 1977).

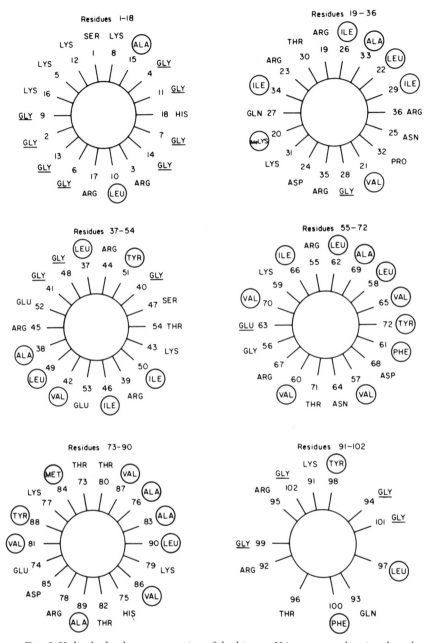

FIG. 6. Helical wheels representation of the histone H4 sequence showing the relative positions of residues in a plane transverse to the helix axis, assuming that the residues lie on an α-helix. Clustering of hydrophobic residues (circled) about the circle indicates a high potential for helix formation (Schiffer and Edmundson, 1967). Helix-destabilizing *glycine* residues are underlined for emphasis. (Reproduced from Bradbury and Crane-Robinson, 1971, with permission.)

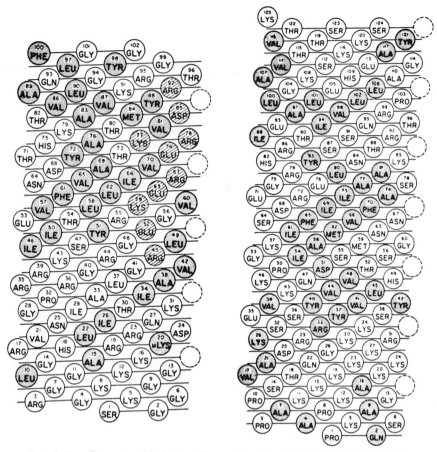

FIG. 7. Two-dimensional helical surfaces of histones H4 (left) and H2B (right). Note the strong tendency for grouping of basic residues in the N-terminal sequence regions, and of apolar residues (shaded) in the middle and C-terminal sequence regions. An unusual alternating arrangement of acidic and basic residues occurs along the hypothetical helical surface of H4 (hatched residues). (Reproduced from Bradbury and Crane-Robinson, 1971, with permission.)

Nuclear magnetic resonance (NMR) studies have provided details of the structure-forming potential of specific histone sequence regions. Interpretation of histone NMR spectra has been based principally on the fact that resonance linewidths are a sensitive function of the configurational mobility of the resonant atom. Atoms with a high degree of configurational mobility exhibit narrow resonances similar in shape to those found for modest-sized random coil polypeptides, whereas those confined by the protein structure will have broadened resonances, with the degree of broadening depending on the degree of rigidity and the tumbling time of the whole structure. Further broadening and apparent loss of peak intensity of such resonances may occur because of shift anisotropy in the structured protein environment. NMR spectra (both ^{1}H and ^{13}C) of all histones characteristically contain sharp resonances superimposed on broad, relatively featureless backgrounds. The linewidths of the sharper resonances are considerably narrower than those found even for small globular proteins, and approach widths expected for random coil polypeptides, providing strong support for the bipartite, random coil plus globular, model of histone conformations. Additionally, computer simulations using known group resonance frequencies, and either narrow or appropriately broad linewidths, reasonably mimic the observed spectra and can be used to pinpoint fairly precisely the boundaries of the random coil and "structured" regions of histone chains.

Histones (excepting H1) aggregate at the concentrations required for NMR measurements (10–100 mg/ml). A basic problem, then, in interpreting these simulated NMR spectra is the question of whether aggregation is solely, or only partially, responsible for structure formation, i.e., immobilization of groups. Although there is some disagreement about the origins of broad resonances observed for histones in low ionic strength solutions, there is general agreement that the basic N-terminal histone tails (and the basic C-terminal tails, where they exist) have close to random coil configurations, while the nearly neutral more hydrophobic central sequence regions are structured and/or aggregated. More specific assignments are summarized in Table II. Addition of simple salts promotes aggregation that does not involve the basic tails (Bradbury *et al.*, 1972, 1975a,b; Bradbury and The Biophysics Group, 1975; Bradbury and Rattle, 1972; Lewis *et al.*, 1975; Lilley *et al.*, 1975; Tancredi *et al.*, 1976; Morris and Lewis, 1977).

Electron microscopic observations (Sperling and Bustin, 1974, 1975, 1976; Sperling and Amos, 1977) have shown that the aggregation of core histones is not random, but instead leads to relatively well-organized, rodlike structures. Although H3 and H4 form aggregates more readily than H2A and H2B, the aggregation pattern of all four histones is similar. As protein concentration or ionic strength is increased, the first aggregates found are 22 Å diameter bent rods, followed by 40–80 Å diameter curly fibrils appearing to consist of two interwound cables, and finally by larger more linear bundles.

TABLE II
STRUCTURED HISTONE REGIONS

Histone	Predicted structure[a]				Structured regions based on NMR data	Reference[c]
	α-Helix regions	β-Sheet regions	Max/min (%) α[b]	Max/min (%) β[b]		
H2A	9–15, 47–66, 81–88, 91–97, 118–127	23–27, 30–34, 76–79, 100–104, 111–116	40/22	34/19	25–113	1
					40–70, 96–116	2
H2B	15–24, 69–82, 93–102, 105–113	39–48, 61–66, 88–90, 117–122	35/21	27/20	31–102	3
					36–48, 55–78, 91–125	2
					50–108 (±5)	4
H3	16–27, 45–53, 58–65, 67–79, 88–98	99–104, 109–113, 117–120, 124–128	39/16	25/15	42–110	5
					21–26, 47–105	2
H4	15–22, 31–39, 57–67	26–30, 46–51, 69–73, 80–90, 96–100	28/11	37/31	33–102	6
					~70–102	7
					25–67 (nucleus for structure)	8
					26–102	2

[a] (Fasman et al., 1976).

[b] The maximum percentage α and minimum percentage β are calculated from the predicted α and β regions listed in the previous columns. The minimum percentage α is calculated from the predicted stable helices which are unaffected by charge repulsion. The maximum percentage β is calculated from the predicted β regions plus new β regions arising from potential α → β transitions (see Fasman, et al., 1976).

[c] Key to references: 1, Bradbury et al., 1975b;   2, Lilley et al., 1975;   3, Bradbury et al., 1972;   4, Tancredi et al., 1976;   5, Bradbury et al., 1973a;   6, Bradbury and Rattle, 1972;   7, Pekary et al., 1975;   8, Lewis et al., 1975.

D. SPECIFIC HISTONE–HISTONE INTERACTIONS

Much early chromatin research was devoted to developing methods for purifying individual histones and for preparing chromatin selectively depleted of specific histones. A recurrent "problem" in these attempts was the coextraction and coelution of histones H3 with H4, and of H2A with H2B. Noted also was the ability of certain histones to inhibit aggregation of others. These and related observations led to the suspicion, now confirmed, that core histones form specific mixed complexes.

Utilizing sedimentation, circular dichroism, and fluorescence methods, Isenberg and co-workers systematically examined interactions between pairs of acid-extracted histones at high dilutions (1–$5 \times 10^{-5}\, M$ total protein) as the ionic strength was increased after mixing of the histones in distilled water. Strong 1:1 association occurred between the pairs H3 + H4, which form $(H3)_2(H4)_2$ tetramers with an association constant of $\sim 10^{21}\, M^{-3}$, and between the pairs H2B + H4, and H2A + H2B, which form dimers with association constants of $\sim 10^6\, M^{-1}$ (D'Anna and Isenberg, 1973, 1974b,d,e; Baker and Isenberg, 1976; Isenberg, 1977). Additional weaker, but detectable complexes formed between all other pairwise combinations of core histones. Similar results were obtained by Sperling and Amos (1977), who examined the association of acid-extracted histones by sedimentation velocity, chemical cross-linking, and electron microscopic methods. A most interesting finding of this latter work is the formation of long bent rods, 40–80 Å across, by the histone pairs (H3, H4) and (H2A, H2B), as well as mixtures of all four histones, at higher ionic strengths and protein concentrations than those utilized by Isenberg and co-workers.

Since the ability of acid-extracted histones to refold properly is open to question, procedures that take advantage of the propensity for coextraction and cochromatography of histone pairs have been utilized for the isolation of H3 + H4 and H2A + H2B mixtures. Sedimentation equilibrium and chemical cross-linking analyses of H3 + H4 and H2A + H2B fractions obtained by Sephadex G-100 chromatography have also demonstrated that the principal complexes formed by these pairs are tetramers and dimers, respectively, and suggest that H2A + H2B dimers may interact more weakly to form higher 1:1 oligomeric complexes (Kornberg and Thomas, 1974; Roark et al., 1974, 1976).

The conformation of the $(H3)_2(H4)_2$ complex is of particular importance with regard to mechanisms of nucleosome assembly and structure, discussed in more detail below. The frictional coefficient of this complex, derived from sedimentation studies, is exceptionally high for a globular protein (f/f_0 = 1.99 in 50 mM acetate, 50 mM bisulfite) and could be accounted for by a high degree of asymmetry, or by a partially globular and partially random coil (i.e., highly hydrated) structure. Strong evidence for the latter structure derives from the observation that the proton magnetic resonance spectrum of the complex is similar to those of partially denatured, globular proteins (Moss et al., 1976). Specifically, several aromatic resonances are clearly perturbed and broadened, while others, most critically the α-carbon proton resonances of glycine (located most frequently in the basic N-terminal histone sequences), are observed with the same chemical shift and with the same linewidth in both the native and denatured (6 M urea) complex. Similar studies of interactions between H3 and H4 polypeptide fragments have indicated that residues 42–120 of H3, and 38–102 of H4, are most important for complex formation (Bohm et al., 1977).

Circular dichroism and infrared measurements have confirmed the existence of partially structured, α-helical, domains containing approximately 29% of the tetramer amino acid residues (Moss et al., 1976). A very similar value was estimated for the α-helix content of the tetramer formed for acid-extracted histones (Baker and Isenberg, 1976). In marked contrast to histone self-aggregates, the H3 + H4 tetramer appears devoid of β-sheet structure.

Within chromatin the conformations of individual histones are influenced by histone–DNA interactions that neutralize basic histone charges and may promote secondary structure formation. Although details of histone conformations in situ are presently experimentally inaccessible, the potential of a given sequence region for structure formation can be estimated in a variety of ways. A more precise approach than the graphic methods illustrated previously takes advantage of statistical information, available from X-ray diffraction data on proteins, on the relative frequencies (probabilities) of occurrence of specific amino acids in α-helical or β-sheet conformations. One expects that a sequence region will be structured if it contains a preponderance of residues often found in structured regions in typical globular proteins. Fasman et al. (1976) have described a method for quantitating these expectations and have constructed conformational probability maps of histones, as illustrated in Fig. 8. Predicted regions of high helix and β-sheet potential of all five histones are included in Table II. In general these predictions are consistent with available experimental data. However, it is interesting to note that regions of high helix potential

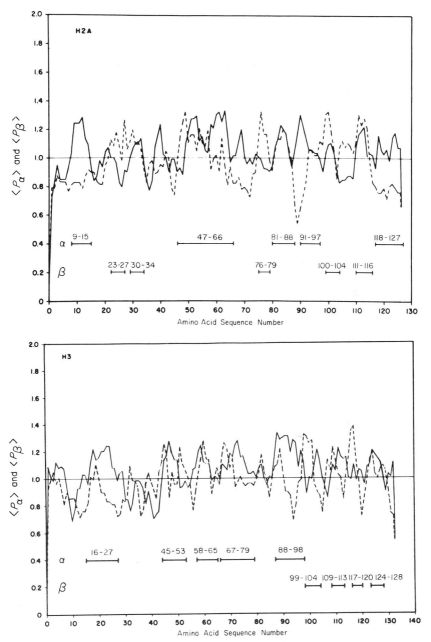

FIG. 8. The conformational probability profiles of histones H2A (top) and H3 (bottom). (——), helical potential $<P_\alpha>$ of tetrapeptide i to $i+3$. (------), β-sheet potential $<P_\beta>$ of tetrapeptide i to $i+3$. Regions with helical and β-sheet-forming potential lie above the cutoff point 1.0, with tetrapeptide breakers falling below 1.0. The predicted α and β regions are underlined. (Reproduced from Fasman *et al.*, 1976, with permission.)

exist in the basic N-terminal tails, as well as in the central regions, of all core histones. This potential could be realized in the presence, though not in the absence, of good charge shielding (e.g., by DNA).

III. Nucleosomes—Discovery, Isolation, and Composition

A. Evidence for a Semiperiodic Arrangement of Histones Along DNA

Chromatin structure is conveniently described in terms of the pattern of distribution of proteins along DNA, and the consequent folding scheme of the nucleoprotein fibril. Over the relatively brief span from late 1973 to early 1975, several independent lines of investigation converged to provide compelling evidence for a repeating structure in chromatin based on the semiperiodic arrangement of specific histone complexes along DNA.

Electron micrographs of chromatin fibrils streaming from osmotically shocked nuclei, centrifuged onto carbon grids, and negatively or positively stained revealed roughly spherical, 60–80 Å diameter particles (Fig. 9), termed "ν-bodies", connected by short, thin filaments (A. L. Olins and Olins, 1973; D. E. Olins and Olins, 1974; Woodcock, 1973). A similar "beads-on-a-string" appearance was observed in micrographs of H1-depleted chromatin (Oudet et al., 1975) and of the SV40 DNA complex with host histones, termed the SV40 "minichromosome" (Griffith, 1975).

Nucleases have proven to be extremely effective probes of chromatin structure because of the strong restricting effect of bound proteins on DNA reactivity (Clark and Felsenfeld, 1971). Slight digestion of chromatin in intact nuclei with endogenous nuclease (Hewish and Burgoyne, 1973) and micrococcal nuclease (Noll, 1974a) yielded relatively narrow distributions of DNA fragments averaging 200 bp, and integer multiples of 200 bp in length (Fig. 10), indicating the presence of well-defined protein arrays located at roughly periodic intervals along DNA. Furthermore, centrifugation of nucleoproteins from such digests on sucrose gradients revealed a polysomelike separation pattern of products (Noll, 1974a). More extensive digestion of chromatin with micrococcal nuclease (Rill and Van Holde, 1973; Sahassrabuddhe and Van Holde, 1974), or DNase II (Oosterhof et al., 1975) yielded predominantly small, compact ($s_{20,w} \cong 11$ S) nucleoprotein particles (originally termed "PS particles", based on their solubility properties) containing 120–140 bp of DNA associated with approximately equimolar amounts of H2A, H2B, H3, and H4. These appeared similar, by electron microscopy, to the repeating particles observed in intact chromatin fibrils (Van Holde et al., 1974a; Olins et al., 1975; Oudet et al., 1975).

Evidence based on X-ray scattering studies for some periodicity in chromatin structure considerably predated the nucleosome model. A characteristic pattern of rings at equivalent Bragg spacings of 110, 55, 37, 27, and 22 Å occurs in the X-ray scattering intensity profile of wet chromatin fibers. Until recently, these rings were attributed solely to DNA scattering, and were interpreted in terms of regular (Pardon et al., 1967; Richards and Pardon, 1970) or irregular (Bram and Ris, 1971) supercoiling of DNA. This interpretation has been altered by elegant neutron diffraction studies in which advantage was taken of the markedly different neutron scattering properties of H_2O and D_2O to vary the neutron scattering of the solvent so as to alternatively contrast match scattering from proteins (computed contrast match at 37.5% D_2O) and from DNA (computed match at 63.5% D_2O) (Baldwin et al., 1975). In 10% D_2O, the composition yielding nearly zero solvent scattering, the X-ray and neutron scattering intensity profiles are nearly identical. The 110 and 37 Å peaks disappear in 30% D_2O and reappear at higher D_2O concentrations, while the 55 and 27 Å peaks weaken progressively with increasing D_2O concentration, nearly disappearing in 100% D_2O. Thus these two pairs of rings arise from distinctly different scattering centers—the former predominantly from proteins, indicating a protein, rather than DNA, repeat array at 110 Å intervals; and the latter predominantly from DNA.

B. DEFINITION OF NUCLEOSOMES

These indications of particulate repeating structural elements in chromatin, coupled with the discovery of specific association complexes of histones, led to the first proposed models from chromatin subunits. Based, in part, on the potentially globular character of the $(H3)_2(H4)_2$ tetramer, the strong conservation of H3 and H4 sequences, and the difficulty of extraction of these histones from chromatin, Kornberg (1974) proposed that this tetramer forms the core of the repeat unit, with much of the ~200 bp of DNA of the repeat following a path on the tetramer. Two (H2A, H2B) pairs were presumed to complete the subunits and to define the path of the remaining DNA, because of the necessity for H2A and H2B for the regeneration of the full native X-ray scattering pattern from reconstituted nucleohistone complexes, and the roughly equimolar stoichiometries of these four histones in native chromatin. Histone H1 was suggested to be associated with, but not an integral part of, the subunit. Van Holde et al. (1974b) pro-

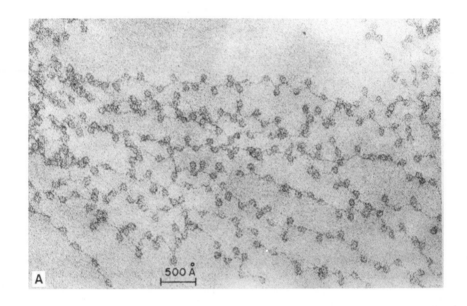

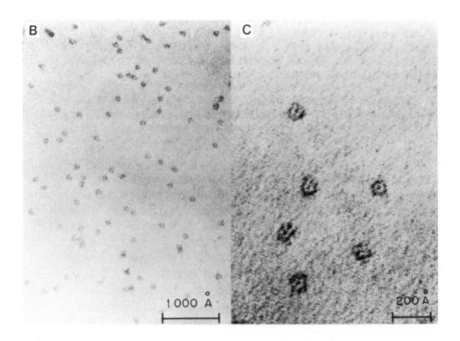

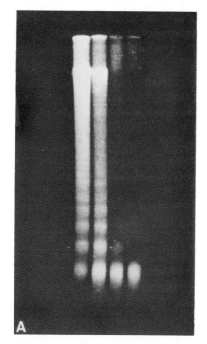
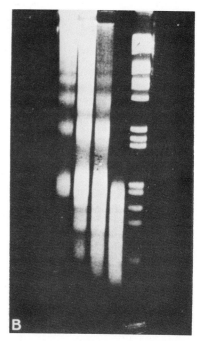

FIG. 10. Electrophoresis, on gels of 1.4% agarose (A) and 3.5% polyacrylamide (B), of DNA from Chinese hamster ovary *nuclei* digested with micrococcal nuclease for 20 seconds, 90 seconds, 10 minutes, and 60 minutes (left to right). Gel on far right contains Endonuclease R *Hae*III fragments of PM2 DNA. The sizes of the 11 smallest PM2 fragments are 642, 592, 498, 322, 288, 263, 160, 145, 117, 94, and 50 bp (Kovacic and Van Holde, 1977). Gels were stained with ethidium bromide and photographed under black light illumination. (Reproduced from Rill *et al.*, 1975, with permission.)

posed a more detailed model for the basic subunit, in which a specific complex between two copies each of histones H2A, H2B, H3, and H4 (formed by interactions between the more hydrophobic C-terminal regions of each chain) provides a hydrophobic core about which ~120 bp (now known to be 140 bp) of DNA is wrapped. The basic N-terminal histone tails were presumed to extend from the histone core,

FIG. 9. (A) High-resolution electron micrograph of a spread chicken erythrocyte nucleus. Note the *ν* bodies (nucleosome cores) with clear internal structure. Many chromatin fibers exhibit a zigzag configuration with the *ν* bodies lying on alternate sides of the connecting strand. Stained with aqueous 0.2% uranyl acetate. (B) Low-resolution and (C) high-resolution electron micrographs of monomer *ν* bodies obtained after micrococcal nuclease digestion and fractionation by sucrose gradient centrifugation. See also Olins *et al.* (1976b). (Reproduced from Olins, 1978, with permission.)

wrap along the DNA, probably in the major groove, and thereby assist in holding DNA about the core.

Although considerable additional information is now available, and more detailed models for chromatin structure have been proposed as described below, these early models have proved correct in essential details and serve to define the basic repeating unit of chromatin, now commonly referred to as the nucleosome. Literature usage of the term "nucleosome" has varied, owing to initial uncertainties in the DNA size of the repeat unit and the relationship between the octamer of core histones and the DNA repeat length, as is reflected by a comparison of the above two models. Closer inspection of the products of nuclease digestion as a function of digestion time have shown that the histone octamers are spaced at intervals averaging 180–210 bp in chromatin from most sources, but are most tightly associated with only 140 bp. The remaining 40–70 bp in the repeat appears associated primarily with lysine-rich histones (Varshavsky et al., 1976a; Whitlock and Simpson, 1976; Noll and Kornberg, 1977) and is probably conformationally distinct from the core-histone associated DNA (see further below). In particular, this spacer DNA is easily extended and is evident as a short, thin fibril connecting globular units in electron micrographs of chromatin from osmotically shocked nuclei or isolated chromatin depleted of histone H1 (i.e., the "string" of the "beads-on-a-string" structure). Although the term nucleosome was originally used to refer to the globular elements of the structure (Oudet et al., 1975), these elements are now commonly referred to as nucleosome cores. Nucleosomes are defined here as the core plus spacer DNA, and any other proteins (e.g., H1) associated in stoichiometric amounts with an adjacent core.

C. MICROCOCCAL NUCLEASE DIGESTION OF CHROMATIN AND THE ISOLATION OF NUCLEOSOMES AND CORES

Micrococcal nuclease digestion of chromatin provided some of the earliest evidence for the nucleosome structure and has continued to be extremely useful for the isolation of nucleosomes and cores and, more recently, of specific nucleosome fragments. The feature that distinguishes this from other nucleases (e.g., DNase I and II) is a strong preference for intercore (spacer) DNA. Thus brief digestion of chromatin in nuclei, particularly at 0°C, yields a characteristically discontinuous distribution of DNA fragments that are unit and integer multiple nucleosome repeat length. The initial scissions of mono- and oligonucleosomes from bulk chromatin are followed by rapid trimming from the ends, most likely because of the exonucleolytic activity of micrococcal nuclease, and by cleavages at specific sites within the core (Figs. 10 and 11) (Noll, 1974a; Axel, 1975; Sollner-Webb and Felsenfeld, 1975; Shaw *et al.*, 1976; Noll and Kornberg, 1977).

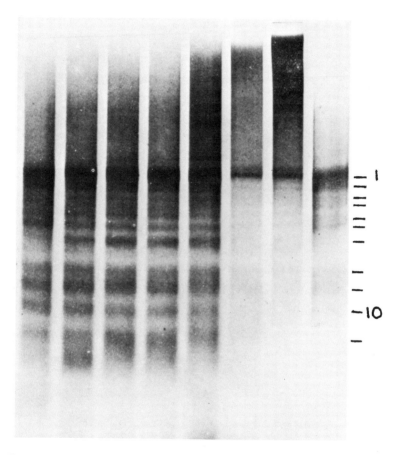

FIG. 11. Kinetics of digestion by micrococcal nuclease of chromatin isolated from chicken erythrocyte nuclei illustrating DNA products of internal digestion of nucleosomes. DNA from samples digested to 8, 16, 21, 30, 31, 35, and 43% acid solubility (second from right to left) was electrophoresed on a 6% polyacrylamide slab gel (stained with Stains-all, Eastman). At the far right is a marker from a limit digest of chromatin (50% acid solubility). Sizes of fragments from the limit digest numbered from 1 to 11, are 157, 138, 126, 118, 106, 99, 88, 68, 58, 48, and 38 bp (Camerini-Otero *et al.*, 1976). A similar distribution of small DNA fragments is produced upon digestion of nuclei, but the appearance of these fragments is delayed (see Fig. 10 and text). Digestion of isolated chromatin generally does not yield bands that are multiples of the nucleosome repeat length. (From R. D. Camerini-Otero *et al.*, 1976, *Cell* **8**, 333–347, reproduced with permission. Copyright © 1978 by MIT Press.)

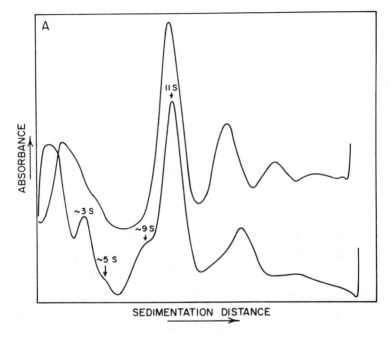

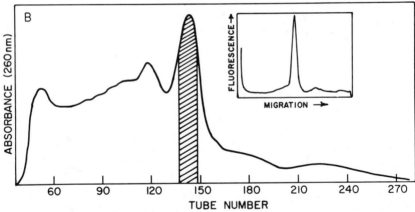

FIG. 12. (A) Fractionation of chromatin digests by centrifugation on sucrose gradients. Upper curve is from chicken erythrocyte nuclei digested to approximately 20% acid solubility with micrococcal nuclease, sedimented on a 5–20% linear sucrose gradient at 36,000 rpm for 13 hr (6°C) in an International SB-283 rotor. Bottom curve is from chicken erythrocyte chromatin (isolated in gel form) digested to approximately 30% acid solubility and sedimented on a 5–20% linear sucrose gradient in the above rotor at 40,000 rpm for 10 hr (6°C). Subnucleosomes with approximate sedimentation coefficients of 3 S, 5 S, and 9 S may be noted in the profile of the chromatin digest. (Reproduced from Rill *et al.*, 1978a, with permission.) (B) Fractionation of chromatin digests by gel filtration. Chromatin stripped of H1 and H5 by the method of Bolund and Johns

Because of their discreteness, mono- and oligonucleosomes are readily separated by a variety of procedures, such as sedimentation in density gradients (Fig. 12) using swinging bucket (Noll, 1974a; Rill *et al.*, 1975) or zonal (Olins *et al.*, 1976a) rotors, chromatography on agarose (Fig. 12) (Shaw *et al.*, 1974), or electrophoresis on polyacrylamide gels (Varshavsky *et al.*, 1976a; Todd and Garrard, 1977). As described in detail elsewhere (Rill *et al.*, 1978a), the composition of the "mononucleosome" fractions ($s_{20,w} = 11 \pm 1$ S) depends considerably on the conditions and extent of digestion, and even on the chromatin source, because of the variability in the relative rates of intercore versus intracore cleavages and end trimming. Typically this fraction contains nearly intact nucleosomes after digestion to only a few percent acid solubility, and contains predominantly cores and trimmed nucleosomes after digestion to 15–20% acid solubility. Internal cleavages occur throughout the digestion, but are most evident in the later stages. Although no systematic study has been reported, the relative rates of these internal cleavages appear particularly sensitive to digestion conditions, and certain tissue variations are evident (D. A. Nelson, R. L. Rill, and D. K. Oosterhof, unpublished observations). Small amounts of quite monodisperse nucleosomes and cores can be obtained by electrophoretic procedures, and larger amounts by other methods with careful control of digestion conditions.

Appreciation of the unique nucleosomal location of H1, and its role in condensing chromatin, has led to improved techniques for isolating very pure nucleosome cores. Nucleosomes retaining H1 (or H5) are precipitated at physiological salt concentrations; thus cores are recovered simply by addition of salt (KCl) to solubilized digests (Olins *et al.*, 1967a; Whitlock and Simpson, 1976) and may be further purified by centrifugation. Alternatively, higher concentrations of salt (0.4–0.6 M KCl) selectively strip H1 and H5 from chromatin (Ohlenbusch *et al.*, 1967), and recovery of the stripped chromatin is facilitated by adsorption of free H1 by a cation exchange resin (Bolund and Johns, 1973). Because of the absence of bound histone, cleavage and trimming of intercore DNA are greatly enhanced, relative to intracore

---

Fig. 12   (*Continued*)
(1973) was digested briefly with micrococcal nuclease, concentrated to $130\,A_{260}$ units/ml, and 20 ml was applied to a 90×5-cm column of Bio-Gel A5m (200–400 mesh; flow rate, 30 ml/hr; fraction volume, 3.2 ml). Elution was performed in a cold room. Inset: Purity of nucleosome core DNA. DNA from the nucleosome core peak (hatched area in B) was electrophoresed on an 8% polyacrylamide gel under nondenaturing conditions. The gel was stained with ethidium bromide and scanned with a Gelman DCD-16 scanning densitometer operated in the fluorescence mode. The median DNA length was 143 ± 5 bp, based on a comparison with endonuclease R *Hae*III fragments of ϕX174 RF DNA.

TABLE III
DNA Lengths in Nucleosomes from Various Sources[a,b]

Cell type	DNA content of nucleosome (bp)	Reference
Aspergillus	154	Morris (1976)
Rabbit cortical neuron	162	Thomas and Thompson (1977)
Yeast	165, 163	Thomas and Farber (1976)
Neurospora		Lohr *et al.* (1977a)
Neurospora	170	Noll (1976)
Physarum	171, 190	Compton *et al.* (1976); Johnson *et al.* (1976)
Cells grown in culture		
CHO	177	Compton *et al.* (1976)
HeLa	183, 188	Lohr *et al.* (1977a)
Hepatoma	188	Compton *et al.* (1976)
Teratoma	188	Compton *et al.* (1976)
P815	188	Compton *et al.* (1976)
Myoblast	189	Compton *et al.* (1976)
CV-1, exponentially growing or confluent	189	Compton *et al.* (1976)
BHK	190	Compton *et al.* (1976)
Rat kidney, primary culture	191	Compton *et al.* (1976)
Myotube	193	Compton *et al.* (1976)
C6, exponentially growing or confluent	198	Compton *et al.* (1976)
Rat bone marrow	192	Compton *et al.* (1976)
Rat fetal liver	193	Compton *et al.* (1976)
Rat liver	198, 196	Noll and Kornberg (1977); Thomas and Farber (1976); Compton *et al.* (1976)
Rat kidney	196	Compton *et al.* (1976)
Syrian hamster liver	196	Compton *et al.* (1976)
Syrian hamster kidney	196	Compton *et al.* (1976)
Chick oviduct	196	Compton *et al.* (1976)
Rabbit cortical glia	197	Thomas and Thompson (1977)
Tetrahymena micro- and macronuclei	199	D. J. Mathis and M. A. Gorovsky, unpublished
Rabbit cerebellar neuron	200	Thomas and Thompson (1977)
Stylonychia micronucleus	202	Lipps and Morris (1977)
Chicken erythrocyte	198, 207, 212	Lohr *et al.* (1977a); Compton *et al.* (1976); Morris (1976)
Sea urchin gastrula	218	Spadafora *et al.* (1976)
Stylonychia macronucleus	220	Lipps and Morris (1977)
Sea urchin sperm	241	Spadafora *et al.* (1976)

[a] Reprinted, by permission from R. D. Kornberg, 1977, *Annu. Rev. Biochem.* **46**, 931–954. Copyright © 1977 by Annual Reviews, Inc.

[b] The values shown are from studies in which standards of known nucleotide sequence and methods for eliminating the effect of degradation were used. [Studies in

cleavage (Noll and Kornberg, 1977), and sedimentation or gel permeation chromatography (Fig. 12) readily yields large quantities of well-defined monomeric cores (Tatchell and Van Holde, 1977; Klevan and Crothers, 1977) plus some dimeric cores that lack intervening DNA (Klevan and Crothers, 1977).

D. DNA COMPOSITION

Originally, considerable uncertainty and controversy developed over the basic repeat length of chromatin subunits. This has largely been dispelled by better understanding of digestion events and the recognition of a distinction between the well-defined nucleosome core and the less understood intercore regions. In addition, adoption of sophisticated gel electrophoresis methods and agreement on calibration standards have made possible precise and rather absolute comparisons of the repeat lengths of chromatin from various sources. Suspected species, and even tissue, variations in repeat lengths have now been confirmed, as illustrated in Table III. Although there appears to be a trend toward smaller repeat lengths in more transcriptionally active cells, there are several important exceptions, and no fully acceptable general statements about the significance of these variations can be made at present. The most commonly observed repeat lengths in higher eukaryotes are 190–200 bp, while lower values (170–180 bp) are more common in lower eukaryotes and certain cultured cell lines.

To complicate matters further, considerable evidence has now accumulated, indicating variations in intercore DNA lengths within chromatin from a single tissue, and even within a single cell line. Careful analyses of the apparent nucleosome repeat length with increasing digestion (Lohr *et al.*, 1977a,b) and of the lengths of thin

---

Table III (*Continued*)

which one or both precautions were not taken give values in approximate agreement with those shown and are useful for extending the range of cell types in which nucleosomes are known to occur (Axel, 1975; McGhee and Engel, 1975; Gorovsky and Keevert, 1975b; Honda *et al.*, 1975; Rill *et al.*, 1975, 1977; Keichline *et al.*, 1976; Woodcock *et al.*, 1976; V. M. Vogt, and Braun, 1976.)] In the case of *Physarum*, where the values shown differ by more than 10%, the smaller was obtained alongside values for rat liver and chicken erythrocyte that agree well with the results from independent studies, so the larger value is probably in error. In other cases where results from two or more studies are available, the discrepancies are 5% or less. Although many of the values may be accurate to even better than 5%, the DNA content of a nucleosome need not be correspondingly precise; the values could represent the means of distributions (see text).

fibers joining the globular elements in electron microscopic images of chromatin fibrils (Johnson *et al.*, 1976) suggested variations in inter-core DNA within chromatin from a single source. This has been confirmed by studies of the products of redigestion of electrophoretically separated polynucleosomes, and by trimming of the ends of isolated dinucleosomes by exonuclease III. Although the major repeat in bovine chromatin is 191 bp, minor repeats exist that are both larger (203 ± 9 bp) and smaller (170 ± 8 and 142 ± 3 bp) than average (Todd and Garrard, 1977). Exonuclease III readily trims polydisperse nucleosome DNA to 140 bp core size, but the length distribution of similarly trimmed dinucleosomes remains fairly disperse, indicating that the original band breadth is a consequence both of partially random initial cleavage and variations in the intercore DNA length (Prunell *et al.*, 1978; Riley and Weintraub, 1978). Particularly intriguing, in view of trend toward smaller (on the average) repeats in highly active tissues, are indications of regions containing repeated core particles without intervening, H1-covered DNA (Varshavsky *et al.*, 1976a; Todd and Garrard, 1977; Rill and Nelson, 1978).

Under appropriately controlled conditions, little background is observed in the multiple-band electrophoretic pattern obtained from mildly digested nuclei (Noll, 1974a; Lohr *et al.*, 1977b), indicating that most chromatin DNA is packaged into nucleosome repeats. More compelling evidence for nearly universal nucleosome packaging has been derived from hybridization experiments demonstrating that the distribution of specific classes of DNA follows the general DNA distribution after nuclease digestion. Numerous experiments of this sort assaying for unique copy DNA (active and inactive), integrated viral DNA, tandemly repeated (e.g., ribosomal) DNA, middle repetitive DNA, and highly repetitive DNA have uncovered no exceptions to the nucleosome theme (Axel *et al.*, 1975; Lacy and Axel, 1975; Bostock *et al.*, 1976; Garel and Axel, 1976; Lipchitz and Axel, 1976; Mathis and Gorovsky, 1976; Piper *et al.*, 1976; Reeves and Jones, 1976; Reeves, 1976; Tien-Kuo *et al.*, 1976; Musich *et al.*, 1977).

The nucleosome core has been defined operationally, in essence, as the fragment of chromatin that is most resistant to micrococcal nuclease. As described previously, the initially disperse distribution of nucleosomal DNA, centered about the repeat size, rapidly narrows first to a doublet with peaks at 160–170 and 140 bp, then to a sharp singlet at 140 bp [with concurrent appearance of small discrete DNA (Fig. 11)]. Despite wide variations in repeat lengths, this accumulation of 140 bp DNA has been noted in all digestion studies reported, indicating that this is either a unique stopping point or recognition site for micrococcal nuclease (since it is both an exo- and endonuclease). The findings that DNase I and DNase II (both endonucleases) do not preferentially yield 140 nucleotide DNA fragments (Noll, 1974b; Altenburger *et al.*, 1976), while exonuclease III rapidly trims nucleosomes to 140 bp, (and much less readily to less than 140 bp), suggests that the accumulation of 140 bp cores in micrococcal nuclease digests is due to exonuclease activity encountering a stopping point (Riley and Weintraub, 1978). This invariant stopping point and the relative resistance of the remaining particle undoubtedly reflect regions of tightest binding of core histones in the nucleosome and the invariance of a large portion of the core histone chains.

E. HISTONE COMPOSITION

The stoichiometry of two copies each of H2A, H2B, H3, and H4, initially suggested by Kornberg (1974), has now been confirmed by direct analysis of isolated nucleosomes, using electrophoretic methods coupled with accurate determinations of staining constants (Olins *et al.*, 1976a; Weintraub, 1978) and specific activities (Weintraub, 1976; Rall *et al.*, 1977). Cross-linking of either intact or partially digested chromatin with certain diimidates preferentially yields histone octamers and multiples of octamers with equimolar amounts of core histones, providing a different sort of evidence for the organization of core histones (Kornberg and Thomas, 1974). Reconstitution experiments have also shown that all four histones are required to regenerate all of the properties of isolated cores (see Section IV,E).

Some uncertainty still exists over the H1 content of nucleosomes. Analyses of whole chromatin and isolated nucleosomes generally indicate an average of one H1 molecule per core, but such determinations are complicated by potential for loss of H1 by proteolysis or dissociation and by uncertainties in staining constants. The discovery of dinucleosomes lacking H1 (Varshavsky *et al.*, 1976a; Klevan and Crothers, 1977) and of mononucleosomes containing High Mobility Group (HMG) nonhistone chromosomal proteins in the place of H1 (Rill and Nelson, 1978; Jackson *et al.*, 1979) raise the possibility of nucleosomes containing two H1's. Such species would be difficult to detect even by electrophoretic methods. Erythrocyte chromatin may be a special case, since Olins *et al.* (1976a) and Weintraub (1978) have shown that chicken erythrocytes contain close to 1 mole each of H1 and H5 per nucleosome.

IV. NUCLEOSOME STRUCTURE

A. GENERAL STRUCTURAL FEATURES AND
PHYSICAL PROPERTIES

Early hydrodynamic and electrooptic measurements on nuclease-resistant particles from chromatin (Rill and Van Holde, 1973; Sahasrabuddhe and Van Holde, 1974), soon correlated with the ~80 Å diameter globular particles observed along chromatin fibrils (Olins and Olins, 1974; Van Holde *et al.*, 1974a; Olins *et al.*, 1975, 1976b; Oudet *et al.*, 1975), indicated a high degree of compaction of DNA within nucleosome cores (ν bodies). Scanning transmission electron microscopy (Langmore and Wolley, 1975) provided some of the first experimental evidence for a disklike, rather than sperical, shape of nucleosomes, and for the peripheral location of DNA as proposed by Kornberg (1974) and Van Holde *et al.* (1974b).

Development of improved methods for isolating large quantities of nucleosome cores made possible the application of the whole gamut of physical techniques typically used to study well-defined macromolecules. X-Ray diffraction and electron microscopic analyses of small crystals of cores, and partially degraded cores, have yielded preliminary information about crystal symmetry ($P2_12_12_1$) and packing (in pseudohexagonally arranged columns) and core dimensions (110 × 110 × 57 Å) (Finch *et al.*, 1977). Electron density maps of the crystal plane (bc) perpendicular to the column axes indicate that the cores are wedge shaped (Fig. 13), as would be the case if DNA were wrapped in more than one, but less than two, superhelical turns about the protein core. From the shape and dimensions of the wedge, Finch *et al.* (1977) have suggested that the DNA is coiled in about one and three-quarters turns, with a pitch and average diameter of 28 and 90 Å, respectively; each full superhelical turn then containing 75–82 bp. The pitch of this helix is sufficiently small to permit easy coupling of one turn with another through histone–histone and histone–DNA interactions. The two halves of the core appear related by mirror symmetry, suggesting the existence of a dyad axis or pseudodyads perpendicular to the superhelix axis. At present resolution (~20 Å) the X-ray data alone do not distinguish between protein and DNA. Unfortunately the occurrence of three cores in the unit cell (MW $\simeq$ 600 000) indicates that high refinement will be extremely difficult.

The relative dispositions of protein and DNA have been established unequivocally by neutron scattering studies of solutions of cores and nucleosomes, using the H_2O–D_2O contrast variation

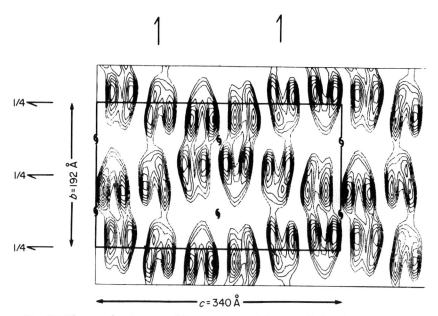

FIG. 13. Electron density map of the projection of the crystal of nucleosome cores in the direction of the *a* axis of the unit cell. (Reproduced from Finch *et al.*, 1977, with permission.)

method described previously. The larger radius of gyration of DNA (~50 Å) compared to protein (~30 Å) clearly locates DNA on the periphery of the core (Pardon *et al.*, 1975, 1977; Hjelm *et al.*, 1977; Suau *et al.*, 1977). Although neutron and X-ray solution scattering measurements do not uniquely determine a three-dimensional structure, the observed angular dependence of the scattering can be compared with that predicted for models differing in shape and relative distribution of protein and DNA scattering densities. The number of possibilities is severely limited by these methods (coupled with other known restrictions on the shape, volume, overall density, etc.), and the best-fit models deduced are in excellent agreement with the X-ray structure and previous electron microscopic measurements (see Table IV). Thus, there is little doubt about the general size and shape of the nucleosome core and the general DNA folding scheme, although questions remain about the precise pattern of DNA folding and conformations of the core histones.

Hydrodynamic data (Table V) are consistent with a compact core configuration, but indicate an unrealistically high degree of asymmetry, if hydration is neglected in interpreting the frictional ratio. Assumption of a spherical shape with a degree of hydration approaching

TABLE IV
DIMENSIONS OF THE NUCLEOSOME CORE[a]

Method	Height (Å)	Diameter (Å)	Reference
Electron microscopy,[b] X-ray diffraction of crystals	57	110	Finch et al. (1977)
Neutron scattering[b]	50–55	110	Pardon et al. (1977)
Neutron scattering	60	110	Suau et al. (1977)

[a] Nucleosome cores contained ~140 bp length DNA.

[b] Electron density maps of crystalline cores indicate a cylindrical slightly wedged shape. Both models deduced from neutron scattering data are flat cylinders, with the DNA located on the outside of a compact histone core. Various types of spherical models do not well fit the neutron scattering curves.

that of ribosomal subunits is in good agreement with other estimates of measurements of particle diameters, and can be reconciled if the particles contain a central depression filled by immobilized water (in a hydrodynamic sense), as indicated by certain models deduced from neutron scattering data and electron microscopic views which show a darkly staining central region that can be interpreted as a stain-filled cavity (Olins et al., 1976b).

TABLE V
HYDRODYNAMIC DATA ON NUCLEOSOME CORES (ν_1)[a]

Parameter	Total ν_1 in 10 mM KCl buffer	KCl-soluble ν_1 in 100 mM KCl buffer
$s^0_{20,w'}$ Svedbergs	10.89 ± 0.28	11.41 ± 0.31
$D^0_{20,w}$ Ficks	3.44 ± 0.13	3.90 ± 0.13
$\bar{v}$ (ml/gm)	0.661 ± 0.006	(0.661 ± 0.006)[b]
$\overline{M}_w$ (S,D)	$226{,}734 \pm 11{,}095$	$209.547 \pm 9{,}634$
$\overline{M}_w$ (Equil)	$230{,}432 \pm 2{,}080$	$215{,}665 \pm 2{,}086$
$f_{20,w'}$ (gm sec^{-1} particle^{-1})	$1.176, 1.195$[c]	$1.037, 1.068$[c]
d (anhydrous sphere) (Å)	$78.0, 78.5$	$76.0, 76.7$
f/f_0	$1.60, 1.61$	$1.44, 1.47$
a/b, prolate axial ratio	$11.1, 11.4$	$8.2, 8.7$
a/b, oblate axial ratio	$13.8, 14.3$	$9.7, 10.3$
d (hydrated sphere) (Å)	$124.5, 126.6$	$109.8, 113.1$
δ_1 (maximum hydration) (gm H$_2$O/gm ν_1)	$2.03, 2.11$	$1.33, 1.45$

[a] From Olins et al., 1976a.

[b] v measured only for total ν_1; it is assumed to be the same for KCl-soluble ν_1.

[c] The first calculation of each pair is based upon $\overline{M}_w$ (S,D); the second, upon $\overline{M}_w$ (Equil).

B. THE CONFORMATIONS OF CORE HISTONES

Rough approximations of the position of total histones within the DNA periphery of the core have been made by subtraction of DNA contributions to the X-ray diffraction pattern (Finch et al., 1977), and by comparing the neutron scattering properties of models with the protein radius of gyration and neutron scattering profiles obtained through the H_2O/D_2O contrast variation method (Pardon et al., 1977; Suau et al., 1977). Both methods place the bulk of the histones ($\geq 75\%$) in the nucleosome center, occupying a volume similar to that of typical globular proteins, but do not provide details of the shape or symmetry of the complex.

Histones in the core are more highly α-helical than individual histones in solution, and even in reconstitution complexes with DNA. Measurements of the amide I$'$ and amide III$'$ vibrations by laser Raman spectroscopy indicate a helix content of $51 + 5\%$ (Thomas et al., 1977), in good agreement with estimates of 44 (Tatchell and Van Holde, 1977) to 50% (Thomas et al., 1977; Mencke and Rill, 1979) based on circular dichroism spectra. β-Sheet content is estimated as $13 + 9\%$ and 5% by the two methods, respectively (Thomas et al., 1977). These values deviate from the predictions by Fasman et al. (1976) of a maximum of 36% helix, 22% β-sheet, but it should be noted that some of the predicted β regions also have high helix potential. Recently we have isolated, from chromatin digests, subnucleosomal particles containing subsets of histones associated with specific less than core size DNA fragments. These have the following histone/DNA stoichiometries: $(H2A)(H2B)(H3)_2(H4)_2/105 \pm 15$ bp, $(H3)(H4)/72 \pm 10$ bp, $(H2A)(H2B)/54 \pm 9$ bp, $(H1)/66 \pm 10$ bp (Nelson et al., 1978b; Mencke and Rill, 1979). The circular dichroism spectra of all of these complexes, except that containing H1, are virtually identical to that of the nucleosome core from 200–240 nm, after subtraction of the DNA contribution, indicating that the helix contents of the H2A + H2B and H3 + H4 pairs in the core are very similar [44–47%, depending on reference spectra used (Mencke and Rill, 1979)].

Interpretations of circular dichroism spectra have been criticized because histones influence the spectrum of DNA from 260–300 nm and presumably also at lower wavelengths. Uncertainties in helix estimates from this source should amount to no more than a few percent, since the intensity of the nucleosome cores at 200–225 nm is more than ten times that of B-form DNA. The spectrum of DNA is only modestly influenced in this region by high concentrations of salt, which mimic the effects of histones on the spectrum from 260–300 nm (Hanlon et al., 1975).

Two extreme models for the histone disposition along DNA can be envisaged: one where the histones lie along DNA (e.g., in a groove) in extended conformations, with chain projections forming a loose network of histone–histone contacts in the nucleosome center; the other where histones form a fully globular central complex about which the DNA wraps (Kornberg, 1977). The former possibility appears to be ruled out by X-ray and neutron studies. A more acceptable alternative, consistent with present conceptions of the conformations of histones in solution, is one where only portions of the histone chains form a central globular complex, while the remaining chain regions (presumably the basic N-terminal 30–40 residues) provide extended "arms" or "tails" that wrap about DNA, neutralizing phosphate charges (Van Holde *et al.*, 1974b; Pardon *et al.*, 1977).

Choice between these latter two arrangements could be aided by studies of appropriate model histone complexes. In fact, such complexes exist. Weintraub *et al.* (1975a) and others have shown that 2 *M* NaCl dissociates DNA from histones, leaving intact complexes that appear mainly to be "heterotypic" tetramers containing each of the core histones. [There is still some unresolved controversy over the stability of the tetramer relative to other higher, i.e., octameric, and lower forms (Thomas, 1978).] Furthermore, cross-linking of histones in 2 *M* salt (pH 9) with dimethyl suberimidate (Thomas and Kornberg, 1975a), or of isolated nucleosome cores with methyl mercaptobutyrimidate followed by dissociation with salt (Stein *et al.*, 1977), yields cross-linked histone octamers. Both the heterotypic tetramer and the cross-linked octamer recombine with DNA upon gradient dialysis out of salt, yielding reconstituted nucleosome cores nearly indistinguishable from native cores (Stein *et al.*, 1977; Tatchell and Van Holde, 1977). The radius of gyration of the heterotypic tetramer (29.7 ± 0.4 Å) is very close to that of the cross-linked octamer (30.2 ± 0.4 Å), which in turn agrees well with the radius of gyration (30.6 ± 2.0 Å) determined for the histone core within intact nucleosome cores (Pardon *et al.*, 1977). Thus, the general conformation of the histone complexes must be similar in the presence and absence (in 2 *M* NaCl) of DNA, and the tetramers must be stacked directly one on top of the other in the octamer. Neutron scattering data are consistent with (but do not uniquely determine) a disklike core histone octamer approximately 70 Å in diameter and 30–40 Å high, with extended, highly hydrated histone projections, e.g., N-terminal tails (Pardon *et al.*, 1977). The low sedimentation coefficients of the tetramer (3.8 S, Weintraub *et al.*, 1975a) and the cross-linked octamer (5.3 S, Stein *et al.*, 1977) are also consistent with a highly hydrated or partially random coil structure.

The primary evidence for chain extensions from the histone core is derived from trypsin digestion and NMR studies. While H1 and H5 in whole chromatin are extensively digested by trypsin, only 20–30 residues are cleaved from the ends of core histones, even after prolonged ("limit") digestion (Weintraub and Van Lente, 1974; Weintraub, 1975; Weintraub *et al.*, 1975a,b). Similar limit polypeptides are obtained upon trypsinization of isolated nucleosome cores and heterotypic tetramers (Weintraub *et al.*, 1975a; Lilley and Tatchell, 1977; Sollner-Webb *et al.*, 1976), indicating that the majority of the core histone chains are involved in a globular state that confers inaccessibility to protease. The cleavable N-terminal tails are presumably located more toward the accessible periphery of the nucleosome, and are in a relatively open configuration in the heterotypic tetramer. However, as pointed out by Smith (1975) and Kornberg (1977), this does not necessarily imply that the complex is not globular, or that the tails are unstructured, since several native, classically "globular" proteins (e.g., ribonuclease) are only partially susceptible to proteases.

Much better evidence for highly mobile portions of histone chains has been obtained from ^{13}C NMR spectra of the heterotypic tetramer, which exhibit C_α and C_β linewidths and spin–spin relaxation times much more typical of denatured proteins than those observed in small native proteins, such as ribonuclease (Lilley *et al.*, 1977). The narrowness of the lysine and arginine carbon resonances places a significant portion of these residues in the mobile regions, while splitting of aromatic resonances in the 270-MHz proton NMR spectrum indicates the presence of tertiary structure placing aromatic groups in close proximity. The latter result is also consistent with the induction of cross-links between aromatic groups in chromatin by tetranitromethane and uv light (Martinson and McCarthy, 1975, 1976; Martinson *et al.*, 1976).

Comparisons of the Raman and circular dichroism spectra of core histones in 2 M salt and in intact nucleosome cores indicate that little, if any, change in histone secondary structure occurs upon loss of DNA, provided there is adequate charge shielding (Thomas *et al.*, 1977; Cotter and Lilley, 1978). If all of this helix is confined to the sequence regions believed to be globular from NMR studies (~60% of the total residues), then the helicity of the globular portion of the histone core would aproach that of hemoglobin (~80%), suggesting a structure based on helix–helix associations between subunits (Thomas *et al.*, 1977). However, though the N-terminal histone tails of the heterotypic tetramer are mobile, they need not be totally unstructured. Short stretches of helix occurring in the N-terminal arms of each of the core histones have been predicted by Fasman *et al.* (1976), and helices containing basic residues can readily be accommodated in the major groove of DNA in a manner appropriate for binding DNA phosphates (Sung and Dixon, 1970). Mobility of the tails could be provided by "hinge" points flanking the short (6–11 residues), potentially helical regions. Hinge points could occur at glycine residues, which are relatively freely rotating and are located within a few residues of these predicted helical regions in all four histones; and particularly by the dipeptide Gly-Gly, which precedes the predicted helical regions in H2A, H3, and H4, and follows these regions in H3 and H4. In addition to allowing for floppy tails, the Gly-Gly sequence has a high probability of involvement in β turns in globular proteins (Fasman *et al.*, 1976). Thus the N-terminal histone tails could be partially structured, yet assume a mobile, partially extended conformation in the absence of DNA, and a conformation appearing similar to totally globular proteins in the presence of DNA. In this event, the major distinctions between models would lie in the configuration of the N-terminal tails and their degree of extension around the DNA.

C. CONFORMATIONAL CONSTRAINTS ON NUCLEOSOMAL DNA

Nucleosomal DNA obviously is constrained to fold or coil back on itself. Packing within the core may yield a 4.4- to 8.5-fold reduction in effective DNA length. (This is based on the length of 140 bp, B-form DNA, and the maximum or minimum dimensions of the core, 110×57 Å. Other methods for calculating the packing ratio are possible, depending on specific assumptions about the internal DNA geometry.) In chromatin, packing will depend on the arrangement of the disklike cores, extremes being precise face-to-face and edge-to-edge arrangements. Estimates from electron microscopic measurements are on the order of 5–7 to 1 [for example, see Griffith (1975), Oudet *et al.* (1975), and Chapter VII]. Limits on the number of required folds or coils are set by the dimensions of the core, the two extremes being approximately 2.7 folds for a hairpin or paper clip arrangement (assuming a DNA diameter of 20 Å) and 1.5–2 coils for a uniform supercoil (Finch *et al.*, 1977; Pardon *et al.*, 1977; Suau *et al.*, 1977). The former arrangement, resembling paracrystalline regions of typical polymers, would yield a relatively uniform DNA distribution through the center of the core that is clearly ruled out by neutron scattering studies; hence the number of folds (coils) is more likely to be 2 or less.

A large variety of configurational variations are consistent with folding of DNA within nucleosome dimensions, with the majority of the DNA density in the outer volume elements. Certain restrictions on the possible DNA configurations have been set by studies of the relationship between nucleosomes and supercoiling of viral DNA, and of the pattern of cleavage of nucleases within nucleosomes.

Covalently closed circular DNA molecules of different superhelix densities are readily separated by electrophoresis on agarose gels (Keller and Wendel, 1974). Since circular superhelical molecules can only differ by integral numbers of superhelical turns [more properly, linking numbers (L), see Fuller (1971, 1972) and Crick (1976)] and there is a Gaussian distribution of superhelical turns in both super-coiled and fully relaxed, closed, circular DNA due to oscillatory fluctuations (Depew and Wang, 1975; Pulleyblank et al., 1975), the number of superhelical turns in any given DNA can be determined by step counting on the ladder pattern generated by electrophoresis of appropriate standards (Keller, 1975).

Native, unrelaxed SV40 DNA contains 24 + 2 superhelical turns, as determined by this method (Keller, 1975), or approximately one negative turn ($L = -1$) per 180–200 bp. This corresponds closely to the number of nucleosomes (21–26) observed on the seemingly relaxed SV40 minichromosome shown in Fig. 14 (Griffith, 1975; Germond et al., 1975). Furthermore, recombination of core histones with SV40 DNA in the presence of nicking–closing enzyme introduces approximately one negative superhelical turn per nucleosome formed (Germond et al., 1975). This does not mean literally that DNA is coiled in one turn about the histone core, nor does it describe the configuration of the DNA, since the linking number is related by the equation $L = W + T$ to both the path of the DNA helix axis (described by the writhing number W) and the winding or twist T of the helix about its own axis [for an informative discussion, see Crick (1976)]. Despite limitations, knowledge that the linking number per nucleosome is about -1 sets strict bounds on possible DNA folding schemes.

Coiling about the core clearly requires significant deformation of DNA from the normal B form found in solution. Such deformation could be spread out uniformly, keeping a constant radius of curvature. Alternatively, the deformation required could be accomplished by a relatively few sharp bends or "kinks" at periodic sites, with retention of the B-form geometry in between. Two sterically and energetically feasible models of kinks have been proposed that are consistent with nucleosome dimensions, if they occur at 20 or 10 bp intervals (Crick and Klug, 1975; Sobell et al., 1976).

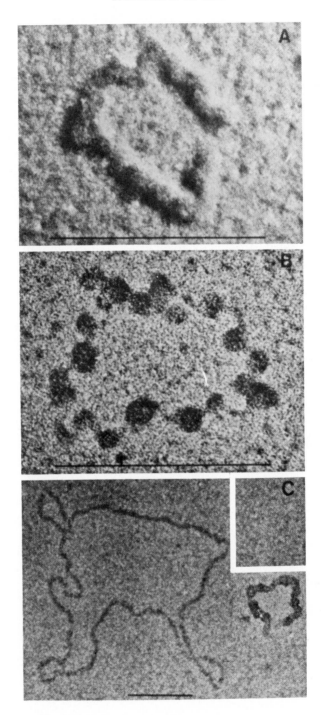

Periodicity within the nucleosome core has been indicated by examination of the products of internal nuclease cleavages. In particular, electrophoresis under denaturing conditions of DNA from nuclei or cores partially digested with DNase I yields a distinctive extended ladder of bands from 10 to 140 nucleotides (nt) in length (and more in the case of nuclei) that are multiples of 10 nt, as shown in Fig. 15 (Noll, 1974b). Very similar patterns have been obtained with DNase II and micrococcal nuclease (Sollner-Webb *et al.*, 1976). [The latter nuclease cleaves across both strands and therefore yields sharp banding patterns on both denaturing and nondenaturing gels (see Sollner-Webb and Felsenfeld, 1975, 1976).] Thus internal cleavage sites are spaced at intervals that are multiples of 10 nt. Certain differences in nuclease specificities are evident; for example, 80 nt fragments are common in DNase I digests, but are nearly absent from micrococcal nuclease digests. To establish the absolute locations of internal cleavages, cores may be labeled on the 5'-end with ^{32}P using polynucleotide kinase and [γ-^{32}P]ATP, and further digested with DNase I. The DNA fragments are then electrophoretically separated on denaturing gels. Autoradiography locates fragments retaining label, which define the distances of cleavages with respect to the ends. Results of several such experiments differ in some details, but generally show that cleavages occur at all or nearly all sites, although with very different frequencies (Simpson and Whitlock, 1976; Lutter, 1978; Noll, 1978). Cleavages 30, 80, and 110 nt from both ends are particularly infrequent. These results demonstrate most directly that DNA must surface and be accessible to nuclease at periodic intervals of 10 nt, and that the constraints of folding and histone binding severely restrict the action of nucleases at other sites.

Several more detailed explanations of these periodic cleavages have been offered. Two are simply related to the above observations,

FIG. 14. SV40 minichromosomes. High-resolution electron micrographs of the SV40 minichromosome in its native (A), beaded (B), and deproteinized (C) state. The example of the native state was taken in the single side-band strioscopy mode to accentuate the appearance of repeating units along the fiber. The example of the beaded state was prepared by diluting native minichromosomes tenfold with distilled water for 1 minute and fixing with glutaraldehyde (0.1%) for 15 minutes at 20°C. Minichromosomes were deproteinized, then relaxed by mild X-ray treatment (C). The scale bars represent 1000 Å; the inset in (C) compares a native minichromosome at an equal magnification with the DNA. (From J. D. Griffith, 1975, *Science* **187**, 1202–1203. Copyright 1975 by the American Association for the Advancement of Science.) An average of 21 ± 1 beads were observed per minichromosome (41 measurements). The lengths of DNA in each bead and interconnecting strand ("bridge") were estimated to average approximately 170 and 40 bp, respectively.

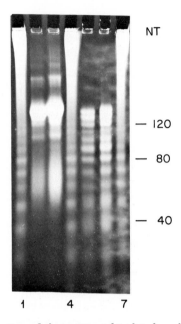

— 120

— 80

— 40

1   4   7

FIG. 15. DNase I digestion of chromatin and isolated nucleosome cores. Chicken erythrocyte nuclei were isolated and digested with pancreatic DNase I as described by Camerini-Otero *et al.* (1976). Nucleosome cores were obtained by Bio-Gel A5m chromatography of a micrococcal nuclease digest of chromatin stripped of H1 and H5 (see Fig. 12). Cores ($20\,A_{260}$ units/ml) were redigested with DNase I (Worthington, $1\,\mu g/A_{260}$ unit, 37°C, 30 min, in 10 mM Tris–HCl, 1 mM MgCl$_2$, pH 8.0). Samples were deproteinized, precipitated with ethanol, dissolved in 98% formamide, and denatured by heating at 100°C for 3 min. Electrophoresis was on a 12% polyacrylamide slab gel containing 7 M urea (Maniatis *et al.*, 1975). Staining was with ethidium bromide. Lanes 1, 4, and 7 are from DNase I-digested nuclei; lanes 2 and 3 are from undigested nucleosome cores; and lanes 5 and 6 are from redigested nucleosome cores. Bands are multiples of 10 nucleotides (NT) in length (Noll, 1974b).

i.e., if DNA is continuously bent about the core and retains 10 bp per turn, then a given strand will surface at 10 bp intervals. Alternatively, strong histone binding sites may be spaced so that accessible regions occur only every 10 bp. (Note that the N-terminal tail of each histone can neutralize 7–9 DNA bp. Obviously these two alternatives are not mutually exclusive and may well be related.) A third alternative relates to the kinky helix hypothesis, that is, cleavages may occur at each kink (Crick and Klug, 1975). Consideration of the single-strand nicking sites on the continuously bent and kinked models leads to distinctly different predictions for the disposition of cuts on opposing strands. In the continuously bent model, while one strand is on the surface, the other is inside. The situation is then reversed a half-twist in either direction. Thus if cleavage occurs precisely from the outside, cuts on opposing strands must be separated by 4 or 6 nt, yielding *double-stranded* fragments with staggered, partially single-stranded ends. Cuts at kinks are expected to occur either directly across strands, or 10 nt away (Sollner-Webb and Felsenfeld, 1977). Recent experiments, in which the DNA polymerase repair of recessed 3'-OH ends of the double-stranded fragments from DNase I digests was quantitated, showed that the 3'-OH and 5'-P ends are recessed by 8 and 2 nt, respectively, in disagreement with both of the above predictions (Sollner-Webb and Felsenfeld, 1977). Additional studies comparing the sites of cleavage of DNase II and micrococcal nuclease to those of DNase I are consistent with a situation in which all three enzymes bind to a common site related to the nucleosome structure, but cleave at different sets of sites related by a dyad axis, as illustrated in Fig. 16 (Camerini-Otero *et al.*, 1978; Felsenfeld, 1978). In this event the pattern of cleavage does not distinguish between the two folding models.

At their present level of sophistication, energy considerations alone do not permit distinction between kinked and uniformly curved structures either. Kinking could occur by bending of DNA toward either the minor (Crick and Klug, 1975) or major (Sobell *et al.*, 1976) grooves, with stereochemically acceptible changes in sugar–phosphate backbone angles, no loss in base pairing, and retention of the B-form geometry between kinks. Although there may be a significant

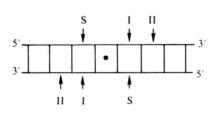

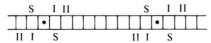

FIG. 16. Relative locations of the cleavage sites on nucleosomal DNA by pancreatic DNase (I), spleen acid DNase (II), and micrococcal (staphylococcal) nuclease (S). The dot represents the local dyad axis of the cuts. The top figure shows the cutting sites in relation to a local dyad axis. The bottom figure shows two sets of local cutting sites separated by 10 nt on each strand. Note the different possible arrangements of staggered ends, depending on the combination of sites cleaved. (Reproduced from Felsenfeld, 1978, with permission.)

activation energy for kinking, the net loss in free energy per kink is due to the unstacking of a single base pair from its neighbor and should amount to only a few kilocalories per mole.

Although DNA is usually thought of as a very stiff chain (which it is), considerable bending occurs in solution on account of normal Brownian motion. Statistically, the behavior of long DNA chains has been described successfully in terms of wormlike coil models that consider the chain to be a continuously bent rod with a characteristic persistence length. The persistence length is an experimentally accessible parameter that in effect measures the degree of stiffness of the chain and corresponds, statistically, to the average length required to achieve a 68.4° bend. Theories of wormlike chain behavior lead to expressions relating persistence length to bending free energies. Measurements of the temperature dependence of the persistence length determine the enthalpy and entropy of bending, and hence can be used to calculate the free energy required to fold DNA into a nucleosomal configuration (Camerini-Otero and Felsenfeld, 1977a; Harrington, 1977; for a general discussion, see Bloomfield et al., 1974). Estimates on this basis range from 20–44 kcal/mole of nucleosomes, depending on the choice of models, and are well within the range expected for kinked structures (Camerini-Otero and Felsenfeld, 1977a; Harrington, 1977; Levitt, 1978; Sussman and Trifonov, 1978). For comparison, the free energy of 140 bp of DNA containing seven kinks would be 28–42 kcal/mole above that of linear DNA, if a free energy of 4–6 kcal/mole of kinks is assumed. Note, however, that the above treatment avoids the basic question of possible kinking of protein-free DNA in solution. As described by the original proponents, kinking offers several unique advantages in considerations of mechanisms of helix packing, intercalation, and other DNA activities. Unfortunately, the expected frequency of kinking in DNA in solution [a minimum of one per 800 bp; see Crick and Klug (1975)] may be too low to permit experimental detection. It is hoped the question of kinking within nucleosomes will be resolved by refinements of X-ray diffraction data.

One additional aspect of DNA in nucleosomes requires mention, and that is its unusual optical activity. Protein-free DNA in solutions of modest ionic strength exhibits a strong positive circular dichroism band in the vicinity of 275 nm. As shown in Fig. 17 this band is suppressed to slightly less than half its original intensity in typical chromatin preparations, and is further suppressed in nucleosome cores, to the point where it resembles spectra recorded for films of DNA in the C form (Tunis-Schneider and Maestre, 1970). DNA assumes the C form in fibers and films at low humidities when lithium is the

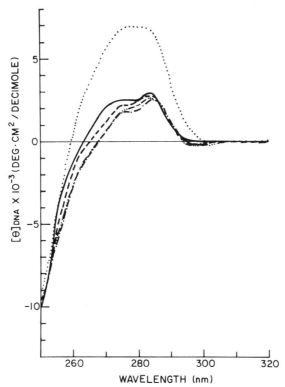

FIG. 17. Near-uv circular dichroism spectra of an unfractionated micrococcal nuclease digest of nuclei (———), nucleosomes (———————), nucleosome cores (—·—··—·—·—),
subnucleosomal particles containing an (H2A)(H2B)(H3)₂(H4)₂ hexamer associated
with 105 bp DNA (—···—····—), and deproteinized DNA (· · · · · ·). Spectra were recorded
in 0.1 mM cacodylic acid, 0.02 mM EDTA, 0.1 mM phenylmenthanesulfonyl fluoride
(pH 7.0). Molecular ellipticities are reported in terms of moles DNA phosphate. (From
Mencke and Rill, 1978.)

counterion. X-Ray diffraction studies have shown that the C form is
similar to the B form, but is more compact and overwound, with a rotation per residue of 39°, rather than 36°, and a translation per residue
along the helix axis of 3.32 Å, rather than 3.38 Å. This overwinding is
accompanied by slight increases in tilt of the bases with respect to the
helix axis, in the dihedral angle between bases, and in the distance of
the bases from the helix axis, thereby deepening and narrowing the
minor groove. [The C form may be considered as an extreme variant of
the B family of configurations, which differ modestly in winding
angles and base orientations and seem to be smoothly interconverti-
ble, depending upon solution conditions. The B family is distinct from

the A family of configurations, which are underwound with respect to the classic B form and otherwise differ significantly in base orientations and sugar configuration. The A form is characteristic of double-stranded RNAs and RNA–DNA hybrids and may be the required form of DNA during transcription. For a tabulated comparison of the various forms, see Bloomfield *et al.* (1974).]

The circular dichroism spectra of DNA in concentrated solutions of monovalent salts, and in certain partially organic solvents (Nelson and Johnson, 1970; Hanlon *et al.*, 1972, 1975; Studdert *et al.*, 1972; Ivanov *et al.*, 1973; Zimmer and Luck, 1973), as well as that of nucleosomes and some histone–DNA complexes, closely mimic that of C-form DNA in films. These spectral similarities naturally suggest a similarity in configuration of the film and solution forms. In fact, the circular dichroism spectra of DNA in chromatin and in concentrated solutions of salt can be explained nicely on the basis of a linear combination of spectral contributions from a mixture of B and C forms (Hanlon *et al.*, 1972, 1976; Johnson *et al.*, 1972). Assignment of the C form to DNA in nucleosomes has been criticized based on the observation of DNase I cleavage sites at 10 nt intervals. This criticism need not apply since there are only 0.7 residues per turn less in C-form than in B-form DNA, and the full overwinding may not occur in nucleosomes. Furthermore, kinking at 10 or 20 bp intervals would not necessarily preclude overwinding of the DNA in between. However, Goodwin and Brahms (1978) have recently reported that the Raman spectrum of B-form DNA in films has a characteristic band at 835 cm^{-1}, probably because of phosphodiester stretching vibrations, that is replaced by a band at 865–870 cm^{-1} for films of C-form DNA. The presence of a prominent band at about 835 cm^{-1} in Raman spectra of mono- and oligonucleosomes (Goodwin and Brahms, 1978) indicates that the "classic" C form does not predominate in nucleosomal DNA.

A different perspective on these matters is obtained if one considers that the common feature of DNA in C-form films, solutions of concentrated salts or organic solvents, and in nucleosomes is a dramatic alteration in the microsolvent environment, i.e., loss or expulsion of water of hydration due to competing ions or histone binding. As shown most recently and elegantly by Eickbush and Moudrianakis (1978), in the absence of sufficient water of hydration, DNA has a natural tendency to compact by forming either left-handed homonemic supercoils (imagine wrapping the DNA double helix about a rod), or paracrystalline arrays (Fig. 18). Observation of left-handed superhelical turns in condensed closed circular DNAs is consistent with an overwinding of DNA about the helix axis in the direction of C-form

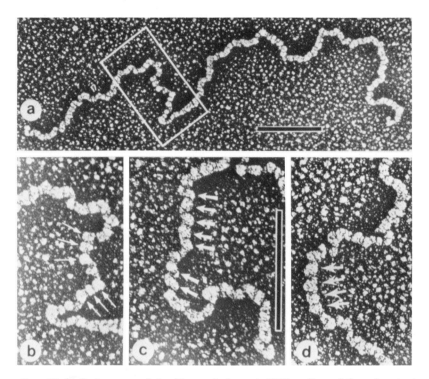

FIG. 18. Coiled nature of the fibers of phage λ DNA collapsed from low ionic strength solution (1 mM Tris, pH 7.5) with 95% ethanol. (a) Low-magnification view of a typical fiber. (b) Higher resolution view of the rectangle in (a) revealing the coiled arrangement of the DNA (see arrows). (c) and (d) Higher resolution views of segments from two other DNA fibers. The left-hand nature of the homonemic supercoil can be ascertained from (c). Scale bars, 2000 Å. (From T. H. Eickbush and E. N. Moudrianakis, 1978, *Cell* **13**, 295–306, reproduced with permission. Copyright © 1978 by MIT Press.)

DNA (Eickbush and Moudrianakis, 1978). However, more than this is required to generate a superhelix, namely, the Cartesian coordinate frame of each base pair must be rotated uniformly in *three* dimensions with respect to that of its neighbor. Rotation of the XY and $X'Y'$ planes of adjacent base pairs about the helix axis (Z) generates only a linear double helix. An additional rotation of Z' (the local double helix axis) with respect to Z is required to generate a superhelix or coiled-coil. The degree of this rotation determines the superhelix pitch (Crick, 1953; Rill, 1972). This extra rotation must involve alterations in both the relative base orientations and sugar–phosphate backbone that are not described precisely by either the classic B- or C-form geometries. Recent modeling studies by Levitt (1978) and Sussman and Trifonov

(1978) provide illustrations of energetically feasible conformations of DNA that is supercoiled without kinking. Insofar as the author is aware, there has been no report of the expected optical properties of highly supercoiled DNA. At present it appears best to regard the C-type circular dichroism spectrum as characteristic of certain partially dehydrated, compacted forms of DNA, without implying a specific geometry. As such it provides a convenient measure of the integrity of nucleosome preparations and of the ability of specific subsets of histones or other factors to mimic the state of DNA in nucleosomes.

D. HISTONE–HISTONE AND HISTONE–DNA INTERACTION SITES IN NUCLEOSOMES

Regardless of the details of the DNA folding scheme, the energies of interaction of specific histone pairs in solution are sufficient to account for nucleosome formation (D'Anna and Isenberg, 1974b,d,e; Sperling and Bustin, 1975; Roark et al., 1974, 1976; Camerini-Otero and Felsenfeld, 1977a; Harrington, 1977). The proximities of histones in chromatin and nucleosomes have been probed by cross-linking with a variety of bifunctional reagents. Products separated electrophoretically have been identified by electrophoresis after cleavage of the cross-link, or by mapping of iodinated peptides. All possible core histone dimers are formed by at least one reagent (Table VI), probably reflecting the ability of long cross-linkers (up to nearly 15 Å for methyl mercaptobutyrimidate) to span a considerable extent of the core. More significant are the relative amounts of different dimers obtained with the same reagent, and the number of reagents of different lengths and specificities able to cross-link a given pair. Analyses of the former type vary with reagent and are difficult to interpret. However, the histone pairs cross-linked in chromatin by the widest variety of reagents, and therefore presumably in the most intimate contact, are the same pairs that interact most strongly in solution. Thus, solution histone interactions appear relevant to nucleosome structure.

Unfortunately, in most cases, these results reveal little about the proximities of specific regions of histone pairs, along DNA or otherwise, because many of the reagents react with abundant lysines that are spread throughout the chains (though not uniformly). Important exceptions are the cross-linking of H2B to H4 by tetranitromethane and of H2B to H2A by uv light (Martinson and McCarthy, 1975, 1976; Martinson et al., 1976). Both of these cross-links are specific and of zero length, involving the coupling of aromatic residues, and occur only in chromatin or properly reconstituted histone–DNA complexes.

TABLE VI
CROSS-LINKING PATTERN OF HISTONES IN NUCLEOSOMES[a]

H3	H4	H2B	H2A	
MMB[b] Sub[c] MMPI[d]	MMB[b] Sub[c] MMPI[d] CH_2O^e CD[f]	MMB[b]	MMB[b]	H3
	MMB[b]	MMB[b] Sub[c] MMPI[d] $CH_2O^{g,h}$ TNM[i,j] GA[h]	MMB[b] Sub[c] MMPI[d]	H4
		MMB[b]	MMB[b] Sub[c] MMPI[d] $CH_2O^{g,h}$ GA[h] uv[j,k]	H2B
			MMB[b]	H2A

[a] Reprinted from Trifonov, 1978, with permission. Abbreviations used: MMB, methyl-4-mercaptobutyrimidate; Sub, dimethyl suberimidate; MMPI, methyl-3-mercaptopropionimidate; CH_2O, formaldehyde; TNM, tetranitromethane; CD, 1-ethyl-3-(3-dimethylamino-propyl)carbodiimide; GA, glutaraldehyde; uv, ultraviolet light.

[b] Hardison et al. (1975).
[c] Thomas and Kornberg (1975a).
[d] Thomas and Kornberg (1975b).
[e] Hyde and Walker (1975).
[f] Bonner and Pollard (1975).
[g] Weintraub et al. (1975).
[h] Van Lente et al. (1975).
[i] Martinson and McCarthy (1975).
[j] Martinson and McCarthy (1976).
[k] Martinson et al. (1976).

Furthermore, H2B can be cross-linked simultaneously to H2A and H4, indicating that association of this trio is an important structural feature of chromatin. Analyses of cyanogen bromide cleavage fragments have shown that the C-terminal half of H2B cross-links near the C-terminus of H4, while the N-terminal half of H2B links to H2A (Martinson and McCarthy, 1976). Less direct, but convincing evidence for the immediate proximity of the Cys-110 residues of the H3 pair in nucleosomes is derived from the observation that nucleosome cores reconstituted from 140 bp DNA and core histones, with or without dimerization of H3 via a disulfide bond, are virtually indistinguishable from native cores (Camerini-Otero and Felsenfeld, 1977b). If the core possesses a dyad axis, then this disulfide bond must lie along it.

Several questions come to mind in considerations of the nature and locations of histone–DNA interactions. Among these are the following. (i) Are the histones located predominantly in the major or minor DNA grooves or neither; (ii) are the binding sites of individual histones located on one or both DNA strands, and are they continuous along the strand(s) or are they interrupted by sites bound by other histones; (iii) where are the locations of the principal binding zones of individual histones along nucleosomal DNA; (iv) are histones symmetrically disposed along DNA; and (v) are histone–DNA interactions primarily electrostatic, or are hydrogen and hydrophobic interactions also important?

Model building studies have indicated that histones can fit into either the major or minor DNA grooves (Zubay and Doty, 1959; Shih and Bonner, 1970; Richards and Pardon, 1970; Sung and Dixon, 1970; Oliver and Chalkley, 1974), and it certainly has been convenient to think that they do so. This question has been approached most recently by examination of the relative rates of methylation of protein-free DNA, chromatin, and polypeptide–DNA complexes (Levina and Mirzabekov, 1975; Mirzabekov et al., 1977). Methylation at the N-7 position of guanine, located in the minor groove, is not affected by bound histones (in chromatin), protamine, polylysine, or polyarginine, in agreement with other evidence against minor groove occupation based on reporter molecule (Simpson, 1970) and antibiotic binding (Melnikova et al., 1975). Yet methylation at the N-3 position of adenine, located in the major groove, is reduced only 14 and 15% by chromatin proteins and polyarginine, respectively, and not at all by polylysine.

In contrast, the minor groove is strongly shielded by binding of the antibiotic netropsin, and the major groove is shielded 31% by glucosyl residues in T4 DNA, in which cytosine is replaced by glycosylated hydroxycytosine (Mirzabekov et al., 1977). Thus, it appears that histones lie more nearly along the DNA backbone, penetrating the major groove modestly, and the minor groove little, if at all.

Determination of the location of specific histones along DNA in nucleosomes has been approached by taking advantage of the ability of nucleases to cleave nucleosomes internally at specific sites spaced by about 10 bp. Internal cleavages should yield two or more nucleoprotein fragments, termed subnucleosomes, each containing a subset of histones associated with a specific length DNA fragment. In principle, the histone compositions of these fragments should indicate which histones are neighbors along the DNA, and the associated DNA lengths should locate these histones with respect to the nucleosome ends, particularly if the ends are labeled with ^{32}P. In certain ways, this approach is analogous to that taken to sequence proteins and nucleic acids. However, there are several complicating factors that must be considered. A major one results from the fact that nuclease cleavage sites may not be defined by the end of the binding region of a given histone; furthermore, a single histone chain may bind at several sites separated by regions of intervening histone binding. In either event, DNA cleavage may not be accompanied by dissociation of the respective subnucleosomal fragments. One must, therefore, distinguish between zones of histone–DNA contact, in which a significant portion of a single histone chain makes intimate contact nearly continuously with a long stretch of DNA, and points of contact by one or a few residues of one histone in a region dominated by the binding of another. One expects that the clustered basic residues in the N-terminal tails of core histones (and the C-terminus of H1) will define zones of coverage, while the remaining portions of the chains may either simply extend these zones or bridge to more distant points of contact.

Electrophoretic methods have proved extremely valuable for detailed examination of the nucleoprotein products of chromatin digestion and of nucleosome processing (Varshavsky *et al.*, 1976a; Bakayev *et al.*, 1977; Todd and Garrard, 1977; Bafus *et al.*, 1978; Rill and Nelson, 1978). Nucleosomes, containing equimolar core histones plus H1 or H5 associated with near repeat length (190 ± 20 bp) DNA, are readily resolved from nucleosome cores with 140–160 bp DNA that lack H1 (Fig. 19), directly demonstrating the location of the major zones of lysine-rich histones on intercore (spacer) DNA (Varshavsky *et al.*, 1976a; Bakayev *et al.*, 1977; Todd and Garrard, 1977; Rill and Nelson, 1978). More detailed studies of the products as a function of extent of digestion have shown that reduction of the nucleosome DNA to 160 bp is sufficient to cause loss of H1, but not of H5 (Todd and Garrard, 1977; Rill *et al.*, 1978b), although digestion of chromatin stripped of H1 and H5 rapidly yields 140 bp cores without a pause of 160 bp (Noll and Kornberg, 1977). These results suggest that H1 may bind up to, or even within, the 140 bp core, but that the primary binding sites are not immediately adjacent to the core.

Only two major subnucleosomes have been detected after digestion to an extent where the *full* spectrum of subnucleosomal DNA

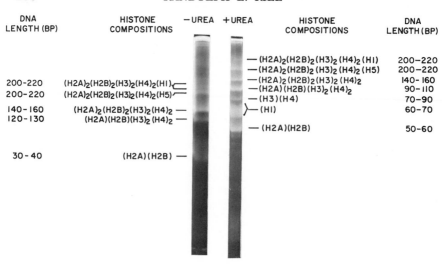

FIG. 19. Electrophoretic resolution of nucleosomes and subnucleosomes from chicken erythrocyte nuclei digested to approximately 18% acid solubility with micrococcal nuclease. Sample on left was electrophoresed on a 6% polyacrylamide tube gel containing 10 mM Tris, 2 mM Na$_2$EDTA, adjusted to pH 7.6 with 3-(N-morpholino)propanesulfonic acid (MOPS from Calbiochem). Sample on right was electrophoresed on a 10% polyacrylamide slab gel containing the same buffer made 3 M in urea. Staining is with ethidium bromide. Histone and DNA compositions of the major bands are as indicated (Rill and Nelson, 1978; Rill et al., 1978b; Nelson et al., 1978b). Note that cleavages 30–40 bp from the end(s) of the nucleosome core cause dissociation of subnucleosomes in the absence of urea, while cleavages 50–60 bp from the end(s) are nondissociative unless 3 M urea is present.

fragments is produced from internal cleavage of 20–25% of all nucleosomes (Fig. 19). These are complementary particles, one containing an H2A + H2B pair associated with 30–40 bp DNA, and the other containing a (H3)$_2$(H4)$_2$(H2A)(H2B) hexamer associated with 110–130 bp DNA, indicating that the primary binding zones of H2A and H2B are located on the terminal 30–40 bp of at least one end of 140–160 bp nucleosome cores (Nelson et al., 1977b; Rill et al., 1978b). Since nucleosomes cleaved at this point freely dissociate into the respective subnucleosomes, it is unlikely that H2A and H2B contacts extend significantly into the center of the core, or that other histones bind significantly to this terminal DNA. Weak point contacts are not ruled out. In contrast, nucleosomes cleaved more near the core center do not dissociate and are recovered with intact nucleosomes and cores. Inclusion of a moderate concentration of urea (3 M) in gels is sufficient to disrupt histone–DNA or histone–histone interactions keeping internally cleaved units intact, without causing histone dissociation or exchange, and permits electrophoretic resolution of four additional major subnu-cleosomes (Fig. 19). Two of these have the same histone compositions as the subnucleosomes described above, but the DNA from the H2A + H2B complex is ~20 bp longer, on the average, and that of the hexamer is correspondingly ~20 bp shorter (Rill and Nelson, 1978; Nelson et al., 1978b). Apparently this extra 20 bp is involved in a his-tone contact, presumably with H3 and H4, that is broken by urea. The remaining two subnucleosomes are single copies of H1 associated with 66 ± 10 bp DNA and (H3)(H4) pairs associated with 74 ± 10 bp DNA. In the case of the H1 complexes it is not clear if urea is required to disrupt histone–DNA interactions, or H1–H1 interactions, since these complexes (but not the others) aggregate upon removal of urea (Mencke and Rill, 1978). Isolation of the H3 + H4 complex indicates that these histones are neighbors along DNA, but the location of this pair within the core cannot presently be specified since the comple-mentary subnucleosome(s) is (are) not uniquely defined.

Little is known about the strand specificity of histone binding, or the locations of point histone–DNA contacts, because of the paucity of methods for conveniently cross-linking proteins to DNA. One method of promise involves methylation of purines with dimethyl sulfate, re-moval of the activated bases by heating, formation of a Schiff's base between the resulting aldehyde and a neighboring amino group, and reduction of the Schiff's base with sodium borohydride to form a sta-ble covalent bond (Levina and Mirzabekov, 1975). Application of this approach to nucleosome cores, labeled on the 5'-termini with ^{32}P, ini-tially indicated that H3 and H4 were both located on the ends of the core (and hence became coupled to labeled nucleotides), but the sig-nificance of this result has been obscured by the discovery of a protein kinase that falsely labels histones in core preparations (Simpson, 1976; 1978a).

Binding of individual histones to only a single DNA strand would appear to be functionally advantagous, since strand separation could occur during replication and transcription without breakage of his-tone–DNA interactions. No direct information is presently available; however, Riley and Weintraub (1978) have recently reported that exo-nuclease III can digest at least 200 nt from one or both 3'-ends of dinu-cleosomes, producing a ladder of bands that differ by 10 nt. Thus, this enzyme, and perhaps polymerases and other nucleic acid processing enzymes, can move along a single strand through the equivalent of an entire nucleosome without destroying, in advance of digestion, the structural features responsible for pauses at 10 nt intervals.

A variety of detailed models for the arrangement of histones have been proposed (D'Anna and Isenberg, 1974d; Li, 1975; Olins *et al.*, 1976a; Weintraub *et al.*, 1976; Rosenberg, 1976; Trifonov, 1978). Since each possesses individual virtues, but there is little experimental basis for choosing between them, a detailed comparison will not be given here. Common features of the models are a symmetric disposition of histones along DNA and twofold symmetry of the folded structure. Twofold symmetry is suggested by the stoichiometry of core histones and is extremely attractive both from theoretical considerations of symmetry in protein complexes and from a functional standpoint (e.g., see Weintraub *et al.*, 1976). At their present resolution, X-ray diffraction and electron microscopic data are consistent with, but do not prove, the existence of a dyad axis. They do indicate a bipartite character of the core. Reported observations of heterotypic histone tetramers in 2 *M* salt (Weintraub *et al.*, 1976), of an increase in the number of spherical bodies on SV40 minichromosomes from 20–24 to 40–50 with decreasing ionic strength (Oudet *et al.*, 1978), and of the digestion of chromatin at 100 bp, rather than 200 bp, intervals by DNase II under certain ionic conditions (Altenburger *et al.*, 1976) all support the concept of a bipartite nucleosome structure capable of unfolding into half-nucleosomes. Such unfolding could be important *in vivo*, permitting retention of the structure of half of the nucleosome while the other half is processed.

The facts that DNase I almost never cleaves positions 30, 80, and 110 nt from the 5'-ends of both strands of nucleosome core (140 bp) DNA and that the smallest major subnucleosomes contain only H2A + H2B pairs argue for a symmetric disposition of histones along DNA, with H2A and H2B occupying most or all of the terminal 30–40 bp. Similar histone binding on each core end is also suggested by the finding that both 3'-ends are readily digested by exonuclease III, with pauses leading to the accumulation of fragments differing by 10 nt (Riley and Weintraub, 1978). A particularly strong pause point is encountered 100 nt from the end, at the same location as a point of high accessibility to DNase I (Simpson and Whitlock, 1976). This pause point may correspond to the beginning of a region of strong H3 + H4 binding, since cleavage at this site by micrococcal nuclease causes release of the subnucleosome containing H2A + H2B, while cleavage 10–20 bp further into the core does not cause release due to binding of neighboring histones, presumably H3 and H4. The high accessibility of the site 40 nt from the end could, in this event, reflect a gap, or weak point, in histone–DNA interactions at a junction between the binding domains of the two homotypic histone classes.

An attractive model for the core structure is obtained if one assumes, as described by Finch *et al.* (1977), that there are 80 nt per turn, so that positions 80 nt apart are in close proximity along the side of the superhelix. These positions could be similarly protected by regions of neighboring histones related by a dyad axis perpendicular to the superhelix axis (Fig. 20). If H2A and H2B are symmetrically located on both core ends, then the positions 60, 70, and 80 nt from the ends, that are only slightly to moderately accessible to DNase I, must be protected by H3 and H4 binding. The symmetry-related pairs of points of low susceptibility to DNase I at positions 30 and 110, and of micrococcal nuclease pause points at positions 0 and 140, may be influenced by either H2A + H2B or H3 + H4 binding. This will depend on the presence or absence of bridging of H3 or H4 to the ends of the core DNA.

Strong electrostatic interaction between histones and DNA is assured by the clustering of basic residues in local sequence regions. Such clustering permits effective competition of histones against small cations for DNA binding, much in the way that chelating agents are able to outcompete monovalent ligands for metal ion binding.

The importance of nonelectrostatic interactions within chromatin also has long been recognized. For example, although H1 has the highest net positive charge of all histones, it is removed much more easily than core histones from chromatin by simple salts, e.g., 0.6 *M*

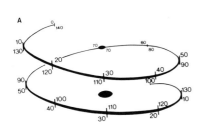
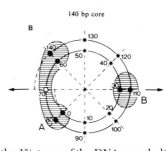

FIG. 20. (A) Diagram, drawn roughly to scale, of the 1³/₄ turns of the DNA superhelix proposed for the 140-bp nucleosome core. The top set of numbers gives the distances, in bases, of the DNase I cutting sites from the 5′-end of one strand, and the bottom set refers to the other strand, related to the first by the dyad axis shown. For definiteness, the number of bases per turn is shown as 80, but the exact number is not established. (B) The 1³/₄ turns of one strand of the DNA are represented here as a spiral to show how the supercoiling brings sites 80 bases apart close together and groups the sites of low or medium cutting frequency by DNase I into the two diametrically opposite areas, A and B, shown shaded. The arrow indicates the dyad. The numbers give the distance in bases from the 5′-end. Asterisks denote high-frequency DNase I cutting; ●, low-frequency cutting (or micrococcal pauses); and ○, medium-frequency cutting. (Reproduced from Finch *et al.*, 1977, with permission.)

NaCl (Ohlenbusch *et al.*, 1967; for a very simple procedure for removing H1, see Bolund and Johns, 1973). High concentrations of urea, a classic disrupter of hydrophobic and hydrogen bonds, cause a dramatic unfolding and thinning of chromatin fibrils (Bartley and Chalkley, 1968, 1972, 1973; Georgiev *et al.*, 1970) without causing histone dissociation, while lower concentrations of urea greatly enhance the extraction of histones with saline solutions and decrease the distinction in extraction between H1 and other histones (Bartley and Chalkley, 1972, 1973). Hydrophobic agents, such as deoxycholate, often alter the order of extraction (Smart and Bonner, 1971).

Since mainly nonionic histone–histone interactions are at least partially responsible for maintaining the folded state of chromatin, these results do not directly indicate nonionic contributions to histone–DNA binding. A comparison of the effects of urea and trypsin on intact, and internally cleaved, nucleosome cores does indicate such interactions. Increasing concentrations of urea differentially affect the conformations of core DNA and histones, causing a gradual, noncooperative increase in core symmetry and a change in DNA circular dichroism toward that of B-form DNA; but a cooperative loss in histone secondary structure (α-helix) that ensues only after the urea concentration approaches $4\ M$ and a considerable increase in core asymmetry have already occurred (Olins *et al.*, 1977). The unfolding of the core at low urea concentrations could be due to disruption of histone–histone or histone–DNA interactions, or both, with retention of histone secondary structure in any case. Since $3\ M$ urea disrupts internally cleaved cores, without releasing free DNA, histone–histone interactions must be broken by urea. However, the much lower than normal protein to DNA ratios of the smallest subnucleosomes indicates that histone–DNA interactions are also broken (Rill and Nelson, 1978; Nelson *et al.*, 1978b). The histone secondary structure in these subnucleosomes is virtually identical to that found in intact cores (Mencke and Rill, 1978). Hence core unfolding must be a consequence of disruption of nonionic interactions of both types.

Trypsin digestion of the N-terminal 20–30 residues of all histones in intact cores is not sufficient to cause histone dissociation or complete unfolding of the core structure (Lilley and Tatchell, 1977). Similar trypsinization of internally cleaved cores causes the release of subnucleosomal fragments, most of which retain bound C-terminal histone sequences (Weintraub, 1975). Hence the C-terminal regions of histones, which possess considerably reduced potential for ionic interactions with DNA, are able to bind both histones and DNA and maintain part of the conformational integrity of the core. Similar conclusions have been reached by examination of the binding of cyanogen bromide cleavage products of histones to DNA (Li and Bonner, 1971) and of the thermal denaturation of subnucleosomes (Mencke and Rill, 1978).

E. NUCLEOSOME ASSEMBLY AND THE ROLES OF HOMOTYPIC HISTONE PAIRS

Nucleosomes are readily reconstituted onto protein-free DNA by direct mixing with histones in 2 M NaCl, followed by removal of salt by gradient dialysis. Such reconstituted complexes appear virtually identical to native chromatin in terms of induction of supercoiling (Germond et al., 1975), X-ray scattering pattern (Boseley et al., 1976), and restricted digestion by a variety of nucleases and proteases (Camerini-Otero et al., 1976; Sollner-Webb et al., 1976). In addition, nucleosome cores indistinguishable from native cores, by several physical and biochemical criteria, have been reconstituted in the same manner from core histones and 140 bp DNA (Tatchell and Van Holde, 1977; Camerini-Otero and Felsenfeld, 1977b).

Nucleosome assembly could occur by concerted binding of individual histones or a preassembled histone octamer to DNA, with concomitant DNA folding, or by a stepwise process where initial binding of a limited subset of individual histones or small histone complexes delineates and partially folds the nucleosomal DNA, and secondary binding of the remaining histones completes the folding and further stabilizes the core structure. In vitro reconstitution protocols have been developed that mimic both of these assembly mechanisms. Surprisingly, nucleosome cores can be reassembled either by combination of 140 bp DNA with cross-linked octamers of core histones (Stein et al., 1977) or by addition of H2A + H2B to 140 bp DNA previously "primed" with H3 + H4 (Camerini-Otero and Felsenfeld, 1977b; Klevan et al., 1978); so neither mechanism for in vivo assembly can be ruled out. One aspect of nucleosome assembly in vivo now seems clear: assembly of new nucleosomes during replication does not result in the mixing of old density labeled (heavy) histones with newly synthesized unlabeled (light) histones, indicating that the formation of new histone octamers is conservative. Furthermore, old and new histone octamers appear to segregate conservatively over two to three generations (Leffak et al., 1977).

Although stepwise histone association may not be operative in vivo, reconstitution experiments, in which specific subsets of histones are combined with DNA, have confirmed the indications from sequence studies that H3 and H4 together are the critical elements in the formation and delineation of nucleosomes. Specifically, H3 and H4 in combination with circular SV40 DNA yield a relaxed beaded circle closely resembling the native SV40 minichromosome. The size, DNA packaging ratio, and degree of DNA supercoiling of these beaded structures are very close to those found for nucleosome cores (Bina-Stein and Simpson, 1977). Beadlike structures similar to nucleosomes are also formed by H3 and H4 in combination with 140 bp DNA (Bina-Stein and Simpson, 1977). Unlike other histone combinations, H3 + H4 complexes with DNA yield X-ray scattering curves very similar to the scattering of whole chromatin (Moss et al., 1977).

Felsenfeld and co-workers (Camerini-Otero et al., 1976; Sollner-Webb et al., 1976) have examined in great detail the products of digestion of nearly all possible histone–DNA combinations with various nucleases and proteases. Only complexes containing H3 and H4 yield specific degradation products resembling those obtained from native chromatin. Of particular significance is the finding that micrococcal nuclease digestion of H3 + H4 complexes with DNA yields, at least transiently, discrete DNA fragments near nucleosome core size (120–130 bp), in addition to several smaller ($\sim$60–80 bp) more stable fragments (Camerini-Otero et al., 1976; Moss et al., 1977). Thus H3 + H4 pairs alone are necessary and sufficient for the formation on DNA of complexes exhibiting many of the structural features typical of nucleosome cores.

Because of contradictory evidence it is not clear if H3 + H4 octamers, or only tetramers, are required to organize and protect core length DNA in this manner. Under certain conditions, combination of H3 + H4 with 140 bp DNA yields compact ($s_{20,w}$ = 9.8 S) particles resembling nucleosome cores, but the histone/DNA stoichiometry of these complexes is unsettled (Bina-Stein and Simpson, 1977; Camerini-Otero et al., 1978). Klevan et al. (1978) have concluded that the complex formed, when $(H3)_2(H4)_2$ tetramers and 140 bp DNA are mixed in equimolar amounts, is highly asymmetric and/or hydrated, based on its low sedimentation coefficient ($s_{20,w}$ = 6.45 S, f/f_0 = 2.1 for $\bar{v}$ = 0.65), low rotational diffusion coefficient, and relatively large linear dichroism. Direct addition of H2A + H2B to this complex caused a transition to a structure very similar to nucleosome cores. Micrococcal nuclease digestion of (H3 + H4)–DNA complexes yielded a nucleoprotein with $\sim$130–140 bp DNA that Moss et al. (1977) have inferred contains a histone octamer, based on the increased yield of this particle with increasing input histone/DNA ratio.

A different sort of indication of relative structural roles of specific histone pairs has been obtained from studies of subnucleosomes isolated directly from chromatin digests. Approximately 50 bp (67%) of the 74 bp DNA fragments found associated with H3 + H4 pairs thermally denature in a sharp transition with a T_m nearly identical to that of intact nucleosome cores, and even the remaining DNA is significantly stabilized ($T_m \cong 62°C$). The circular dichroism of this DNA, from 260 to 300 nm, is also significantly supressed by H3 and H4. In contrast, the circular dichroism of the 54 bp fragments associated with H2A + H2B pairs is similar to that of free DNA, and only 25% (~ 14 bp) of this DNA is extremely stable to denaturation, while the remainder denatures over a wide range with a $T_m \cong 62°C$ (Mencke and Rill, 1978). Thus a single arginine-rich histone *pair* is able to interact strongly with half-nucleosome length (~ 70 bp) DNA—much more DNA than their slightly lysine-rich counterparts, and much more DNA than can be accounted for in terms of purely electrostatic interactions. The degree of folding of 70 bp DNA by the arginine-rich pair is not known.

V. CONCLUDING REMARKS

The primary structure of eukaryotic chromatin is now firmly established as based on the semiperiodic wrapping of DNA about octamers of histones H2A, H2B, H3, and H4, forming repeating structural subunits or "nucleosomes." This knowledge and our present information about the general features of nucleosome structure and assembly have gone far in explaining many physical features of chromatin fibrils and in relating the chemical and physical properties of histones to their roles as the dominant structural organizers of chromatin, thus providing a necessary conceptual framework for understanding molecular aspects of gene function. Yet the picture of nucleosomes that has emerged is rather static and sterically constrained, raising interesting questions about accessibilities of specific gene sequences and the fate of nucleosomes in active processes. Insights are needed into the structural distinctions between chromatin regions that are permanently repressed, actively engaged in transcription, capable of future activation and expression, or are of other functional significance.

The evolutionary stability of core histones guarantees a monotony of structural repetition consistent with the inactivity of most DNA in a typical eukaryotic cell. Enormous potential exists for the relief of this monotony in functionally distinctive regions through the devices of posttranslational histone modifications, programmed changes in histone sequence variant populations, interactions with nonhistone proteins, and conformational fluctuations. The unique sequence variations and special structural role of histone H1 are likely also to be of considerable importance. Although some progress has been made, relatively little is known about the relationships between these factors and nucleosome structure–function. Most exciting in this regard are recent reports that the preferential sensitivity of active genes in chromatin to DNase I digestion (Weintraub and Groudine, 1976; Garel and Axel, 1976) may be related to histone acetylation (Nelson *et al.*, 1978a; Simpson,1978b; Vidali *et al.*, 1978). It is hoped the functional significance of nucleosomes will become clear as more correlations of this type are made and affirmed, and as the internal architecture and key interactions of nucleosomes are identified.

ACKNOWLEDGMENTS

The author wishes to thank D. Nelson, A. Mencke, J. Jackson, S. Chambers, and D. Oosterhof for their contributions to the research efforts of this laboratory over the past few years. This research was supported by funds from the National Institutes of Health and the Department of Energy. The author is a recipient of a USPHS Career Development Award.

REFERENCES

Altenburger, W., Horz, W., and Zachau, H. G. (1976). *Nature (London)* **264**, 517–522.
Axel, R. (1975). *Biochemistry* **14**, 2921–2925.
Axel, R., Cedar, H., and Felsenfeld, G. (1975). *Biochemistry* **14**, 2489–2495.
Bafus, N. L., Albright, S. C., Todd, R. D., and Garrard, W. T. (1978). *J. Biol. Chem.* **253**, 2568–2574.
Bailey, G. S., and Dixon, G. H. (1973). *J. Biol. Chem.* **248**, 5463–5472.
Bakayev, V. V., Bakayeva, T. G., and Varshavsky, A. J. (1977). *Cell* **11**, 619–629.
Baker, C. C., and Isenberg, I. (1976). *Biochemistry* **15**, 629.
Baldwin, J. P., Boseley, P. G., Bradbury, E. M., and Ibel, K. (1975). *Nature (London)* **253**, 245–249.
Bartley, J. A., and Chalkley, R. G. (1968). *Biochim. Biophys. Acta* **160**, 224.
Bartley, J. A., and Chalkley, R. (1972). *J. Biol. Chem.* **247**, 3647–3655.
Bartley, J. A., and Chalkley, R. (1973). *Biochemistry* **12**, 468–474.
Bina-Stein, M., and Simpson, R. T. (1977). *Cell* **11**, 609–618.
Blankstein, L. A., and Levy, S. B. (1976). *Nature (London)* **260**, 638–640.
Bloomfield, V. A., Crothers, D. M., and Tinoco, I. (1974). *In* "Physical Chemistry of Nucleic Acids." Harper, New York.
Bohm, L., Hayashi, H., Cary, P. D., Moss, T., Crane-Robinson, C., and Bradbury, E. M. (1977). *Eur. J. Biochem.* **77**, 484–494.
Bolund, L. A., and Johns, E. W. (1973). *Eur. J. Biochem.* **35**, 546–553.
Bonner, W. M., and Pollard, H. B. (1975). *Biochem. Biophys. Res. Commun.* **64**, 282–288.

Bosely, P. G., Bradbury, E. M., Butler-Browne, G. S., Carpenter, B. G., and Stephens, R. M. (1976). *Eur. J. Biochem.* **62**, 21–31.
Bostock, C. J., Christie, S., and Hatch, F. T. (1976). *Nature (London)* **262**, 516–519.
Bradbury, E. M. (1975). *Ciba Found. Symp.* No. 28 (New Ser.), pp. 1–4.
Bradbury, E. M., and Crane-Robinson, C. (1971). *In* "Histones and Nucleohistones" (D. M. P. Phillips, ed.), pp. 85–134. Plenum, New York.
Bradbury, E. M., and Rattle, H. W. E. (1972). *Eur. J. Biochem.* **27**, 270–281.
Bradbury, E. M., Cary, P. D., Crane-Robinson, C., Riches, P. L., and Johns, E. W. (1972). *Eur. J. Biochem.* **26**, 482–489.
Bradbury, E. M., Inglis, R. J., Matthews, H. R., and Sarner, N. (1973a). *Eur. J. Biochem.* **33**, 131.
Bradbury, E. M., Cary, P. D., Crane-Robinson, C., and Rattle, H. W. E. (1973b). *Ann. N.Y. Acad. Sci.* **222**, 226.
Bradbury, E. M., and The Biophysics Group. (1975). *Ciba Found. Symp.* No. 28 (New Ser.), pp. 131–148.
Bradbury, E. M., Cary, P. D., Chapman, G. E., Crane-Robinson, C., Danby, S. E., Rattle, H. W. E., Boublik, M., Palau, J., and Auiler, F. J. (1975a). *Eur. J. Biochem.* **52**, 605–613.
Bradbury, E. M., Cary, P. D., Crane-Robinson, C., Rattle, H. W. E., Boublik, M., and Sautiere, P. (1975b). *Biochemistry* **14**, 1876–1885.
Bram, S., and Ris, H. (1971). *J. Mol. Biol.* **55**, 325–336.
Brandt, W. F., and von Holt, C. (1972). *FEBS Lett.* **23**, 357–360.
Brandt, W. F., and von Holt, C. (1974a). *Eur. J. Biochem.* **46**, 407–417.
Brandt, W. F., and von Holt, C. (1974b). *Eur. J. Biochem.* **46**, 419–429.
Brandt, W. F., Strickland, W. F., and von Holt, C. (1974a). *FEBS Lett.* **40**, 167–172.
Brandt, W. F., Strickland, W. F., and von Holt, C. (1974b). *FEBS Lett.* **40**, 349–352.
Camerini-Otero, R. D., and Felsenfeld, G. (1977a). *Nucleic Acids Res.* **4**, 1159–1181.
Camerini-Otero, R. D., and Felsenfeld, G. (1977b). *Proc. Natl. Acad. Sci. U.S.A.* **74**, 5519–5523.
Camerini-Otero, R. D., Sollner-Webb, B., and Felsenfeld, G. (1976). *Cell* **8**, 333–347.
Camerini-Otero, R. D., Sollner-Webb, B., Simon, R. H., Williamson, P., Zasloff, M., and Felsenfeld, G. (1978). *Cold Spring Harbor Symp. Quant. Biol.* **42**, 57–75.
Candido, E. P. M., and Dixon, G. H. (1972). *Proc. Natl. Acad. Sci. U.S.A.* **69**, 2015–2019.
Clark, R. J., and Felsenfeld, G. (1971). *Nature (London), New Biol.* **229**, 101–106.
Cohen, L. H., Newrock, K. M., and Zweider, A. (1975). *Science* **190**, 994–997.
Compton, J. L., Bellard, M., and Chambon, P. (1976). *Proc. Natl. Acad. Sci. U.S.A.* **73**, 4382–4386.
Cotter, R. I., and Lilley, D. M. J. (1977). *FEBS Lett.* **82**, 63–68.
Crick, F. H. C. (1953). *Acta Crystallogr.* **6**, 685.
Crick, F. H. C. (1976). *Proc. Natl. Acad. Sci. U.S.A.* **73**, 2639–2643.
Crick, F. H. C., and Klug, A. (1975). *Nature (London)* **255**, 530–533.
D'Anna, J. A., Jr., and Isenberg, I. (1972). *Biochemistry* **11**, 4017.
D'Anna, J. A., Jr., and Isenberg, I. (1973). *Biochemistry* **12**, 1035–1043.
D'Anna, J. A., Jr., and Isenberg, I. (1974a). *Biochemistry* **12**, 2093–2097.
D'Anna, J. A., Jr., and Isenberg, I. (1974b). *Biochemistry* **13**, 2098–2104.
D'Anna, J. A., Jr., and Isenberg, I. (1974c). *Biochemistry* **13**, 4987–4991.
D'Anna, J. A., Jr., and Isenberg, I. (1974d). *Biochemistry* **13**, 4992–4997.
D'Anna, J. A., Jr., and Isenberg, I. (1974e). *Biochem. Biophys. Res. Commun.* **61**, 343–347.

Dayhoff, M. O., ed. (1972). "Atlas of Protein Sequence and Structure," Vol. 5. Natl. Biomed. Res. Found., Silver Spring, Maryland.

DeLange, R. J., and Smith, E. L. (1975). *Struct. Funct. Chromatin, Ciba Found. Symp.* No. 28 (New Ser.), pp. 59–69.

DeLange, R. J., Fambrough, D. M., Smith, E., and Bonner, J. (1969).*J. Biol. Chem.* **244,** 319–334 and 5669–5679.

DeLange, R. J., Hooper, J. A., and Smith, E. L. (1972). *Proc. Natl. Acad. Sci. U.S.A.* **69,** 882–884.

DeLange, R. J., Hooper, J. A., and Smith, E. L. (1973). *J. Biol. Chem.* **248,** 3261–3274.

Depew, R. E., and Wang, J. C. (1975). *Proc. Natl. Acad. Sci. U.S.A.* **72,** 4275–4279.

Desai, L., Ogawa, Y., Mauritzen, C. M., Taylor, C. W., and Starbuck, W. C. (1969). *Biochim. Biophys. Acta* **181,** 146–153.

Dixon, G. H., Candido, E. P. M., Honda, B. M., Louie, A. J., Macleod, A. R., and Sung, M. T. (1975). *Struct. Funct. Chromatin, Ciba Found. Symp.* No. 28 (New Ser.), pp. 229–249.

Eickbush, T. H., and Moudrianakis, E. N. (1978). *Cell* **13,** 295–306.

Elgin, S., and Weintraub, H. (1975). *Annu. Rev. Biochem.* **44,** 726–773.

Fasman, G. D., Chou, P. Y., and Adler, A. J. (1976). *Biophys. J.* **16,** 1201–1238.

Felden, A. R., Sanders, M. M., and Morris, N. R. (1976). *J. Cell Biol.* **68,** 430–439.

Felsenfeld, G. (1975). *Nature (London)* **257,** 177–178.

Felsenfeld, G. (1978). *Nature (London)* **271,** 115–122.

Finch, J. T., Lutter, L. C., Rhodes, D., Brown, R. S., Rushton, R., Levitt, M., and Klug, A. (1977). *Nature (London)* **269,** 29–36.

Franklin, S. G., and Zweidler, A. (1977). *Nature (London)* **266,** 273–275.

Frearson, P. M., and Crawford, L. V. (1972). *J. Gen. Virol.* **14,** 141–155.

Fuller, F. B. (1971). *Proc. Natl. Acad. Sci. U.S.A.* **68,** 815–819.

Fuller, F. B. (1972). *Rev. Roum. Math. Pures Appl.* **17,** 1329–1334.

Garel, A., and Axel, R. (1976). *Proc. Natl. Acad. Sci. U.S.A.* **73,** 3966–3970.

Georgiev, G. P., Ilyin, Y. V., Tikhonenko, A. S., and Stelmashchok, V. Y. (1970). *Mol. Biol. (Moscow)* **4,** 246–255.

Germond, J. E., Hirt, B., Oudet, P., Gross-Bellard, M., and Chambon, P. (1975). *Proc. Natl. Acad. Sci. U.S.A.* **72,** 1843–1847.

Goff, C. G. (1976). *J. Biol. Chem.* **251,** 4131–4138.

Goodwin, D. C., and Brahms, J. (1978). *Nucleic Acids Res.* **5,** 835–850.

Gorovsky, M. A., and Keevert, J. G. (1975a). *Proc. Natl. Acad. Sci. U.S.A.* **72,** 2672–2676.

Gorovsky, M. A., and Keevert, J. B. (1975b). *Proc. Natl. Acad. Sci. U.S.A.* **72,** 3536–3540.

Gorovsky, M. A., Pleger, G. L., Keevert, J. B., and Johmann, C. A. (1973). *J. Cell Biol.* **57,** 773–781.

Griffith, J. D. (1975). *Science* **187,** 1202–1203.

Hanlon, S., Johnson, R., Wolf, B., and Chan, A. (1972). *Proc. Natl. Acad. Sci. U.S.A.* **69,** 3263.

Hanlon, S., Brudno, S., Wu, T. T., and Wolf, B. (1975). *Biochemistry* **14,** 1648–1660.

Hanlon, S., Glonek, T., and Chan, A. (1976). *Biochemistry* **15,** 3869–3875.

Hardison, R., and Chalkley, R. (1977). *Methods Cell Biol.* **17,** 235–252.

Hardison, R., Eichner, M. E., and Chalkley, R. (1975). *Nucleic Acids Res.* **2,** 1751–1759.

Harrington, R. E. (1977). *Nucleic Acids Res.* **4,** 3519–3535.

Hayashi, H., Iwai, K., Johnson, J. D., and Bonner, J. (1977).*J. Biochem. (Tokyo)* **82,** 503–510.

Hewish, D. R., and Burgoyne, L. A. (1973). *Biochem. Biophys. Res. Commun.* **52,** 504–510.

Hjelm, R. P., Kneale, G. G., Suau, P., Baldwin, J. P., and Bradbury, E. M. (1977). *Cell* **10**, 139–151.

Honda, B. M., Baillie, D. L., and Candido, E. P. M. (1975). *J. Biol. Chem.* **250**, 4643–4647.

Hooper, J. A., Smith, E. L., Sommer, K. R., and Chalkley, R. (1973). *J. Biol. Chem.* **248**, 3275–3279.

Hyde, J. E., and Walker, I. O. (1975). *FEBS Lett.* **50**, 150–154.

Isenberg, I. (1977). *In* "Search and Discovery" (B. Kaminer, ed.), pp. 195–215. Academic Press, New York.

Ivanov, V. I., Minchenkova, L. E., Schyolkima, A. K., and Poletayev, A. I. (1973). *Biopolymers* **12**, 89.

Iwai, K., Hayashi, H., and Ishikawa, K. (1972). *J. Biochem. (Tokyo)* **72**, 357–367.

Jackson, J. B., Pollock, J. M., and Rill, R. L. (1979). *Biochemistry* (in press).

Joffe, J., Keene, M., and Weintraub, H. (1977). *Biochemistry* **16**, 1236–1238.

Johns, E. W. (1971). *In* "Histones and Nucleohistones" (P. M. P. Phillips, ed.), pp. 1–45. Plenum, New York.

Johnson, E. M., Littau, V. C., Allfrey, V. G., Bradbury, E. M., and Matthews, H. R. (1976). *Nucleic Acids Res.* **3**, 3313–3329.

Johnson, R. S., Chan, A., and Hanlon, S. (1972). *Biochemistry* **11**, 4347.

Keichline, L. D., Villee, C. A., and Wasserman, P. M. (1976). *Biochim. Biophys. Acta* **425**, 84–94.

Keller, W. (1975). *Proc. Natl. Acad. Sci. U.S.A.* **72**, 4876–4880.

Keller, W., and Wendel, I. (1974). *Cold Spring Harbor Symp. Quant. Biol.* **39**, 199–208.

Klevan, L., and Crothers, D. M. (1977). *Nucleic Acids Res.* **4**, 4077–4089.

Klevan, L., Dattagupta, N., Hogan, M., and Crothers, D. M. (1978). *Biochemistry* **17**, 4533–4539.

Kootstra, A., and Bailey, G. S. (1976). *FEBS Lett.* **68**, 76–78.

Kornberg, R. (1974). *Science* **184**, 868–871.

Kornberg, R. D. (1977). *Annu. Rev. Biochem.* **46**, 931–954.

Kornberg, R. D., and Thomas, J. O. (1974). *Science* **184**, 865–868.

Kovacic, R. T., and Van Holde, K. E. (1977). *Biochemistry* **16**, 1490–1498.

Lacy, E., and Axel, R. (1975). *Proc. Natl. Acad. Sci. U.S.A.* **72**, 3978–3982.

Laine, B., Sautiere, P., and Biserte, G. (1976). *Biochemistry* **15**, 1640–1645.

Lake, R. S., Barban, S., and Salzman, N. P. (1973). *Biochem. Biophys. Res. Commun.* **54**, 640–647.

Langmore, J. P., and Wooley, J. C. (1975). *Proc. Natl. Acad. Sci. U.S.A.* **72**, 2691–2695.

Leffak, I. M., Grainger, R., and Weintraub, H. (1977). *Cell* **12**, 837–846.

Levina, E. S., and Mirzabekov, A. D. (1975). *Dokl. Akad. Nauk SSSR* **221**, 1222–1225.

Levitt, M. (1978). *Proc. Natl. Acad. Sci. U.S.A.* **75**, 640–644.

Lewis, P. N., Bradbury, E. M., and Crane-Robinson, C. (1975). *Biochemistry* **14**, 3391–3400.

Li, H. J. (1975). *Nucleic Acids Res.* **2**, 1275–1290.

Li, H. J., and Bonner, J. (1971). *Biochemistry* **10**, 1461–1470.

Li, H. J., Wickett, R., Craig, A. M., and Isenberg, I. (1972). *Biopolymers* **11**, 375.

Lilley, D. M. J., and Tatchell, K. (1977). *Nucleic Acids Res.* **4**, 2039–2056.

Lilley, D. M. J., Howarth, O. W., Clark, V. M., Pardon, J. F., and Richards, B. M. (1975). *Biochemistry* **14**, 4590–4600.

Lilley, D. M. J., Pardon, J. F., and Richards, B. M. (1977). *Biochemistry* **16**, 2853–2860.

Lipchitz, L., and Axel, R. (1976). *Cell* **9**, 355–364.

Lipps, H. J., and Morris, N. R. (1977). *Biochem. Biophys. Res. Commun.* **74**, 230–234.

Lohr, D., Corden, J., Tatchell, K., Kovacic, R. T., and Van Holde, K. E. (1977a). *Proc. Natl. Acad. Sci. U.S.A.* **74,** 79–83.

Lohr, D., Tatchell, K., and Van Holde, K. E. (1977b). *Cell* **12,** 829–836.

Luck, J. M. (1964). In "The Nucleohistones" (J. Bonner and P. O. P. Ts'o, eds.), pp. 3–12. Holden-Day, San Francisco, California.

Lutter, L. C. (1978). *Cold Spring Harbor Symp. Quant. Biol.* **42,** 137–147.

McGhee, J. D., and Engel, J. D. (1975). *Nature (London)* **254,** 449–450.

Maniatis, T., Jeffrey, A., and van de Sande, H. (1975). *Biochemistry* **14,** 3787–3794.

Martinson, H. G., and McCarthy, B. J. (1975). *Biochemistry* **14,** 1073–1078.

Martinson, H. G., and McCarthy, B. J. (1976). *Biochemistry* **15,** 4126–4431.

Martinson, H. G., Shetlar, M. D., and McCarthy, B. J. (1976). *Biochemistry* **15,** 2002–2007.

Mathis, D. J., and Gorovsky, M. A. (1976). *Biochemistry* **15,** 750–755.

Melnikova, A. F., Zasedatelev, A. S., Gursky, G. V., Kolchinsky, A. M., Grochovsky, S. L., Zhuze, A. L., and Mirzabekov, A. D. (1975). *Mol. Biol. Rep.* **2,** 135–142.

Mencke, A. J., and Rill, R. L. (1978). In preparation.

Mirzabekov, A. D., Sanko, D. F., Kolchinsky, A. M., and Melnikova, A. F. (1977). *Eur. J. Biochem.* **75,** 379–390.

Morris, G., and Lewis, P. M. (1977). *Eur. J. Biochem.* **77,** 471–477.

Morris, N. R. (1976). *Cell* **8,** 357–363.

Moss, T., Cary, P. D., Crane-Robinson, C., and Bradbury, E. M. (1976). *Biochemistry* **15,** 2261–2267.

Moss, T., Stephens, R. M., Crane-Robinson, C., and Bradbury, E. M. (1977). *Nucleic Acids Res.* **4,** 2477–2486.

Musich, P. R., Brown, F. L., and Maio, J. J. (1977). *Proc. Natl. Acad. Sci. U.S.A.* **74,** 3297–3301.

Nelson, D. A., Beltz, W. R., and Rill, R. L. (1977a). *Proc. Natl. Acad. Sci. U.S.A.* **74,** 1343–1347.

Nelson, D. A., Oosterhof, D. K., and Rill, R. L. (1977b). *Nucleic Acids Res.* **4,** 4223–4234.

Nelson, D. A., Perry, W. M., and Chalkley, R. (1978a). *Biochem. Biophys. Res. Commun.* **82,** 356–363.

Nelson, D. A., Oosterhof, D. K., Mencke, A. J., and Rill, R. L. (1978b). Submitted for publication.

Nelson, R. G., and Johnson, W. C., Jr. (1970). *Biochem. Biophys. Res. Commun.* **41,** 211.

Noll, M. (1974a). *Nature (London)* **251,** 249–251.

Noll, M. (1974b). *Nucleic Acids Res.* **1,** 1573–1578.

Noll, M. (1976). *Cell* **8,** 349–355.

Noll, M. (1978). *Cold Spring Harbor Symp. Quant. Biol.* **42,** 77–85.

Noll, M., and Kornberg, R. D. (1977). *J. Mol. Biol.* **109,** 393–404.

Ogawa, Y., Quagliarotti, G., Jordan, J., Taylor, C. W., Starbuck, W. C., and Busch, H. (1969). *J. Biol. Chem.* **244,** 4387–4392.

Ohlenbusch, H. H., Olivera, B. M., Tuan, D., and Davidson, N. (1967). *J. Mol. Biol.* **25,** 299–310.

Olins, A. L. (1978). *Methods Cell Biol.* **18,** 61–68.

Olins, A. L., and Olins, D. E. (1973). *J. Cell Biol.* **59,** 252a.

Olins, A. L., Carlson, R. D., and Olins, D. E. (1975). *J. Cell Biol.* **64,** 528–537.

Olins, A. L., Carlson, R. D., Wright, E. B., and Olins, D. E. (1976a). *Nucleic Acids Res.* **3,** 3271–3291.

Olins, A. L., Senior, M. B., and Olins, D. E. (1976b). *J. Cell Biol.* **68,** 787–793.

Olins, D. E., and Olins, A. L. (1974). *Science* **183**, 330–332.
Olins, D. E., Bryan, P. M., Harrington, R. E., Hill, W. E., and Olins, A. L. (1977). *Nucleic Acids Res.* **4**, 1911–1931.
Oliver, D., and Chalkley, R. (1974). *Biochemistry* **13**, 5093–5097.
Olson, M. O. J., Jordan, J., and Busch, H. (1972). *Biochem. Biophys. Res. Commun.* **46**, 50–55.
Oosterhof, D. K., Hozier, J. C., and Rill, R. L. (1975). *Proc. Natl. Acad. Sci. U.S.A.* **72**, 633–637.
Ord, M. G., and Stocken, L. A. (1975). *Struct. Funct. Chromatin, Ciba Found. Symp.* No. 28 (New Ser.), pp. 259–265.
Oudet, P., Gross-Bellard, M., and Chambon, P. (1975). *Cell* **4**, 281–300.
Oudet, P., Spadafora, C., and Chambon, P. (1978). *Cold Spring Harbor Symp. Quant. Biol.* **42**, 301–312.
Palau, J., and Padros, E. (1972). *FEBS Lett.* **27**, 157–160.
Panyim, S., Bilek, D., and Chalkley, R. (1971). *J. Biol. Chem.* **246**, 4206–4215.
Pardon, J. F., Wilkins, M. H. F., and Richards, B. M. (1967). *Nature (London)* **215**, 508–509.
Pardon, J. F., Worcester, D. L., Wooley, J. C., Tatchell, K., Van Holde, K. E., and Richards, B. M. (1975). *Nucleic Acids Res.* **2**, 2163–2176.
Pardon, J. F., Worcester, D. L., Wooley, J. C., Cotter, R. I., Lilley, D. M. J., and Richards, B. M. (1977). *Nucleic Acids Res.* **4**, 3199–3214.
Patthy, L., Smith, E. L., and Johnson, J. (1973). *J. Biol. Chem.* **248**, 6834–6840.
Pekary, A. E., Li, H. J., Chan, S. I., Hsu, C. J., and Wagner, T. E. (1975). *Biochemistry* **14**, 1177–1183.
Piper, P. W., Celis, J., Kaltoft, K., Leer, J. C., Nielsen, O. F., and Westergaard, O. (1976). *Nucleic Acids Res.* **3**, 493–505.
Prunell, A., Baer, B., and Kornberg, R. D. (1978). *Cold Spring Harbor Symp. Quant. Biol.* **42**, 103–108.
Pulleybank, D. E., Shure, M., Tang, D., Vinograd, J., and Vosberg, H. P. (1975). *Proc. Natl. Acad. Sci. U.S.A.* **72**, 4280–4284.
Rall, S. C., Okinaka, R. T., and Strniste, G. F. (1977). *Biochemistry* **16**, 4940–4943.
Reeves, R. (1976). *Science* **194**, 529–532.
Reeves, R., and Jones, A. (1976). *Nature (London)* **260**, 495–500.
Richards, B. M., and Pardon, J. F. (1970). *Exp. Cell Res.* **62**, 184–196.
Riley, D., and Weintraub, H. (1978). *Cell* **13**, 281–293.
Rill, R. L. (1972). *Biopolymers* **11**, 1929.
Rill, R. L., and Nelson, D. A. (1978). *Cold Spring Harbor Symp. Quant. Biol.* **42**, 475–482.
Rill, R. L., and Van Holde, K. E. (1973). *J. Biol. Chem.* **248**, 1080–1083.
Rill, R. L., Oosterhof, D. K., Hozier, J. C., and Nelson, D. A. (1975). *Nucleic Acids Res.* **2**, 1525–1538.
Rill, R. L., Nelson, D. A., Oosterhof, D. K., and Hozier, J. C. (1977). *Nucleic Acids Res.* **4**, 771–790.
Rill, R. L., Shaw, B. R., and Van Holde, K. E. (1978a). *Methods Cell Biol.* **18**, 69–103.
Rill, R. L., Nelson, D. A., and Oosterhof, D. K. (1978b). Submitted for publication.
Rizzo, P. J., and Nooden, L. D. (1972). *Science* **176**, 796–797.
Roark, D. E., Geoghegan, T. E., and Keller, G. H. (1974). *Biochem. Biophys. Res. Commun.* **59**, 542–547.
Roark, D. E., Geoghegan, T. E., Keller, G. H., Matter, K. V., and Engle, R. L. (1976). *Biochemistry* **15**, 3019–3025.

Roblin, R., Harle, E., and Dulbecco, R. (1971). *Virology* **45**, 555–566.

Rosenberg, B. H. (1976). *Biochem. Biophys. Res. Commun.* **72**, 1384–1391.

Ruiz-Carrillo, A., Waugh, L. J., and Allfrey, U. G. (1975). *Arch. Biochem. Biophys.* **174**, 273–290.

Sahasrabuddhe, C. G., and Van Holde, K. E. (1974). *J. Biol. Chem.* **249**, 152–156.

Sautiere, P. (1975). *Struct. Funct. Chromatin, Ciba Found. Symp.* No. 28 (New Ser.), pp. 77–88.

Sautiere, P., Tyrou, D., Moschetto, Y., and Biserte, G. (1971a). *Biochimie* **53**, 479–483.

Sautiere, P., Lambelin-Breynaert, M. D., Moschetto, Y., and Biserte, G. (1971b). *Biochimie* **53**, 711–715.

Sautiere, P., Tyrou, D., Laine, B., Mizon, J., Ruffin, P., and Biserte, G. (1974). *Eur. J. Biochem.* **41**, 563–576.

Schiffer, M., and Edmundson, A. B. (1967). *Biophys. J.* **7**, 121.

Shaw, B. R., Corden, J. L., Sahasrabuddhe, C. G., and Van Holde, K. E. (1974). *Biochem. Biophys. Res. Commun.* **61**, 1193–1198.

Shaw, B. R., Herman, T. M., Kovacic, R. T., Beaudreau, G. S., and Van Holde, K. E. (1976). *Proc. Natl. Acad. Sci. U.S.A.* **73**, 505–509.

Shih, T. Y., and Bonner, J. (1970). *J. Mol. Biol.* **48**, 469–487.

Shmatchenko, V. V., and Varshavsky, A. J. (1978). *Anal. Biochem.* **85**, 42–46.

Simpson, R. T. (1970). *Biochemistry* **9**, 4814–4818.

Simpson, R. T. (1976). *Proc. Nat. Acad. Sci. U.S.A.* **73**, 4400–4404.

Simpson, R. T. (1978a). *Nucleic Acids Res.* **5**, 1109–1119.

Simpson, R. T. (1978b). *Cell* **13**, 691–699.

Simpson, R. T., and Whitlock, J. P. (1976). *Cell* **9**, 347–354.

Small, E. W., Craig, A. M., and Isenberg, I. (1973). *Biopolymers* **12**, 1149.

Smart, J. E., and Bonner, J. (1971). *J. Mol. Biol.* **58**, 651–659.

Smerdon, M. J., and Isenberg, I. (1973). *Biochem. Biophys. Res. Commun.* **55**, 1029.

Smerdon, M. J., and Isenberg, I. (1974). *Biochemistry* **13**, 4046.

Smith, E. L. (1975). *Struct. Funct. Chromatin, Ciba Found. Symp.* No. 28 (New Ser.), pp. 307–308.

Sobell, H. M., Tsai, C. C., Gilbert, S. G., Jain, S. C., and Sakore, T. D. (1976). *Proc. Natl. Acad. Sci. U.S.A.* **73**, 3068–3072.

Sollner-Webb, B., and Felsenfeld, G. (1975). *Biochemistry* **14**, 2915–2920.

Sollner-Webb, B., and Felsenfeld, G. (1977). *Cell* **10**, 537–547.

Sollner-Webb, B., Camerini-Otero, R. D., and Felsenfeld, G. (1976). *Cell* **9**, 179–193.

Spadafora, C., Bellard, M., Compton, J. L., and Chambon, P. (1976). *FEBS Lett.* **69**, 281–285.

Sperling, R., and Amos, L. A. (1977). *Proc. Natl. Acad. Sci. U.S.A.* **74**, 3772–3776.

Sperling, R., and Bustin, M. (1974). *Proc. Natl. Acad. Sci. U.S.A.* **71**, 4625–4629.

Sperling, R., and Bustin, M. (1975). *Biochemistry* **14**, 3322–3331.

Sperling, R., and Bustin, M. (1976). *Nucleic Acids Res.* **3**, 1263–1275.

Stein, A., Bina-Stein, M., and Simpson, R. T. (1977). *Proc. Natl. Acad. Sci. U.S.A.* **74**, 2780–2784.

Strickland, M., Strickland, W. N., and Brandt, W. F. (1974). *FEBS Lett.* **40**, 346–348.

Studdert, D. S., Patroni, M., and Davis, R. C. (1972). *Biopolymers* **11**, 761.

Suau, P., Kneale, G. G., Braddock, G. W., Baldwin, T. P., and Bradbury, E. M. (1977). *Nucleic Acids Res.* **4**, 3769.

Sung, M. T., and Dixon, G. H. (1970). *Proc. Natl. Acad. Sci. U.S.A.* **67**, 1616–1623.

Sussman, J. L., and Trifonov, E. N. (1978). *Proc. Natl. Acad. Sci. U.S.A.* **75**, 103–107.

Tancredi, T., Temussi, P. A., Paolillo, L., Trivellone, E., Crane-Robinson, C., and Bradbury, E. M. (1976). *Eur. J. Biochem.* **70**, 403–408.

Tatchell, K., and Van Holde, K. E. (1977). *Biochemistry* **16**, 5295–5303.
Thomas, G. J., Prescott, B., and Olins, D. (1977). *Science* **197**, 385–388.
Thomas, J. (1978). *Cold Spring Harbor Symp. Quant. Biol.* **42**, 119–125.
Thomas, J. O., and Farber, V. (1976). *FEBS Lett.* **66**, 274–279.
Thomas, J. O., and Kornberg, R. D. (1975a). *Proc. Natl. Acad. Sci. U.S.A.* **72**, 2626–2630.
Thomas, J. O., and Kornberg, R. D. (1975b). *FEBS Lett.* **58**, 353–358.
Thomas, J. O., and Thompson, R. J. (1977). *Cell* **10**, 633–640.
Tien-Kuo, M., Sahasrabuddhe, C. G., and Saunders, G. F. (1976). *Proc. Natl. Acad. Sci. U.S.A.* **73**, 1572–1575.
Todd, R. D., and Garrard, W. T. (1977). *J. Biol. Chem.* **252**, 4729–4738.
Trifonov, E. (1978). *Nucleic Acids Res.* **5**, 1371–1380.
Tunis-Schneider, M. J. B., and Maestre, M. F. (1970). *J. Mol. Biol.* **52**, 521–541.
van der Westhuyzen, D. R., and von Holt, C. (1971). *FEBS Lett.* **14**, 333.
Van Holde, K. E., and Isenberg, I. (1975). *Acc. Chem. Res.* **8**, 327.
Van Holde, K. E., Sahasrabuddhe, C. G., Shaw, B. R., van Bruggen, E. F. J., and Arnberg, A. C. (1974a). *Biochem. Biophys. Res. Commun.* **60**, 1365–1370.
Van Holde, K. E., Sahasrabuddhe, C. G., and Shaw, B. R. (1974b). *Nucleic Acids Res.* **1**, 1597.
Van Lente, F., Jackson, J. F., and Weintraub, H. (1975). *Cell* **5**, 45–50.
Varshavsky, A. J., Bakayev, V. V., and Georgiev, G. P. (1976a). *Nucleic Acids Res.* **3**, 477–491.
Vidali, G., Boffa, L. C., Bradbury, E. M., and Allfrey, V. G. (1978). *Proc. Natl. Acad. Sci. U.S.A.* **75**, 2239–2243.
Vogt, V. M., and Braun, R. (1976). *FEBS Lett.* **64**, 190–192.
Weintraub, H. (1975). *Proc. Natl. Acad. Sci. U.S.A.* **72**, 1212–1216.
Weintraub, H. (1976). *Cell* **9**, 419–422.
Weintraub, H. (1978). *Nucleic Acids Res.* **5**, 1179–1188.
Weintraub, H., and Groudine, M. (1976). *Science* **193**, 848–856.
Weintraub, H., and Van Lente, F. (1974). *Proc. Natl. Acad. Sci. U.S.A.* **71**, 4249–4253.
Weintraub, H., Palter, K., and Van Lente, F. (1975a). *Cell* **6**, 85–100.
Weintraub, H., Van Lente, F., and Blumenthal, R. (1975b). *Struct. Funct. Chromatin, Ciba Found. Symp.* No. 28 (New Ser.), pp. 291–306.
Weintraub, H., Worcel, A., and Alberts, B. (1976). *Cell* **9**, 409–417.
Whitlock, J. P., and Simpson, R. T. (1976). *Biochemistry* **15**, 3307–3314.
Wickett, R., Li, H. J., and Isenberg, I. (1972). *Biochemistry* **11**, 2952.
Woodcock, C. L. F. (1973). *J. Cell Biol.* **59**, 368a.
Woodcock, C. L. F., Safer, J. P., and Stanchfield, J. E. (1976). *Exp. Cell Res.* **97**, 101–110.
Yeoman, L. C., Olson, M. D. J., Sugano, N., Jordan, J. J., Taylor, C. W., Starbuck, W. C., and Busch, H. (1972). *J. Biol. Chem.* **247**, 6018–6023.
Zimmer, C., and Luck, G. (1973). *Biochim. Biophys. Acta* **312**, 215.
Zubay, G., and Doty, P. (1959). *J. Mol. Biol.* **1**, 1–20.

Chapter VII

Nucleosomes and Higher Levels of Chromosomal Organization

JOHN C. HOZIER

I. INTRODUCTION

Eukaryotic chromosome morphology at the light microscopic level was an active and productive field of basic research long before much was known about the molecular components of chromosomes. Most of this work was concerned with cells in mitosis when chromosomes are recognizable entities and led to the concept of chromosomes as packaging units for the segregation of genetic information during cell divi-

315

MOLECULAR GENETICS, PART III

sion. Less can be learned from studies of interphase nuclei by light microscopy when the chromosomes are dispersed and in a functional state. Important exceptions are specialized forms, including polytene chromosomes found in dipteran species and lampbrush chromosomes in *Xenopus laevis* and other species, where the chromosomes can be seen and their structure related to their functions, particularly RNA transcription.

One of the main principles that has emerged from cytological studies is that chromosome morphology is highly variable among different species, and among different tissues and stages of the cell cycle within the same species. However, electron microscopic and biochemical studies of chromosomes have demonstrated a remarkable similarity of the components among different species and among chromosomes with different morphology. This chapter will deal with those molecular aspects of chromosome structure which are common to many species and will show how the various structural components may be arranged in chromosomes of different morphology.

Central to such a discussion is the concept of the fibrous nature of chromosomes. That chromosomes are composed of fibers has been well established by electron microscopic and biochemical investigations of mitotic chromosomes and interphase nuclei. The dimensions, composition, and molecular organization of chromosome fibers are being vigorously studied today. A strong impetus to this research has been provided by the nucleosomal concept of DNA–histone interaction, in which groups of histones are associated with DNA in discrete units. (The composition and substructure of nucleosomes have been covered in detail in Chapter VI). This discontinuous structure is superimposed on a very long continuous DNA fiber. The folding of DNA in nucleosomes and in higher-order structures finally produces the native chromosome in its various forms.

Here we will discuss three levels of fiber folding. The first involves the DNA and the nucleosomal histones to form the "beads-on-a-string" configuration recently described for isolated chromatin, disrupted interphase nuclei, and mitotic chromosomes. The second level is the folding of chains of nucleosomes to form the native chromosome fiber. Here histone H1 (which is not an integral part of the nucleosome) appears to be particularly important. The third level involves the organization of fibers in both interphase nuclei and during mitosis. The first two levels of fiber organization have many features in common for a variety of species, cell types, and stages of the cell cycle, while the great diversity in chromosome morphology is associated with the third level of fiber organization. For this reason, the word "chromosome" will be used generally to describe the different fiber arrangements at mitosis, meiosis, and during interphase.

II. THE CONCEPT OF THE CHROMOSOME FIBER

A. CONTINUITY OF DNA IN CHROMOSOMES

1. *DNA as the Fiber "Backbone"*

The belief that chromosomes are fibrous structures has its origin in studies of the structure of the DNA molecule within them. The major question concerns the number of DNA molecules in chromosomes. A metaphase chromatid may contain one continuous DNA double helix (a unineme model) or, at the other extreme, it may contain a multi-stranded complex of many small molecules (a multineme model). The question is posed in terms of the mitotic chromosome, since the first convincing answer was found for this case in experiments designed to investigate the mode of replication of the DNA double helix in eu-karyotic chromosomes (Taylor *et al.*, 1957; Taylor, 1963). In these studies, DNA replication was allowed to proceed in the presence of tritiated thymidine, and the distribution of the labeled DNA was ob-served by autoradiography of chromosomes in the first and second successive mitoses. In the first postlabeling mitosis, both chromatids were labeled to the same extent. But at the second postlabeling mi-tosis, only one chromatid per chromosome was labeled. Figure 1

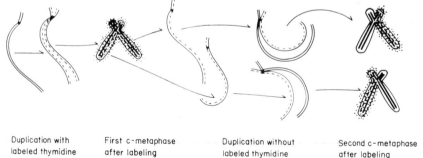

Duplication with   First c-metaphase   Duplication without   Second c-metaphase
labeled thymidine   after labeling   labeled thymidine   after labeling

FIG. 1. Diagrammatic representation of the labeling pattern at mitosis in a DNA au-toradiography experiment. Solid lines represent unlabeled DNA strands. Broken lines represent labeled DNA. Dots (over metaphase figures) demonstrate the patterns of au-toradiographic grains at the first and second postlabeling metaphases. (From Taylor *et al.*, 1957, reproduced by permission.)

shows an interpretation of the autoradiographic labeling pattern based on the now well-established concept of semiconservative replication of DNA. The important result of these experiments to this discussion is that the DNA within each chromatid functions as a unit structure; that is, the complementary chains separate during replication with little or no fragmentation. Therefore, the DNA double helix within each chromosome must consist of one or, at most, only a few very long continuous molecules.

Other evidence based on autoradiography of disrupted interphase chromosomes indicates that DNA also exists in interphase nuclei as very long molecules. In a typical experiment, DNA is labeled with tritiated thymidine to a high specific activity followed by lysis of cells and dispersal of the labeled DNA. After exposure and development of a photographic emulsion, individual DNA fibers are revealed by their autoradiographic "tracks." In some experiments, individual DNA fibers can be followed for several millimeters (Cairns, 1966; Huberman and Riggs, 1966). Here again the primary emphasis of the experiments was not on the length of DNA but rather on the pattern of DNA replication. However, the results support the argument that each chromosome consists of one or only a few long molecules of DNA.

The measurement of viscoelastic recoil of very large DNA molecules in solution has added considerable strength to the concept of DNA fiber continuity in chromosomes. This new development deserves to be described in some detail, since it offers the greatest potential for the measurement of DNA molecules that might be expected from a unineme chromosome model. The technique makes use of a concentric cylinder viscometer modified to follow the viscoelastic recoil of DNA that has been subjected to a shear stress (Fig. 2). The shear stress is applied by rotating the inner cylinder relative to the outer. The DNA random coil, solubilized from cells lysed in a very gentle manner, is distorted in the shear gradient. When the shear

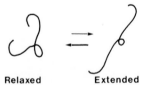

Relaxed   Extended

FIG. 2. Schematic diagram of the viscoelastic recoil of DNA. The DNA is first stretched out in a shear gradient and then allowed to return to the random coil configuration. The relaxation of the DNA in solution is measured in a viscoelastometer, which records the angular rotation of the inner rotor of a Couette viscometer in response to the DNA in solution. (From Kavenoff *et al.*, 1974, reproduced by permission.)

stress is released, the molecules return to a random coil configuration, causing the rotor to revolve in response to their movement. The rotor recoil follows essentially an exponential decay. The longest time constant of this decay is the principal retardation time τ_1. It is a strong function of the molecular weight of the longest DNA molecules in solution. In practice, τ_1 is related to molecular weight by calibration with DNA species of known size.

Measurement of viscoelastic recoil for molecular weight determinations has several advantages over other techniques. It is relatively insensitive to degradation of a fraction of the DNA (Kavenoff and Zimm, 1973), which is quite likely to happen in even the most gentle cell lysis and DNA solubilization procedure. It is insensitive to shear stress and shear-dependent anomalies (Klotz and Zimm, 1972), which are particular problems in viscometric and sedimentation velocity measurements. It does not require radioactive labeling of the DNA, which may produce toxic effects and degradation of the DNA.

One of the most interesting applications of the technique has been the measurement of DNA lengths in various *Drosophila* species (Kavenoff and Zimm, 1973). Proteolytic enzymes, EDTA, and heat were used to inactivate nucleases and to release most proteins associated with the DNA. Table I shows viscoelastic recoil data for various *Drosophila* species and mutants with differing chromosome morphologies. The value of τ_1 for *Drosophila melanogaster* wild type corresponds (on the basis of extrapolation of the empirically derived relationship between τ_1 and molecular weight) to a largest DNA molecule of 4.1×10^{10}. This corresponds remarkably well with the cytologically determined DNA content of the largest chromosome (No. 3) in *D. melanogaster* wild type of 4.3×10^{10}. In addition, the value of τ_1 for a translocation mutant of *D. melanogaster* yields a molecular weight of the longest DNA (5.8×10^{10}) very close to the amount in the largest chromosome as established cytogenetically. Other species and mutants give values of τ_1 predictable on the basis of chromosome morphology. For chromosomes of similar size, the length of the DNA appears to be independent of the chromosome arm length ratio, indicating that the DNA is continuous through the centromere. Information extracted from the recoil profiles also indicates that only one or a few molecules of the length corresponding to the longest chromosome exist per cell. These data can be taken as strong evidence for "chromosome-sized" DNA molecules in *Drosophila*.

Studies on other eukaryotes using this technique also suggest chromosomal length DNA molecules. For DNA released from spheroplasts of the yeast *Saccharomyces cerevisiae*, a molecular weight of

TABLE I

COMPARISON OF CHROMOSOME SIZE WITH RESULTS OF VISCOELASTIC MEASUREMENTS OF DROSOPHILA LYSATES[a]

Drosophila	Chromosomes	τ, hr (Number of measurements)	Molecular weight of largest DNA	DNA content of largest chromosome[b]
melanogaster				
wild-type		1.67 ± 0.23 (44)	$41 \pm 3 \times 10^9$	43×10^9
inversion		1.77 ± 0.30 (9)	$42 \pm 4 \times 10^9$	43×10^9
translocation		2.98 ± 0.54 (10)	$58 \pm 6 \times 10^9$	59×10^9
hydei				
wild-type		1.62 ± 0.27 (8)	$40 \pm 4 \times 10^9$	—
deletion		0.69 ± 0.17 (8)	$24 \pm 4 \times 10^9$	—
virilis				
		2.12 ± 0.28 (5)	$47 \pm 4 \times 10^9$	—
americana				
		$5.00 + 1.1$ (3)	$79 \pm 10 \times 10^9$	—

[a] For each *Drosophila*, the idealized appearance of the chromosomes at metaphase is shown in the second column, with the largest chromosomes emphasized. The third column lists the values of τ, corrected to 50°C and 0.2 M Na$^+$, with the average deviation of the measurements. The fourth column gives the values of M calculated from the values of τ; the precision measure reflects average deviation only. For further details see Kavenoff and Zimm (1973). (Reproduced from Kavenoff *et al.*, 1974, by permission.)

[b] Based on values of Rudkin (1965).

2×10^9 (assuming an open linear DNA configuration) or 4.3×10^9 (assuming a circular DNA configuration) has been determined by Lauer and Klotz (1975). A surprising aspect of their results is that in relation to estimates of the total size of the yeast genome, this piece of DNA could consist of one-fourth to all of the haploid DNA content, in spite of genetic evidence that *S. cerevisiae* has 17 linkage groups (Mortimer and Hawthorne, 1973) which segregate independently. Lauer and Klotz (1975) suggest that at least in some phases of the cell cycle the DNA may be continuous, or that there may be one very large chromosome containing about half of the nuclear DNA in a continuous piece, and 16 smaller ones. It was previously suggested (DuPraw, 1970) from electron microscopy that DNA may be continuous between chromosomes in interphase and that mitotic chromosomes represent discrete regions of condensation with interconnecting DNA molecules. Micromanipulation experiments also demonstrate possible interchromosomal connectives (Hoskins, 1968; Diacumakos *et al.*, 1971) that may contain DNA. Consistent with interchromosomal DNA are the results of flow microfluorometric measurements of DNA content of cell populations, coupled with studies of chromosome number and morphology (Kraemer *et al.*, 1974) which demonstrate DNA constancy in spite of variability in chromosome number and morphology. How these features of chromosomal DNA structure may relate to Kavenoff and Zimm's (1973) data indicating discrete chromosomal length DNA in *Drosophila* is not yet known. It is possible that interchromosomal DNA connectives are very fragile or do not exist at all times in the cell cycle or in all species.

Evidence from viscoelastic measurements on mammalian DNA is sparse, although one preliminary study (Lange, 1975) reports molecular weights of approximately 1.6×10^{10} from mouse leukemia cells. Part of the problem in mammalian studies derives from the extreme instrumental stability required to measure retardation times of 10 hours or more expected from chromosomal length DNA.

All these studies show that chromosomal DNA has the form of a very long continuous fiber in eukaryotic chromosomes and may thus serve as a "backbone" upon which to build up the various levels of chromosome structure and as the basis for the concept of the chromosome fiber. Very long or chromosomal length DNA would in effect provide structural continuity through the length of such chromosome fibers. In fact, there is cytological evidence that this is the case. In the oocytes of amphibians, chromosomes are found in a greatly elongated form during the diplotene stage of meiosis. The longitudinal axis of these so-called lampbrush chromosomes may reach 500–800 μm in length and possess many lateral chromatin loops. Callan and MacGregor (1958) have shown that treatment with DNase will sever and fragment the lateral loops, indicating that DNA passes through and maintains the integrity of the loops. Even though RNA and protein are in the loops, neither RNase or proteases can fragment them. By studying the kinetics of fragmentation caused by DNase, Gall (1963) was able to show that the lampbrush loops are composed of a single DNA double helix, while the axial elements are digested with the kinetics expected from a parallel pair of DNA double helices. This is consistent with the interpretation of paired chromatids, each containing a single DNA double helix in which the loops represent single elements (chromatids) having DNA as their backbone.

2. *Structural Units in the DNA Backbone*

As well as its role as backbone, there may be other features of DNA important at the various levels of chromosome organization. Only those structural features of chromosomal DNA that may be interesting in the discussion of chromosome fiber organization to follow will be discussed here. These include the possible existence of non-DNA linkers along the double helix, the distribution of single-strand discontinuities, and other possibly nonrandom structural features of DNA.

The experiments described in Section II,A,1, for measurements of double-stranded length of chromosomal DNA provide only limited information about the overall continuity of the DNA double-helical backbone. For instance, proteolytic enzymes have been used in both DNA strand autoradiographic and viscoelastometric measurements of double-stranded DNA. Such treatments do not appear to fragment chromosomal DNA as a double-stranded structure, making it unlikely that double-strand interruptions in the form of protein linkers are present in the DNA. Any single-strand discontinuities introduced by such treatments would not be detected by either technique. The distribution of possible "linker" substances and single-strand discontinuities has been approached more successfully by the technique of sedimentation velocity of DNA under denaturing conditions. However, problems of artifacts have plagued these experiments and have probably led to large errors in the estimates of single-strand DNA sizes. The major problems that must be overcome include aggregation (interstrand interference) of single-strand DNA, shear breakage, and anomalous rotor speed effects on the determination of sedimentation coefficients. Shear breakage effects were the first to be recognized (Davison, 1959) and corrected, and probably have not been a major problem in more recent studies. Interstrand interference and rotor speed effects have only recently been recognized and considered in measurements of single-strand DNA lengths.

Rotor speed effects have been demonstrated for both prokaryotic (Kavenoff, 1972; Hozier and Taylor, 1975) and for eukaryotic DNA (McBurney et al., 1971; Ormerod and Lehmann, 1971; Hozier and Taylor, 1975) for neutral and alkaline sedimentation of DNA. In general, this effect involves a decrease in the apparent sedimentation coefficient of both double- and single-stranded DNA with increasing DNA size and rotor speed. A theoretical basis of this anomaly has been proposed by Zimm (see Kavenoff, 1972), which explains the effect in terms of molecular distortions in high centrifugal fields. In any case, past and future studies on sedimentation properties of very large DNA must be evaluated on an individual basis to determine if rotor speed anomalies may be important in interpreting the results.

A potentially even more serious problem was brought to light when it was discovered that under the usual conditions of alkaline lysis and sedimentation, strand separation may not be complete for very large DNA (Simpson et al., 1973; Jolley and Ormerod, 1974; Cleaver, 1974). This problem most likely involves entanglement of very long complementary single-strand DNA during unwinding and actual failure of some regions of DNA duplex to become completely denatured. Persistence of double-stranded DNA under denaturing conditions requires great caution in interpreting results in alkaline sedimentation studies of single-strand DNA. However, this difficulty has recently been overcome in a study of Chinese hamster DNA in which special precautions were taken to ensure that complete strand separation occurred upon alkaline denaturation (Hozier and Taylor, 1975). Bromodeoxyuridine was incorporated into one strand of the DNA double helix, and advantage was taken of the increased sensitivity to ultraviolet radiation for the production of single-strand breaks in bromodeoxyuridine-substituted DNA. After irradiation, the substituted strand could be separated from its complementary strand for alkaline sedimentation analysis without damage to the unsubstituted strand. Under these conditions, and after correcting for speed effects, the number average molecular weight of unsubstituted single-strand DNA was determined to be 1.7×10^8. Further analysis of the sedimentation profiles suggested that the distribution of discontinuities between these DNA subunits is nonrandom. Larger "complexes" and proposed single-strand subunits of DNA structure (Elkind and Kamper, 1970; Lett et al., 1970) must also be shown to be free of possible double-strand DNA or incompletely separated strands.

Even after all these precautions, there remains the problem of determining the nature of the site of the single-strand DNA discontinuity and the role of such sites in chromosome structure. Additional information can be obtained by varying the conditions of lysis and incubation. For example, Hozier and Taylor (1975) found that a protease lysis procedure revealed single-strand DNA discontinuities at approximately 6.0×10^7 dalton intervals (or two breaks in each of the 1.7×10^8 dalton subunits) that also appear to be stable and nonrandom under these conditions of incubation. A similar size single-strand unit (about 4.5×10^7 daltons) has been observed in mouse L cells (Fujiwara, 1975).

Another important structural feature of chromosomal DNA has been revealed by electron microscopic analysis of partially denatured DNA of Chinese hamster cells and chick fibroblasts (Evenson *et al.*, 1972). When the DNA was heat denatured to approximately a 10% level and stabilized with formaldehyde, melted regions were found at 0.4- to 0.5-μm intervals throughout. One possible interpretation is that regions containing a relatively high A + T content, and therefore less stable to thermal denaturation, are present at approximately 0.5-μm intervals in chromosomal DNA.

B. ELECTRON MICROSCOPIC ANALYSIS OF CHROMOSOME FIBERS

Having established that chromosomal DNA has the properties expected of a backbone structure for fibers, it is also important to examine the morphology of interphase and mitotic chromosomes to demonstrate their fibrous nature. The discussion of electron microscopic morphology will be confined to results obtained by standard techniques including thin sectioning, surface spreading, and freeze etching. In thin sectioning tissue, cells or isolated nuclei are fixed and stained (usually with a combination of glutaraldehyde and osmium tetroxide), dehydrated in ethanol, and embedded in a solid electron transparent matrix. Ultrathin sections are cut, attached to electron microscope grids, and poststained (usually with a combination of uranyl acetate and lead citrate). In surface spreading, cells or isolated nuclei are broken open by surface tension when they are applied to an air–water interface. The released chromosome fibers are picked up on electron microscope grids, dehydrated in ethanol, and finally transferred to pure amyl acetate. The grids are then placed in a pressure chamber and flooded with liquid carbon dioxide (which is miscible with amyl acetate) at room temperature. The temperature of the closed chamber is raised so that the carbon dioxide passes through its critical point to the gas phase. The gaseous CO_2 is bled off to dry the specimen. This procedure effectively eliminates the damage to the specimen caused by drying from the liquid phase. The third, less frequently used technique of freeze etching is included because it allows observation of specimens that have not been exposed to drying, fixation, or staining. Typically, specimens are prepared by plunging tissue into liquid nitrogen. The frozen tissue is fractured *in vacuo*, coated with carbon, and shadowed with heavy metals while still in the frozen state. The metal replica is then released from the specimen after thawing and mounted on an electron microscope grid.

1. Interphase Chromosomes

In thin sections of interphase nuclei there is usually a well-defined division of chromatin into condensed and diffuse regions. Condensed chromatin is densely packed and fiber diameters appear larger on the average than in diffuse chromatin. Surveys of the literature of fiber diameters of thin sectioned nuclei reveal a large variability among tissues and species (Davies, 1968; Zirkin and Wolfe, 1972; Solari, 1974). It is difficult to say how much of the variability is real and how much is due to differences in tissue preparation, measurement techniques, or other experimental causes. Some of the variation may be due to differences in the relative amounts of condensed and diffuse chromatin. However, most measurements have been in the range of 150–250 Å, at least for condensed chromatin, which makes up the bulk of chromatin in many cases. Smaller diameters, on the order of 100 Å, are often seen in diffuse chromatin (Frenster, 1974). Several studies that have taken advantage of special features of chromosome organization of certain species and of the properties of chromatin in isolated nuclei will be described, since they give a good overall view of chromosome morphology by thin sectioning.

Davies and his colleagues have shown that in many species, the condensed chromatin displays a specialized organization in the vicinity of the nuclear envelope (Davies, 1968; Davies and Small, 1968; Davies and Haynes, 1976). Apparently contact with the nuclear envelope induces a layering effect in such a way that parallel arrays of very distinct chromosome fibers are seen (Fig. 3). Center-to-center spacing of the fibers is about 280 Å, while densely staining fibers have an average diameter of 170 Å. Farther from the nuclear periphery, the organization into layers becomes less and less distinct so that the overall effect is a more granular appearance. Similar arrangements have been observed in studies of the organization of the nuclear envelope (Zentgraf *et al.*, 1975). Chromosome fibers can also be observed when con-

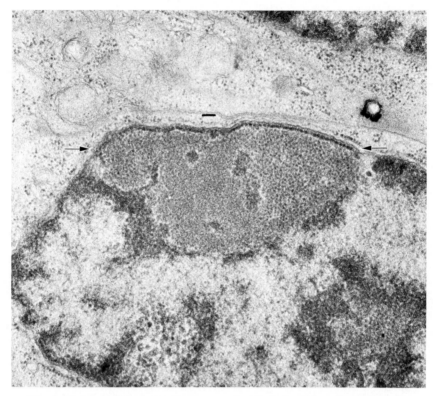

FIG. 3. Electron micrograph of an intestinal epithelial cell from the goldfish *Carassius auratus* displaying a line of chromosome fibers (shown by arrows) between the interior of the nuclear envelope and the nucleolus. The average diameter of the fibers in this section is about 170 Å. Scale line represents 1000 Å. (From Davies and Haynes, 1976, reproduced by permission.)

densed chromatin is dispersed in certain pathological states. Figure 4 shows thin sections of lymphoplasmablasts from a patient with Waldenström disease (Bouteille *et al.*, 1974). Here chromosome fibers 200–300 Å in diameter can be seen at high magnification. When nuclei are isolated from frog erythrocytes and chromatin is dispersed by treatment with low salt concentrations 200 Å fibers are also seen throughout (Ris, 1975). Thus, thin section studies reveal that condensed chromatin, especially when viewed in the vicinity of the nuclear envelope or when it is decondensed in diseases or by chemical treatment of isolated nuclei, is composed of 200–300 Å fibers. Diffuse chromatin regions of the nucleus appear to contain fibers of more variable thickness than those of condensed chromatin.

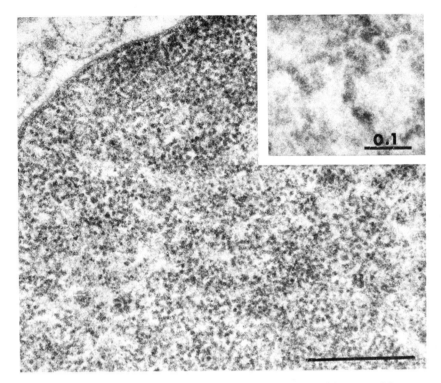

FIG. 4. Thin section electron micrograph of a lymphoplasmablast in Waldenström disease. Fibers of irregular morphology are seen at high magnification in the insert, with average diameters in the range of 200–300 Å. Scale, main, 1 μm; insert, 0.1 μm. (From Boutielle *et al.*, 1974, reproduced by permission.)

By the very nature of the technique, thin sectioning does not allow one to follow the path of a single fiber for any great distance, since the sections are only two or three times as thick as a fiber and the fibers are rather closely packed. Therefore surface spreading techniques, in which the fibers are laid out on the plane of the electron microscope grid, offer a great advantage over thin sections for studying individual fibers. Figure 5 shows chromosome fibers prepared by surface spreading and critical point drying. The consensus of such studies is in very good agreement with thin section studies in that fiber diameters are usually in the 200–300 Å range (DuPraw, 1970; Bahr and Golomb, 1974; Ris, 1975; Bahr, 1977). At the same time, it is possible to follow individual fibers for considerable lengths and to gain some impression of overall features of fiber morphology. For example, in most preparations, fibers have a bumpy or "knobby" appearance (see Fig. 5). They

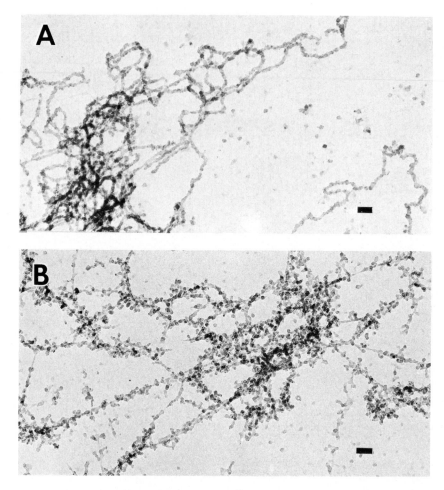

FIG. 5. Electron micrographs of surface spread chromosome fibers from frog (*Rana pipiens*) erythrocytes (A) and yeast (*Saccharomyces cerevisiae*) spheroplasts (B). Fibers are about 200 Å thick. Scale lines represent 1000 Å. (From Ris, 1975, reproduced by permission.)

appear to be quite flexible; in some places the radius of curvature is not much larger than the fiber thickness. It is likely that the fibers seen in surface-spread preparations come mostly from condensed chromatin regions of the nucleus, since this makes up the bulk of the material. Only occasionally are fibers seen that are significantly smaller than the 200–300 Å average.

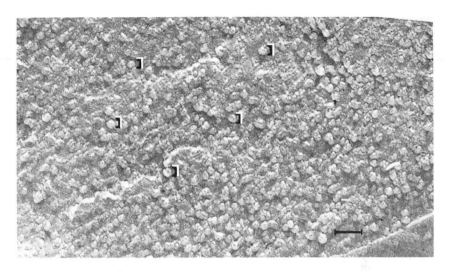

FIG. 6. Electron micrograph of a freeze-etched salamander spleen cell. The light disks are interpreted as broken chromosome fibers. The average diameter of the disks (in brackets) is about 250 Å. Scale line represents 1000 Å. (From Ris, 1969, reproduced by permission.)

In thin sections or in surface-spread preparations, one may be biased in making fiber diameter measurements depending upon what type of fibers are available for convenient measurement. In thin sections, diffuse chromatin fibers are more easily measured, while in surface-spread preparations, condensed chromatin fibers predominate after their dispersal. This may contribute, along with differences in specimen preparation, to differences in average fiber diameters obtained by the two techniques. An additional criticism has been leveled at surface-spread preparations, namely, that fibers may become severely contaminated with cytoplasmic or nonchromosomal nuclear proteins during spreading (Wolfe, 1968; Solari, 1971). To circumvent these various criticisms, freeze etching has been used to study the fibers of interphase chromosomes without fixing, staining, dying, or spreading. Both honey bee embryonic tissue (DuPraw, 1974) and salamander spleen (Fig. 6) typically show 200–300 Å diameter fibers. The fibrous nature of interphase chromosomes has been particularly well preserved in spermatids of the prosobranch mollusk *Goniobasis* (Henley, 1973), where bundles of fiber can be seen lying in parallel. Although these are sperm precursor cells whose fiber structure may not be typical of other tissues, the fibrous nature is quite evident.

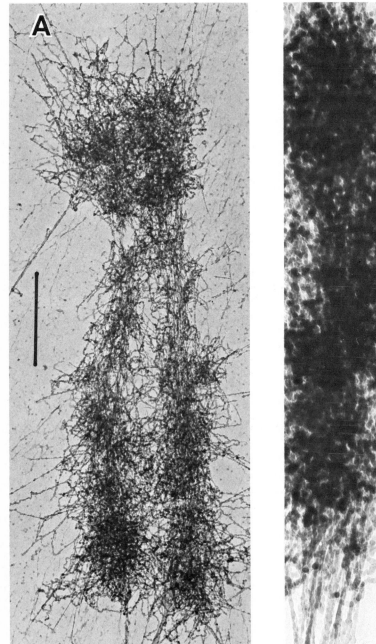
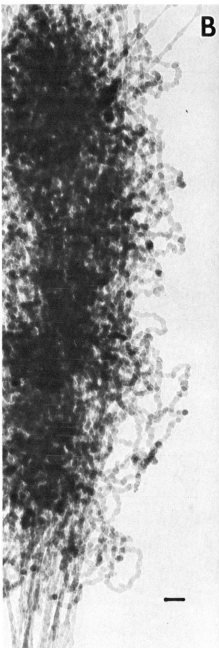

2. *Mitotic Chromosomes*

Although thin sectioning has revealed much about the orientation and movement of chromosomes and arrangement of the mitotic apparatus, very little can be learned about the structure of the chromosome fibers at mitosis. Chromosome fibers appear so tightly packed that even serial sections have not been very helpful in clearly showing structural details. However, recent developments in high-voltage electron microscopy may soon make it possible to view stereoscopically sections that are thick enough to contain entire chromosomes (Ris, 1975, 1978).

In the study of mitotic fibers, surface spreading has been by far the most successful technique. Mitotic cells burst on an air–water interface release chromosomes whose overall morphology is strikingly similar to light microscopic preparations (Fig. 7). Also, the fibrous nature of mitotic chromosomes is clearly defined, and the average fiber diameter from many sources is virtually the same as in interphase nuclei. Some early studies reported large diameters, but this was subsequently shown to be due to the effects of high concentrations of colchicine in the preparations (Bahr, 1975). The similarity in fiber diameter and typical "knobby" morphology opens the possibility that the molecular organization of fibers at interphase and mitosis is similar. This view is supported by data which will be presented in a discussion of nucleosome organization in fibers. In any case, it is quite apparent that the arrangement of fibers at mitosis is quite different from interphase. Fibers appear to pass in and out of the chromatid longitudinal axis in a looping fashion. Some fibers appear to run parallel to the longitudinal axis (Fig. 7), but it is not clear whether they actually do or result from stretching of the chromatids in a longitudinal direction (Comings, 1974a).

The condensation of chromosomes from interphase to metaphase has also been visualized at the electron microscopic level by the surface spreading technique (Bahr *et al.*, 1973). The tendency is for long

---

FIG. 7. Electron micrographs of surface spread and critical point dried human metaphase chromosomes. (A) Low magnification of a whole metaphase chromosome. Fibers can be seen running parallel to the chromosome axis, while others loop out transversely. Transverse fibers appear to be clustered along the chromatid axis, giving the impression of chromosome "banding" or "coiling." Scale line represents 1 μm. (B) Higher magnification of a condensed metaphase chromosome showing typical knobby 200 Å chromosome fibers. Scale line represents 1000 Å. (From Bahr *et al.*, 1973, reproduced by permission.)

thin chromatids of late prophase to thicken and become shorter at metaphase. However, in the absence of colchicine, the morphology of the
individual chromosome fibers does not appear to change. Therefore,
changes in the fiber morphology do not seem to be important in the
consensation of chromosomes at mitosis.

C. PHYSICAL AND CHEMICAL CONSTRAINTS ON FIBER ORGANIZATION

1. Chemical Composition of Chromosomes

We have established that both interphase and mitotic chromosomes are made up of fibers and that the typical fiber has a knobby
appearance and an average diameter of 200–300 Å. Very long DNA
molecules probably form the backbone upon which the continuity of
the fibers depends. In addition to DNA, chromosomes are composed of
several classes of proteins and smaller amounts of other components,
including RNA. Analyses of relative amounts of the different components in chromosome fibers has been the subject of many studies [for a
review, see Johnson et al. (1974)], including those in which mitotic
chromosomes are isolated followed by detailed chemical analyses
(Sadgopal and Bonner, 1970; Mohberg and Rusch, 1970).

One general principle that seems to hold for many species and for
all phases of the cell cycle is that the histone proteins are always present in approximately a one-to-one weight ratio with DNA (see Chapter
VI). Also, the major histone fractions in most mammalian species (H1,
H2A, H2B, H3, and H4) occur in a fixed overall molar ratio. This is
part of the basis of the nucleosomal concept of fiber structure to be
presented below.

The other major class of chromosomal proteins, the nonhistones, is
much more variable in type and relative amount. Studies concerning
nonhistones as bound components of chromosomes have produced
highly variable results. The content of nonhistone chromosomal proteins seems especially sensitive to the method of isolation, with both
pH and ionic strength playing important roles. In most careful studies
of metabolically active tissues there appears to be at least as much
nonhistone as histone in the chromosome fraction [for a review, see
Elgin and Weintraub (1975)]. This is not to say that this much nonhistone protein is involved in chromosome fiber structure. Certainly
many of the nuclear nonhistones are enzymes and structural proteins
of the nucleolus. Others are involved with DNA replication and nonribosomal RNA synthesis. A large proportion may take part in the further processing of primary RNA transcripts and their packaging into

ribonucleoprotein particles (Georgiev, 1974). A general regulatory role for nonhistones has also been proposed (Stein and Kleinsmith, 1975). However, it seems likely that some nonhistones may play a vital role in chromosome fiber organization.

The amount and types of RNA in interphase and mitotic chromosomes are also highly variable, again partially because of differences in the methods of isolation. Most of the chromosome-bound RNAs are, of course, primary transcripts. However, there is growing evidence for a structural role for certain types of RNA (Benyajati and Worcel, 1976; Cook *et al.*, 1976).

2. *Packing Ratio of DNA in the Chromosome Fiber*

The fiber packing ratio refers to the length of DNA per unit length of chromosome fiber. It provides a measure of the efficiency of DNA packaging caused by the chromosomal proteins. The limits on the degree of packing of DNA in the chromosome fiber are set by the total length of the fiber, the relative amounts of the major fiber components, and the closeness of fit of the fiber components. Starting with a variety of assumptions and independent measurements of these fiber parameters, a surprising level of agreement has emerged about the packing ratio.

DuPraw and Bahr (1969) used quantitative electron microscopy of interphase and mitotic chromosomes of human lymphocytes to determine total fiber mass and mass per unit length of chromosome fibers. Thus, knowing the total length of DNA in nuclei and individual chromosomes, they could arrive at estimates of fiber packing ratios from their electron microscopic measurements. Packing ratios of 150 : 1 were determined for mitotic chromosomes. However, the colchicine exposure used in the preparation of mitotic chromosomes caused a variable degree of shortening and thickening of the fibers (Bahr, 1975), with fiber diameters up to 450 Å being recorded. Only fibers this thick could accommodate a packing ratio of 150 : 1 (Bahr, 1970). When colchicine is omitted, metaphase chromosomes have an average diameter of 200 Å (Bahr and Golomb, 1974), and measurements on unstimulated and stimulated lymphocytes in all stages of the cell cycle give the same packing ratio of 28.3 : 1.

Comings (1972), starting from the point of view of folding of a primary nucleohistone structure, or from estimates of total fiber length and DNA length in metaphase chromosomes, has arrived at values for the packing ratio in chromosome fibers of 8.4 : 1 to 30 : 1. The lower estimate was based on interpretations of X-ray diffraction data on fiber structure that no longer appear to be valid (see below), so that the higher estimate based on maximum level of packing in a native fiber seems more reasonable. As will be shown below, new estimates of packing ratio based on the nucleosome model of chromatin structure are also in the range of 25:1 to 40:1.

Thus the estimates of packing ratio in the native chromosome fiber appear to converge at about 30:1. This figure will be used in discussions of higher-order folding of the native chromosome fiber in interphase and mitotic chromosomes. The similarity in packing ratio at mitosis and interphase further supports the idea of constancy and stability of the native chromosome fiber structure.

III. Components of Chromosome Fibers

Biochemistry and high resolution electron microscopy have yielded valuable information on the nature of chromosome fiber components. Biochemical analysis is usually carried out on "chromatin," which is a solubilized version of chromosome fibers. Chromatin is prepared from isolated nuclei by swelling and repeated washing in low ionic strength buffers. This produces a chromatin "gel" which is often sheared to produce a less viscous solution. Electron microscopy is also quite often performed on nuclear material that has been treated in a similar manner. Together, these techniques serve well in defining the fundamental interactions of fiber components.

A. NUCLEOSOMES: DIMENSIONS AND DISTRIBUTION ALONG THE DNA FIBER

The structure and molecular organization of nucleosomes have been treated in detail in Chapter VI. However, a brief review emphasizing those aspects of structure important to higher-order organization of chromosomes will be presented here. The current view holds that nucleosomes are composed of two each of histones H2A, H2B, H3, and H4 tightly bound to about 140 base pairs (bp) of DNA (Van Holde et al., 1974; Kornberg, 1974). These discrete structural units are connected by DNA in a beads-on-a-string arrangement, with an overall repeat unit every 200 bp of DNA. The most important evidence for such a structural arrangement of DNA and histones comes from high-resolution electron microscopy of chromatin (Olins and Olins, 1974; Oudet et al., 1975) which shows globular 60–100 Å structures (ν bodies or nucleosomes) connected in tandem by thinner threads (presumably DNA) and from biochemical investigations of nuclease action on chromatin, which reveal alternating nuclease-sensitive and nuclease-resistant regions (Noll, 1974; Oosterhof et al., 1975).

Limited nuclease digestion produces DNA fragments that are integral multiples of a basic length, estimated to be 140–200 bp for a variety of species and tissues. Individual monomer nucleoprotein units isolated after nuclease digestion and prepared for electron microscopy have the appearance of nucleosomes. This structural pattern, as probed by electron microscopy and biochemical analysis, seems to be due to the folding of DNA and protection against nuclease attack afforded by the histones and is present in all histone-containing organisms thus far studied.

If one wishes to reconcile this mode of organization of nucleoprotein with the native structure of the chromosome fiber, the first problem is that the nucleosome fiber is too thin (<100 Å in diameter) and too loosely packed (maximum packing ratio $7:1$) to represent the native chromosome fiber as described previously (i.e., about 200 Å in diameter, with a packing ratio of $30:1$). One can take the view that one of these chromosome fiber configurations is "right" and the other is "wrong" or that they may reflect different levels of fiber organization. The latter view will be argued here, following the presentation of other features of the nucleosomal model that bear on higher-order structure.

1. *The Size and Shape of Nucloesomes*

The first measurements of nucleosome diameters came from electron microscopic examination of nuclei disupted in hypotonic solutions and treated with formaldehyde (Olins and Olins, 1974) and chromatin depleted of histone H1 (Oudet *et al.*, 1975). Average diameters ranged from about 60 Å for the formaldehyde-fixed preparations to about 125 Å for the H1-depleted chromatin. Such a large difference in average diameters may reflect some real differences in nucleosome structure, but more likely are a result of differences in technique. For example, the fixation and staining procedures of Olins and Olins (1974) would be expected to cause shrinkage of the particles and smaller apparent dimensions, while the heavy metal shadowing technique of Oudet *et al.* (1975) would make the particles appear larger. In our own studies (Fig. 8) on isolated chromosome fibers disrupted by treatment with EDTA followed by fixation, negatively stained preparations have an average diameter of about 80 Å (Hozier *et al*, 1977; Renz *et al.*, 1978).

In any case, dried and stained specimens alone probably should not be relied upon to give an accurate value for nucleosome dimensions. Other techniques in which the particles can be studied in the hydrated state are preferable. Sedimentation analysis of isolated parti-

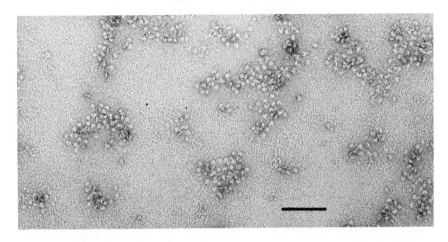

FIG. 8. Electron micrograph of nucleosomes from bovine lymphocytes. Lymphocyte nuclei were mildly digested with micrococcal nuclease, followed by lysis in 0.2 mM EDTA and fixation with glutaraldehyde. Specimens were picked up on carbon-coated grids and stained with 1% aqueous uranyl acetate. Scale line represents 1000 Å.

cles would seem to be the most likely approach, but to date only general features of size and shape have been revealed. Isolated monomer particles have a molecular weight of 200,000–300,000 in sedimentation studies, which is compatible with a complex of 140–200 bp of DNA having one pair each of histones H2A, H2B, H3, and H4 (Senior et al., 1975; Olins et al., 1976). They exhibit a frictional coefficient compatible with a roughly spherical particle (Rill and Van Holde, 1973; Olins et al., 1976).

More detailed information has come from X-ray and neutron scattering studies of chromatin and isolated nucleosomes. Low-angle X-ray diffraction of isolated monomer nucleosomes and chromatin at high concentrations shows a first-order maximum at about the 110 Å position for wet material (Kornberg and Thomas, 1974; Olins et al., 1976). This may correspond to the diameter of the hydrated sphere at high concentration and reflect the true diameter of the nucleosome more nearly than measurements obtained from electron microscopy. In fact, the first maximum for dried nucleosome preparations is in the 76–84 Å range (Olins et al., 1976) and may be due to the close packing of dry particles having an average diameter closer to that determined by electron microscopy.

Neutron scattering offers some specific advantages over X-ray diffraction when working with two-component systems such as nucleosomes. It is possible to "contrast match" one of the components with the solvent so that the neutron atomic scattering factor of that component and the solvent are the same. Then the scattering properties of the other component can be studied. For nucleosomes, both components (DNA and protein) have average neutron scatter values within the range of various mixtures of H_2O and D_2O. Thus contrast matching occurs in 37% D_2O to 63% H_2O volume ratio for histones and 63% D_2O to 37% H_2O for DNA (Bradbury, 1975). Scattering data obtained in this way have yielded information not only on the size of nucleosomes but also on their shape and the distribution of components in nucleosomes. Most studies of isolated nucleosome monomers conclude that a histone core is surrounded by a DNA-rich shell (Pardon *et al.*, 1975; Hjelm *et al.*, 1977), although the exact shape of the particle and the dimensions of the protein core and surrounding DNA coil are still in dispute. Contrast matching experiments indicate an average particle diameter of 104 Å (Hjelm *et al.*, 1978), which is in good agreement with the results from X-ray diffraction, and an average histone core diameter of 64 Å. In addition, comparison of scatter curves of monomer nucleosomes to theoretical scatter curves for ellipsoids shows that the ratio of the shortest to the longest dimension for a monomer-equivalent ellipsoid should not be smaller than 0.5.

The picture that emerges from electron microscopic and biophysical data is of a rather flat cylindrical particle, about 100 Å in diameter in the hydrated form. The DNA surrounding the protein core is very much open to the solvent, suggesting that interactions among nucleosomes, including those responsible for higher-order structure, may depend strongly upon the ionic conditions within the nucleus. The large amount of solvent within the particle (Hjelm *et al.*, 1978) may also be responsible for its apparent shrinkage in the dry state as measured by electron microscopy.

2. *Arrangement of Nucleosomes on DNA*

The question of nucleosome spacing arises both from electron microscopic studies, in which different preparative techniques show different average distances between nucleosomes, and from biochemical studies which demonstrate considerable tissue and species variability in the DNA repeat length in spite of similarly sized monomer structures.

The earlier electron microscope pictures of strings of nucleosomes (Olins and Olins, 1974) showed most nucleosomes separated by thin strands that appeared to be DNA, and were consistent with an average of about 60 bp of DNA separating nucleosomes. To what degree this spacing was affected by stretching of the fibers is still not known. Subsequent studies of nucleosome chains have produced highly variable results concerning nucleosome spacing along DNA (Oudet *et al.*, 1975; Finch *et al.*, 1975; Woodcock *et al.*, 1976b). However, the separation of particles in electron microscopic preparations, even though possibly artifactual, led in some part to the establishment of the nucleosomal concept since the separated particles were easy to visualize. Much of the concern over the electron microscopic morphology of nucleosome chains is misplaced, since in almost every case the preparative conditions were far from those that should preserve the chromosome fiber in its native state.

A more meaningful analysis of the distribution of nucleosomes along DNA has involved nuclease digestion of isolated nuclei. Detailed study of the digestion kinetics and sizes of the nucleosome monomer and multimer DNA species has led to a reasonably unified view of the distribution of DNA in and between nucleosomes. In a variety of tissues and species, a rather uniform monomer "core" particle appears to consist of pairs of each of the nucleosomal histones and 140 bp of DNA, even though the repeat length of the DNA (measured, for instance, as the average difference in DNA length between consecutive nucleosome multimers at intermediate levels of digestion) may vary from 140 bp to well over 200 bp in different tissues (Compton *et al.*, 1976a; Rill *et al.*, 1977). Therefore nuclease-resistant core structures appear to be separated by variable lengths of nuclease-sensitive DNA "spacers." How this spacer DNA may be situated between adjacent nucleosomes is not known. Possibly, it interacts with nonnucleosomal components of the chromosome fiber in the formation of higher order structures.

B. HISTONE H1

Histone H1 is treated separately since it does not participate with the other histones in the formation of nucleosomes. This can be shown by electron microscopy of histone H1-depleted fibers (Oudet *et al.*, 1975; Hozier *et al.*, 1977) which show nucleosomes, often with greater clarity and separation than when H1 is present. In addition, when nucleosome chains are sedimented in sucrose gradients following nuclease digestion, the monomer fraction may be depleted in H1 (Noll and Kornberg, 1977; Renz *et al.*, 1977a).

There are several properties of H1 that set it apart from the nucleosomal histones. Its molecular weight (23,000) is greater than the others. It has a very high proportion of lysine residues (27%) and a high lysine to arginine ratio. It also has a high proline content, which appears to reduce the amount of α-helix in its structure. Histone H1 shows more sequence diversity than the other histones. There is sequence microheterogeneity even within individual tissues (Bustin and Cole, 1968). It is also the histone least tightly bound to DNA and most easily dissociated from chromatin by raising the salt concentration. This property of H1 has led to some important insights into its structural role in chromatin, since it facilitates nuclear magnetic resonance (NMR) studies of the mode of binding of H1 to chromatin and to DNA. Nuclear magnetic resonance is an especially sensitive technique for the study of molecular interactions. The loose binding of H1 relative to the other histones allows it to dominate the NMR spectra of chromatin. Based on NMR and other spectroscopic studies of H1, and its known amino acid sequence (Bustin and Cole, 1968), a schematic representation of H1 structure has been developed (Hartman *et al.*, 1977) (Fig. 9), and much has been learned about the interaction of its structural domains with chromatin and DNA. H1 consists of a random coil "nose" in the N-terminal region up to residue 34. There is a great deal of variation among subfractions in this region. The next 80 residues form a globular structure under physiological conditions of ionic strength and pH. This segment, the "head," has a highly conserved amino acid sequence and a high concentration of hydrophobic residues. The rest of the molecule, from residue 120 to the C-terminal end forms the random coil lysine-rich "tail."

The head, nose, and tail segments of H1 may be enzymatically cleaved and studied separately. The head region is capable of forming a globular structure in the absence of a nose or tail. The nose and head segments together bind DNA very weakly. Chapman *et al.* (1976) have suggested that this region interacts with nucleosomes or with nonhistone proteins. The tail region alone binds DNA as strongly as

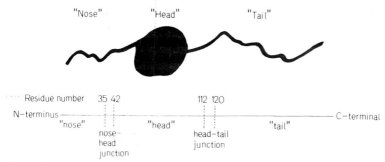

FIG. 9. Schematic representation of the three structural domains of histone H1. The limits of each structural region are shown below the diagram. (From Hartman *et al.*, 1977, reproduced by permission.)

the whole molecule. Nuclear magnetic resonance studies on this segment show that its lysine residues interact with DNA in quite the same manner as whole histone H1.

The NMR spectra of chromatin and DNA–H1 complexes have been studied in detail in the ionic strength range from 0–$0.5\,M$, which includes the ionic strength conditions under which chromatin is usually studied (low salt), conditions of physiological ionic strength (about $0.15\,M$), and the limit above which histone H1 dissociates from nucleohistone ($> 0.45\,M$). Bradbury et al. (1975) have shown that over this entire range the behavior of H1 is identical in chromatin and in DNA–H1 complexes, implying that H1 interacts predominately with DNA in chromatin. The macroscopic effect of H1 is to contract and cross-link both DNA and chromatin under physiological ionic strength conditions to form a precipitate. Sodium chloride was used in most of these experiments. Although monovalent cations other than sodium give similar results, divalent cations are much more effective in causing contraction. For example, maximum contraction of chromatin gels can be achieved in 2 mM $MgCl_2$. The NMR spectroscopic behavior of chromatin and DNA–H1 complexes in the presence of divalent cations parallels the results with monovalent ions and again indicates that H1 interacts primarily with DNA in chromatin.

Since the strongest binding of H1 is to DNA rather than to other histones, the study of DNA–H1 complexes in the absence of the other histones may be instructive. For instance, when a DNA–H1 mixture in high salt is dialyzed to $0.15\,M$ NaCl, a precipitate is formed from H1 and a portion of the DNA. With DNA from a variety of sources, including calf thymus, crab, and several bacterial species, H1 always selects the portion with the highest $A + T$ content (Sponar and Sormova, 1972).

Filter retention experiments with DNA–H1 complexes reveal another kind of binding specificity. Properly treated nitrocellulose filters pass DNA readily in $0.15\,M$ NaCl but retain histone H1 and DNA–H1 complexes. Renz (1975) has used the nitrocellulose filter system to study the cooperative and selective nature of DNA–H1 interations. He found a strong preference of lymphocyte H1 for lymphocyte DNA over E. coli DNA when the DNA is larger than about 10^6 daltons. This selectivity is seen in competition experiments in which both kinds of DNA are mixed in equal amounts, along with a limiting amount of H1. Under these conditions, selectivity can exceed 15 molecules of lymphocyte DNA bound per molecule of E. coli DNA. This effect appears to be too large to be explained only in terms of the 7% higher $A + T$ content of lymphocyte DNA over E. coli DNA. Even more interesting is the finding that for lymphocyte DNA there is a strong size preference of H1 binding, with larger DNA preferentially selected over smaller DNA. In these experiments, competition for a limiting amount of H1 between DNA of about 1×10^6 daltons and equal weights of DNA of various sizes was measured in a filter retention assay (Fig. 10). There is a remarkable increase in the H1 preferential binding capacity between 0.3 and 2×10^6 daltons. *Escherichia coli* DNA does not behave this way, but only shows a gradual increase in H1 binding capacity with increasing size. Renz (1975) has interpreted these results as indicating the presence of cooperative H1 binding sites spaced at approximately 10^6 dalton intervals (or every 1600 bp) along lymphocyte DNA. At these sites, H1 binds cooperatively, so that in equal weight mixtures of small and large DNA, more large molecules would be retained on filters, since a larger proportion of them possess preferential binding sights. By shearing lymphocyte DNA to small fragments and mixing them with a limiting amount of H1, the preferential H1 binding sites could be enriched as that fraction retained by filters. DNA isolated from the retained fraction did indeed bind H1 two to three times more efficiently than the originally unbound fraction. The fraction enriched in H1 binding sites has a melt-

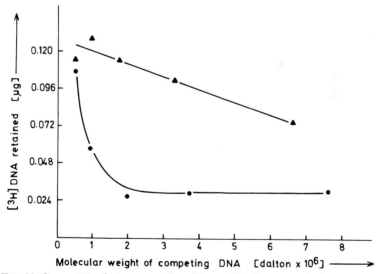

FIG. 10. Competition between small and large DNA fragments for histone H1. Histone H1 (0.05 μg) was added to 1 μg of [3H]-labeled DNA with an average molecular weight of 10^6 and 1 μg of unlabeled lymphocyte DNA (●) or 1 μg of *E. coli* DNA (▲) of various molecular weights. Filter retention was measured in 1 ml of a buffer containing 0.15 M NaCl. (From Renz, 1975, reproduced by permission.)

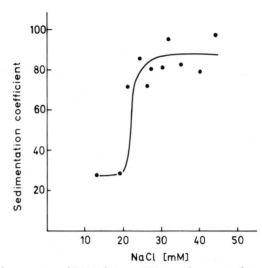

FIG. 11. Sedimentation of DNA–histone H1 complexes as a function of NaCl concentration. Sedimentation coefficients for mixtures of DNA and H1 were determined by analytical ultracentrifugation. Above 20 mM NaCl the DNA sediments as two subpopulations. Only the component with the higher sedimentation coefficient is shown here. The proportion of DNA sedimenting as the fast component (the DNA–H1 complex) increases from 23 to 55% in the range of 21 to 44 mM NaCl, while that of the slow component (free DNA, sedimentation coefficient = 20 S) decreases from 68 to 25% in this concentration range. (From Renz et al., 1978, reproduced by permission.)

ing temperature 5°C lower than bulk DNA, indicating a high A + T content.

As mentioned previously, Evenson et al. (1972) found sites along Chinese hamster and chicken erythrocyte DNA that melted before bulk DNA. These sites are most likely rich in A + T and are spaced at approximately 1500 bp intervals, in remarkable agreement with the spacing of the proposed A + T-rich binding sites for histone H1.

Renz and Day (1976) studied the ionic strength dependence of the DNA–H1 interaction for DNA larger than 1500 bp and found that a transition from noncooperative binding to cooperative and selective binding occurs between 20 and 40 mM NaCl (Fig. 11). Below the transition range, H1 binds to all of the DNA and causes it to sediment slightly faster than DNA alone. At NaCl concentrations above the transition range, H1 binds to a subpopulation of the DNA, leaving the rest free. The H1-bound DNA has a much higher sedimentation coefficient (~80 S) than unbound DNA. The increase in sedimentation coefficient is probably due to intramolecular folding rather than intermolecular aggregation, since the 80 S component is soluble, and only at much higher salt concentrations are aggregates with very high sedimentation coefficients formed.

The initiating events that lead to the accumulation of H1 on certain favored DNA regions (cooperativity) is still not known. However, the spacing of these regions and their absence on bacterial DNA, together with the striking change in sedimentation characteristics of bound DNA, strongly implicate them in the specific folding of the DNA fiber by H1. The possibility that chains of nucleosomes may interact with H1 in an analogous manner will be discussed below.

C. NONHISTONE CHROMOSOMAL PROTEINS

Nonhistone chromosomal proteins may be defined as the proteins other than histones that coisolate with DNA in purified chromatin. The relative amount of nonhistone in chromatin is much more variable than the amount of histone. As a class, nonhistones are much more complex, with perhaps a hundred or more separate proteins resolvable by polyacrylamide gel electrophoreses in a molecular weight range from 10,000 to over 200,000. However, only 15 to 20 of these make up 50 to 70% of the total amount (Elgin and Bonner, 1972). Although tissue specificity has been demonstrated for some nonhistones and significant quantitative variations have led to the belief that some nonhistones are important in the regulation of nuclear functions, most of the nonhistones are common among many tissues and may have structural roles in the nucleus.

Very little is known about the role of nonhistones in the structure of the chromosome fiber. Probably most of them are associated with active regions, where DNA replication and transcription are taking place, since nuclear enzymatic activities are included in this class. Other nonhistones in actively transcribed regions may not be directly associated with the chromosome fiber, but rather with the newly transcribed RNA in the form of ribonuclear proteins. It is possible to separate condensed chromatin from diffuse chromatin in nuclei and to show that the diffuse chromatin contains a higher amount of nonhistones. This is consistent with an "active" role for many of the nonhistones. Perhaps some of the nonhistones in diffuse chromatin are responsible for the diffuse state.

Recent observations on the structure of the eukaryotic nucleus implicate some nonhistones in the higher level folding of chromosome fibers. When nuclei are isolated in the presence of nonionic detergents, most of the nuclear membrane is removed while leaving the nuclear contents intact. Most of the chromatin may then be removed by nuclease digestion to reveal a network of protein containing a small amount of DNA. The residual structure is known as the nuclear matrix (Berezney and Coffey, 1974; Aaronson and Blobel, 1975; Riley et al., 1975). The protein composition seems to be quite simple, with only three or four major species resolved by acrylamide gel electrophoresis. The matrix may serve as a framework for the nucleus, while the DNA content suggests strong association with chromosome fibers. That is, the proteinaceous matrix may provide a skeleton on which to hang the chromosome fibers.

IV. MOLECULAR ORGANIZATION OF CHROMOSOME FIBERS

A. THE NUCLEOSOME AS A UNIVERSAL STRUCTURAL ELEMENT

There appear to be no exceptions to the rule that any tissue which has histones associated with its nuclear DNA contains DNA–histone complexes in the form of nucleosomes. Furthermore, nucleosome structure shows a great deal of evolutionary stability. One example is the marked similarity between the nucleosome core structures in yeast and humans. Each contains the four nucleosomal histones in the same molar ratios and about 140 bp of DNA. The evolutionary stability of the structure is also reflected in the high degree of histone amino acid sequence conservation.

Although interphase nuclei were studied first, it has become clear that nucleosomes persist throughout the cell cycle, including mitosis. Mitotic chromosomes of species as diverse as the slime mold *Physarum polycephalum* (Hozier and Kaus, 1976) and the Chinese hamster (Compton et al., 1976b) have been studied using biochemical techniques and electron microscopy. Mitotic chromosomes of *Physarum* display the same pattern of nucleosome monomer and higher-order multimer DNA species on polyacrylamide gels as do interphase nuclei. Mitotic chromosomes of Chinese hamster also display the characteristic DNA digestion pattern with micrococcal nuclease. In addition, electron microscope studies of hamster and mouse mitotic chromosomes show nucleosomes (Fig. 12) when the chromosome fibers are well dispersed (Rattner et al., 1975; Howze et al., 1976).

Within the interphase nucleus, all types of chromatin appear to contain nucleosomes. Centromeric (constitutive) heterochromatin appears to have the same nucleosome pattern as bulk chromatin (Lipchitz and Axel, 1976; Musich et al., 1977). Even regions of the chromosome fiber that are actively engaged in DNA replication and transcription may contain nucleosomes (Seale, 1975; Weintraub and Groudine, 1976).

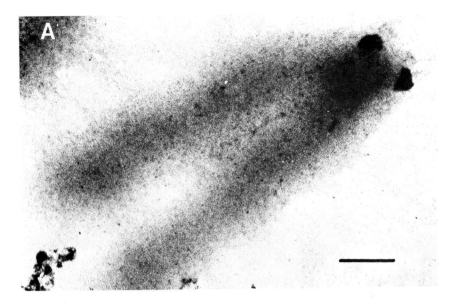

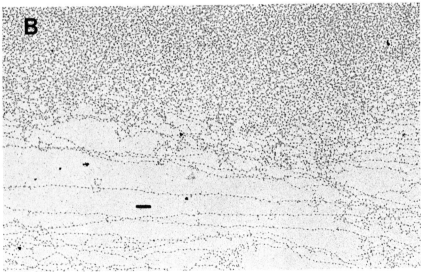

FIG. 12. Electron micrographs of metaphase mouse L 929 cells showing nucleo-somes. (A) Cells lysed in 0.7% Triton X-100 at pH 9.5 were deposited on a carbon-coated grid by centrifugation through sucrose. After drying, chromosomes were post-fixed with 10% formalin and stained with 1% ethanolic phosphotungstic acid. Scale line represents 1 μm. (B) Higher magnification of nucleosomes. More thorough dispersal of fibers was obtained by pretreatment with 10% formalin prior to centrifugation. Stained with 1% ethanolic phosphotungstic acid. Scale line represents 1000 Å. (From Rattner *et al.*, 1975, reproduced by permission.)

Thus the nucleosome seems to be the universal structural element in the eukaryotic chromosome regardless of the stage in the cell cycle or the degree of chromosome condensation. Any proposal for the structure of the native chromosome fiber must take into account the nucleosome as the first level of organization upon which higher levels may be superimposed.

B. RELATIONSHIP BETWEEN THE NATIVE FIBER AND THE CHAIN OF NUCLEOSOMES

The nucleosome chain may be folded under *in vivo* conditions to form a thicker fiber with a higher packing ratio. It is not likely that the thicker fiber seen in surface-spread preparations and thin sections is caused by embedding the nucleosome chain in additional protein, such as nonhistones, for although this would increase the fiber diameter, it would not increase the packing ratio. More likely, the ionic environment of the fiber determines its conformation in the native state.

Ris (1969) has shown that 200 Å fibers do indeed become thinner when exposed to low ionic strength buffers containing chelating agents, such as phosphate or EDTA, which bind divalent cations. The thinner fiber described by Ris is about 100 Å in diameter and appears quite knobby in surface-spread and critical point dried preparations. Some micrographs reveal obvious nucleosomal structures on the thin fiber. Clearly, nucleosome chains undergo a conformational transition to produce the thicker chromosome fiber.

In a recent series of papers (Renz *et al.*, 1977; Hozier *et al.*, 1977; Rentz *et al.*, 1978) the factors that determine whether the nucleosome chain is in a folded or unfolded state have been studied in detail. As in the case of nucleosomes, both electron microscopic and biochemical approaches have been used to determine the important features of fiber folding. Some of the biochemical studies were prompted by earlier results with DNA–H1 complexes which indicated that histone H1 might be capable of folding DNA. If H1 were responsible for the folding of nucleosome chains *in vivo*, previous experience with DNA–H1 complexes would indicate a cooperative, ionic strength-dependent process (Renz and Day, 1976). Therefore, the first approach was to isolate large chromosome fiber fragments at relatively high ionic strength, so that the isolated material would be in the native conformation. It was found that very large chromosome fiber fragments could be obtained from bovine lymphocyte nuclei by mild micrococcal nuclease digestion at 0°C in the presence of at least 1 mM $CaCl_2$, followed by lysis in buffers containing EDTA and at least 40 mM

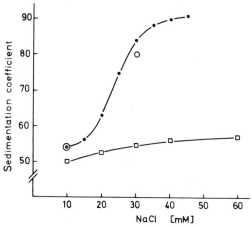

FIG. 13. Average sedimentation coefficients of chromosome fiber fragments at different NaCl concentrations. Aliquots of isolated fiber fragments produced by mild nuclease digestion of bovine lymphocyte nuclei were dialyzed against 1 mM sodium phosphate and 0.2 mM EDTA (pH 6.8), and various NaCl concentrations between 10 and 60 mM and sedimented through sucrose gradients in the same buffers. Average sedimentation coefficients (●) are plotted against NaCl concentration. (□) Histone H1 was removed after isolation of fibers and the samples were dialyzed to the appropriate buffer. (○) Histone H1 (20% by weight relative to DNA) was added back to H1-depleted fiber fragments before centrifugation. (From Renz et al., 1977, reproduced by permission.)

NaCl. High yields of very large fiber fragments (average sedimentation coefficient greater than 90 S) could be obtained in this way.

When the sedimentation coefficient of the isolated fiber fragments was studied as a function of NaCl concentration, a sharp decrease occurred in the range of 20–40 mM (Fig. 13), precisely the range in which DNA–H1 complexes show a transition in binding properties. In addition, when H1 was removed from isolated fiber fragments, the sedimentation coefficient was greatly reduced, and only a small change in sedimentation coefficient is seen in the same salt range. When H1 was added back in the proper molar ratio, the salt-dependent sedimentation behavior was restored.

This behavior is indicative of a histone H1-mediated transition in chromosome fiber structure. Electron microscopy of isolated chromosome fibers above and below the salt transition range confirms the structural basis of the transition to be the conversion of a 200 Å native fiber at high ionic strength to an open chain of nucleosomes at low ionic strength (Fig. 14). This structural transition is probably the same as the one observed by Ris (1975) for divalent cations. Most likely, a minimum monovalent or divalent cation concentration is required to

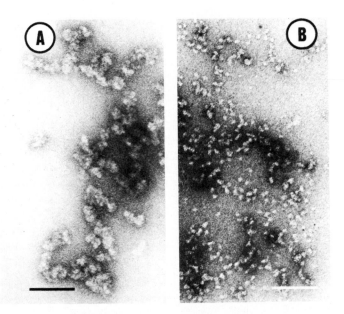

FIG. 14. Electron micrographs of chromosome fiber fragments (A) above (70 mM NaCl) and (B) below (10 mM NaCl) the structural transition range. Fiber fragments were fixed with glutaraldehyde in high and low salt, attached to carbon-coated electron microscope grids, and stained with 1% uranyl acetate. Scale lines represent 1000 Å. (From Renz et al., 1977, reproduced by permission.)

maintain the thick fiber configuration. This is in accordance with the NMR results of Bradbury et al. (1975), which showed that the mode of binding of H1 to chromatin (or DNA) is the same in the presence of monovalent or divalent ions.

This striking similarity between the salt-induced transition of chromosome fibers and DNA–H1 complexes is accompanied by other similarities indicating a cooperative binding mode for H1 in chromosome fibers (Renz et al., 1977). This is best shown in filter retention experiments in which isolated chains of nucleosomes of various lengths, from which H1 has been removed, are used instead of DNA (Fig. 15). In 40 mM NaCl, nucleosome chains pass through nitrocellulose filters but are retained after binding H1. When equal weights of trinucleosomes and nucleosome chains of various lengths compete for H1, the longer nucleosome chains compete more effectively, up to the level of octomers. In terms of the length of DNA contained in the nucleosomes, this behavior is almost identical to that of DNA–H1 complexes (Renz, 1975). A similar effect is seen when more extensively digested chromosome fiber fragments are fractionated on sucrose gra-

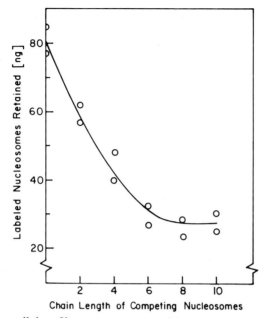

FIG. 15. Nitrocellulose filter retention assay of the competition between large and small histone H1-depleted nucleosome chains for histone H1. Histone H1 (40 ng) was added to 150 ng of ^{3}H-labeled nucleosomes (chain length = 3) and 150 ng of nucleosome chains of various lengths in 0.5 ml of 40 mM NaCl, 0.2 mM EDTA, and 5 mM Tris–HCl (pH 7.5). (From Renz *et al.*, 1977, reproduced by permission.)

dients in 80 mM NaCl (i.e., above the salt-dependent transition range). Monomers and low multimer chains have a full complement of nucleosomal histones, but a reduced amount of H1 relative to higher multimers. Thus, H1 prefers to bind to longer nucleosome chains. This size preference of H1 is abolished when nucleosome chains are sedimented in salt concentrations below the structural transition range (Renz *et al.*, 1977).

The structural role of H1 in the maintenance of the native fiber is unambiguous, and the salt-dependent structural transition may be responsible for some of the variability in the reports on chromosome fiber morphology.

C. FIBER FOLDING: X-RAY AND NEUTRON DIFFRACTION STUDIES

In the brief description of chromatin given at the beginning of Section III, it was stated that low ionic strength is necessary for solublization. In fact, biochemical studies on chromatin are usually done at ionic strengths well below the chromosome fiber structural transition range, where the open chain of nucleosomes exists. This is also true of most X-ray diffraction studies of fiber structure, upon which models of fiber structure have been built. Recent neutron diffraction studies using contrast matching techniques, along with the realization that salt concentration affects fiber structure, have led to the reevaluation of older X-ray diffraction studies that were made before the discovery of the nucleosome.

Chromatin displays a characteristic series of low angle X-ray diffraction rings at 110, 55, 37, 27, and 22 Å (Wilkins *et al.*, 1959; Luzzati and Nicolaieff, 1959). These maxima increase in intensity with increasing chromatin concentration, but finally disappear in dried specimens, to be replaced by a new set of maxima. (This effect has already been discussed in terms of specimen drying for electron microscopy.) The pattern of maxima in wet chromatin was interpreted in terms of a uniform supercoil of DNA with a diameter of 100–130 Å and a pitch of 120 Å (Pardon and Wilkins, 1972). The 110 Å reflection was attributed to the supercoil spacing of the DNA with the other reflections being due to higher orders of the spacing. The concentration dependence of the diffraction pattern was explained by the packing together of superhelical fibers.

Neutron diffraction studies of chromatin using contrast matching techniques have shown that the diffraction rings on chromatin do not behave in the manner expected from this model. If the uniform DNA supercoil model were correct, then the whole series of rings should go to a minimum when the DNA is contrast matched. It was found instead that the diffraction rings at 55 and 27 Å are due primarily to the DNA component of chromatin, while the rings at 110 and 37 Å are due to the protein component (Baldwin *et al.*, 1975). Therefore the DNA and protein have a different repeating arrangement, and the complete series of X-ray maxima cannot be due to a uniform DNA supercoil. This new information on the diffraction pattern of chromatin is more satisfactorily interpreted in terms of the nucleosomal model of fiber structure. The 110 Å ring would originate from the repeat spacing of the histone core of the nucleosome and the 37 Å ring from the packing of histones within the core. The 55 and 27 Å rings would come from the coiling of the DNA on the outside of the particle.

The chromatin used in all of these studies was prepared in very low ionic strength buffers, so there is no reason to believe that the arrangement of nucleosomes is the same as in the native chromosome fiber. However, Olins *et al.* (1975) have shown that isolated nuclei in buffers of nearly physiological ionic strength show the same set of X-ray diffraction rings as isolated chromatin. It seems likely that the native fiber can give the characteristic diffraction pattern, even though the fiber diameter is much greater than that of a simple nucleosome chain. Thus, the repeat spacing of nucleosomes in native fibers may be similar to that in chromatin in spite of the difference in fiber morphology. Carlson and Olins (1976) have calculated theoretical small angle X-ray scattering curves for a variety of nucleosome packing patterns, including the simple chain of nucleosomes expected in chromatin and other patterns that would yield a 200 to 250 Å chromosome fiber. Several of the thicker fiber models give the observed series of X-ray maxima and also give packing ratios expected of nucleosome chains and native fibers. This is consistent with the view that the thick chromosome fiber is built up from nucleosome chains.

D. MODELS OF MOLECULAR ORGANIZATION OF CHROMOSOME FIBERS

Three molecular models for the structure of the native chromosome fiber are presented here, each starting from the point of view that nucleosomes are the basic structural unit. They also attempt to take into account the influence of ionic strength on fiber structure.

1. *The Simple Chain of Nucleosomes*

This is the class of structures in which each nucleosome is in contact with only two other nucleosomes, the ones immediately adjacent to it on the DNA backbone. The simple chain of nucleosomes is almost always seen by electron microscopy at very low ionic strength (Olins and Olins, 1974) or after removal of histone H1 (Oudet *et al.*, 1975), but there are only a few cases in which the simple chain of nucleosomes can be visualized under conditions of ionic strength that would favor a native structure. For instance, Griffith (1975a) has demonstrated 110 Å fibers that appeared to be composed of closely spaced nucleosomes in cultured monkey cells disrupted in the presence of 0.15 M NaCl. Treatment of *E. coli* cells in a similar manner also produces 120 Å knobby fibers. This is particularly intriguing, since bacteria do not contain histones.

The most interesting example of the simple chain of nucleosomes is the simian virus 40 (SV40) minichromosome (Griffith, 1975b). Monkey cells lytically infected with SV40 accumulate a pool of viral DNA molecules which are complexed with cellular histones. When prepared in 0.15 M NaCl, the viral nucleohistone complex is a circular fiber (Fig. 16). The fiber width is about 100 Å and its contour length is

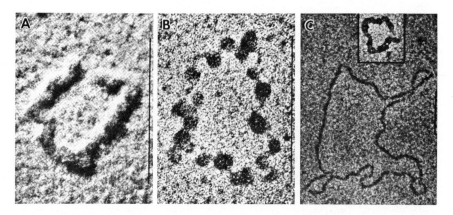

FIG. 16. SV40 minichromosomes. Electron micrographs of SV40 minichromosomes in (A) 150 mM NaCl, (B) 15 mM NaCl, and (C) after deproteinization and relaxation by X-ray treatment. The insert in (C) is to the same scale as the deproteinized DNA to show the sevenfold packing of DNA in a "native" minichromosome. All samples were prepared by absorption onto a carbon-coated grid, washing, dehydrating, and staining by vacuum tungsten decoration. Scale lines represent 1000 Å. (From J. D. Griffith, 1975, *Science* **187**, 1202–1203. Copyright 1975 by the American Association for the Advancement of Science.)

about 2100 Å. When the salt concentration is reduced tenfold, an average of 21 nucleosomes can be seen in the SV40 minichromosome. It is possible that the minichromosomes isolated in this study do not contain histone H1, which would account for the thin fiber morphology. The presence of H1 on SV40 chromatin *in vivo* is a matter of controversy. Other extraction techniques yield SV40 nucleohistone with H1 bound (Varshavsky *et al.*, 1976). Müller *et al.* (1978) have studied the structure of H1-containing SV40 chromatin under nearly physiological ionic strength conditions and found them to have a more compact conformation with globular subunits 190 Å in diameter. These large subunits unfold to form chains of nucleosomes when the ionic strength is lowered.

In any case, the SV40 minichromosome reveals several important aspects of nucleosome fiber organization. From the known length of the circular DNA molecule in SV40 (14,800 Å) and the contour length of the minichromosome, a packing ratio of about 7:1 is determined, which is close to the original estimate for a simple chain of nucleosomes based on X-ray diffraction and nuclease digestion data. Furthermore, a striking relationship exists between the number of nucleosomes in the minichromosome and the superhelical content of

SV40 DNA. Germond *et al.* (1975) isolated SV40 DNA in its native co-valently closed circular form, called DNA 1, which contains approximately 20 negative superhelical twists. The superhelical turns were removed with an enzymatic untwisting extract from mammalian cells (Champoux and Dulbecco, 1972) which forms covalently closed, relaxed circular DNA (DNA 1r). When histones were reassociated with DNA 1r to form nucleosomes, the circular nucleosome chain appeared twisted under electron microscopic examination [as opposed to the "relaxed" minichromosomes seen by Griffith (1975b) (Fig. 16)]. When these "twisted" minichromosomes were treated with untwisting extract, they became relaxed, but DNA isolated from the relaxed minichromosomes was again supercoiled in the same sense as DNA 1. Apparently, nucleosomes are responsible for the supercoiling of SV40 DNA *in vitro*. A careful analysis of the degree of supercoiling relative to the number of nucleosomes on SV40 DNA shows that each nucleosome is responsible for one superhelical turn.

Thus the minichromosome may be very valuable as a model for the structure of the mammalian chromosome fiber, but the lack of histone H1 makes it doubtful that it represents the conformation of the average native fiber. H1-containing minichromosomes (Müller *et al.*, 1978) are probably more closely related structurally to the average mammalian chromosome fiber.

2. A Regular Helix of Nucleosomes

A simple helical model of nucleosomal organization to form a high-order structure has been proposed by several laboratories (Finch and Klug, 1976; Alberts *et al.*, 1977; Carpenter *et al.*, 1976). Finch and Klug (1976) have presented electron micrographs of chromosome fibers mildly digested with micrococcal nuclease and dispersed in 0.2 m*M* EDTA. Fibers negatively stained with uranyl acetate are about 100 Å in diameter, but show little contrast variation along the fiber axis that would indicate nucleosomes. They attribute this picture to very close packing of nucleosomes in the 100 Å fiber, which they call a "nucleofilament." The addition of $MgCl_2$ to nucleofilaments causes them to condense and form thick fibers with diameters in the range of 300–500 Å. In favorable views the nucleofilament appears to be coiled up to form the thicker fiber, which is called a "solenoid." Formation of solenoids depends upon the presence of histone H1. H1-depleted nucleofilaments have the appearance of open chains of nucleosomes even in the presence of $MgCl_2$. Finch and Klug (1976) propose that the 110 Å ring in X-ray diffraction studies arises from the spacing between

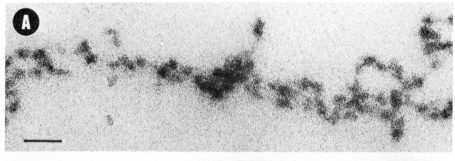

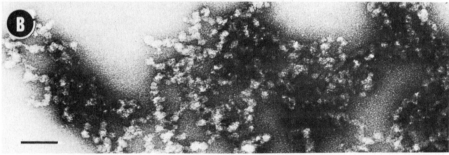

FIG. 17. Electron micrographs of spread preparations of chromosome fibers from (A and B) bovine lymphocytes and (C) rat liver cells. (A) Fibers from nuclei spread on a water surface and prepared for electron microscopy by the Ficoll drying technique. (B) Nuclei fixed with glutaraldehyde in 140 mM NaCl, 2.5 mM MgCl₂, 10 mM triethanolamine (pH 7.5), and then dialyzed to 1.0 mM EDTA for lysis. Fibers were prepared for electron microscopy by negative staining with 1% uranyl acetate. Scale lines represent 1000 Å. (From Hozier *et al.*, 1977.) (C) Chromosome fibers from murine sarcoma cells that were dispersed, spread, positively stained with ethanolic phosphotungstic acid, and shadowed with platinum:palladium. Insert shows higher magnification of fibers. Scale lines represent 1 μm (main micrograph) and 0.1 μm (insert). [(C) is from Franke *et al.*, 1976, reproduced by permission.)] In (A) and (B) average knob diameters are about 200–220 Å. In (C) they are about 260 Å.

nucleofilaments in the regular helical higher order structure. They estimate about six nucleosomes per turns of the helix to produce an overall packing ratio of 40:1 along the fiber axis.

A similar model is proposed by Alberts *et al.* (1977), where, instead of histone H1, crystal packing forces generated by direct interactions among the nucleosomes along the chain would stabilize the helix. Such a scheme might explain the extreme amino acid sequence stability of histones during evolution, since even the outer surfaces of nucleosomes which do not interact with DNA would be important in interactions with adjacent nucleosomes. Their model also takes into

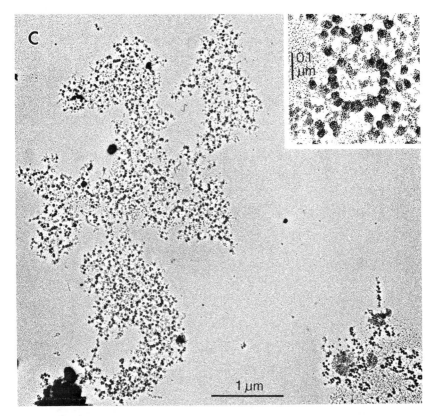

FIG. 17. (*Continued*)

account functional aspects of higher-order structure. They suggest that modification of histones would lead to slightly different helical arrays of nucleosomes which might correspond to different genetic units. These different microcrystalline arrays would respond differently to regulatory proteins. Any response in a receptive genetic unit (that is, a "target" for the regulatory element) is likely to be highly cooperative, since a local modification of the chromosome fiber by protein binding would lead to a destabilization of the microcrystalline array of nucleosomes. In this model gene structure is very closely tied to gene expression.

Carpenter *et al.* (1976) have based their helical model primarily on neutron diffraction studies of oriented chromatin fibers. In pulled and dried fibers from which histone H1 has been removed, the 100 Å reflection is slightly off the meridian. The angular displacement from the meridian is about 8° or 9°, which, together with a 100 Å pitch, is consistent with a helix of nucleosomes with an outside diameter of 300 Å and six nucleosomes per turn. The splitting of the 100 Å arc at the meridian is probably too small to be detected with standard X-ray diffraction apparatus. The advanced neutron detection system and the large specimen to detector distances available with neutron instruments are probably responsible for the improved detail in the 100 Å ring. It is interesting that these studies were done with H1-depleted chromatin precisely because removal of H1 results in improved definition in the low angle diffraction rings (Bradbury *et al.*, 1972). It is important to find out if H1 causes any significant change in the orientation of the 110 Å arc.

3. *The Superbead Model*

As stated previously, chromosome fiber fragments isolated so that they are not exposed to conditions of low ionic strength are about 200 Å in diameter. Bovine lymphocyte fibers isolated in this way have a distinctly knobby appearance (Figs. 17A and 17B). Kiryanov *et al.* (1976) and Franke *et al.* (1976) (Fig. 17C) have also observed 200 Å knobs on chromosome fibers. Hozier *et al.* (1977) (Figs. 17A and 17B) have observed knobby 200 Å fibers using several electron microscopic techniques and have analyzed the isolated chromosome fibers in detail biochemically. Fibers were always isolated in at least 40 mM NaCl. Lower ionic strength resulted in drastic changes in fiber morphology, including loss of knobby character.

The major biochemical finding is that apparent monomer "knobs" can be released from the chromosome fiber after digestion with micrococcal nuclease and can be isolated as a 40 S peak in sucrose gradients. The 40 S position appears to be a plateau region in the nuclease digestion kinetics of chromosome fibers. When the average sedimentation coefficient of isolated fiber fragments is plotted against digestion time, there is a definite leveling off in the average sedimentation rate at 40 S (Fig. 18). The 40 S plateau is absent when histone H1 is removed, and digestion rapidly produces a peak in the 11 S nucleosome monomer position.

From nucleosome multimer peaks superimposed on the 40 S peak and from analysis of DNA in the 40 S region by agarose gel electrophoresis, each isolated 40 S component appears to contain six to ten nucleosomes with few internal breaks. Structures with the dimensions of individual knobs can also be isolated by agarose gel chromatography of nuclease-digested chromosome fiber fragments. Frequency histograms of fiber fragment lengths for early eluting fractions show peaks at approximately integral multiples of 220 Å.

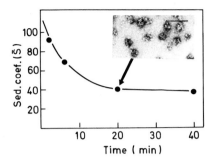

FIG. 18. Sedimentation coefficients of fiber fragments as a function of micrococcal nuclease digestion time in bovine lymphocyte nuclei. Average sedimentation coefficients ($\bar{S}$) for fiber fragments were measured in sucrose gradients containing 40 mM NaCl. Insert: Electron micrograph of individual 40 S fiber fragments isolated from sucrose gradients after a 20-minute micrococcal nuclease digestion. Fragments were fixed in glutaraldehyde and negatively stained with uranyl acetate. Average "knob" diameter is about 240 Å. Scale line represents 1000 Å. (From Hozier et al., 1977, reproduced by permission.)

The electron microscopic and biochemical observations are all consistent with a discontinuous superstructure in the native chromosome fiber, in which the DNA and chromosomal proteins are not distributed uniformly along the fiber axis. Instead, discrete assemblies of nucleosomes are present. The term "superbead" has been proposed to describe the individual 200 Å unit (Hozier et al., 1977). "Super" denotes the superstructural character, and "bead" refers both to the shape of the superstructural unit and to its nucleosomal makeup. In this model, trandem arrangements of superbeads in the native chromosome fiber are relatively resistant to micrococcal nuclease attack, while the DNA connecting them is sensitive. Thus 40 S "monomer" superbeads can be isolated which contain on the average eight nucleosomes. Reducing the ionic strength or removing H1 from the native chromosome fiber disrupts the superbead structure. Using an average of eight nucleosomes per superbead with a repeat length of 180 bp of DNA, one can calculate a fiber packing ratio of about 25:1 for a tandem arrangement of 200 Å superbeads.

Although very little is understood concerning the detailed structure of superbeads, it has recently been suggested that they may contain limited domains of nucleosome supercoiling (Worcel, 1978). Various aspects of chromosome function related to the superbead level of chromosome structural organization will be discussed in following sections.

V. Organization of Fibers into Chromosomes

A. structure of chromosomes in the interphase nucleus

1. *Chromosome Fibers in Interphase Mammalian Nuclei*

Thin section electron microscopy typically shows condensed chromatin localized at the nuclear periphery and diffuse chromatin in the interior region. One or more nucleoli are also usually in contact with clumps of condensed chromatin. Nuclear pores seem to have some structural significance in the organization of interphase chromosomes, since condensed chromatin masses are interrupted in the vicinity of pore complexes (Franke and Scheer, 1974). However, with the exception of the rather special arrangement of chromosome fibers near the inner nuclear membrane, very little is known about the organization of chromosome fibers at interphase.

Several studies have indicated an attachment of chromosome fibers at the nuclear periphery. For example, Brasch *et al.* (1971) have shown that after hypotonic swelling of chicken erythrocyte nuclei, chromatin fibers can be seen streaming from the nuclear envelope inward toward the center of the nucleus. A similar effect can be seen when living cells are subjected to an electric potential (intranuclear electrophoresis) in which fibers are pulled to one side of the nucleus, revealing connections to the nuclear envelope at the other side (Skaer *et al.*, 1976). Evidence that connections at the nuclear periphery involve the nuclear pore complex comes from electron microscopy of surface-spread nuclei (DuPraw, 1965; Comings and Okada, 1970). In these studies, 200 Å chromosome fibers can be seen converging at pore complexes. Usually more than one fiber is connected to each pore complex.

These observations suggest the possibility that the nuclear periphery provides points of anchorage for chromosome fibers at intervals along their length. An important and as yet unanswered question is whether the anchoring points are positioned nonrandomly and thereby introduce some order in the distribution of chromosome fibers at interphase. Arguments for such an orderly arrangement of chromosome fibers in interphase have been presented elsewhere (Comings, 1968; Vogel and Schroeder, 1974). Instead, the possible effects such anchoring points may have as topological constraints on fiber organization will be discussed here.

Biochemical as well as electron microscopical evidence is available indicating a topological constraint on the DNA in interphase nuclei. Cook and Brazell in a series of articles have shown that the sedimentation properties of isolated nuclei depleted of histones have a marked dependence on the concentration of ethidium bromide and other DNA intercalating agents. Nuclei prepared by the lysis of cells in detergent–EDTA mixtures containing at least 1 *M* NaCl sediment as intact structures (nucleoids) in 15–30% sucrose gradients. When ethidium bromide is added to the gradients, the sedimentation rate of the nuclear structures decreases markedly and then increases again as the ethidium concentration is increased further (Fig. 19). If the nucleoids are X-irradiated to introduce single-strand nicks into the DNA before sedimentation, the ethidium bromide concentration has little effect on sedimentation rate, and the structures sediment at the minimum rate for unirradiated nucleoids.

This is precisely the behavior expected of negatively supercoiled DNA in which strand rotation is inhibited (Wang, 1974). The interca-

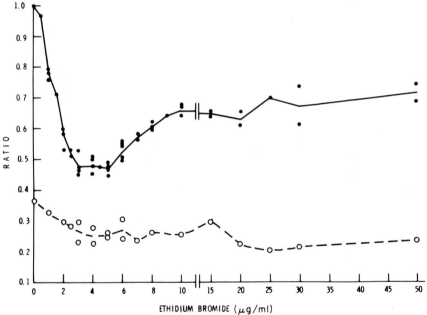

FIG. 19. Effect of ethidium bromide on the sedimentation of HeLa nucleoids. The distance sedimented by unirradiated (●) and irradiated (○) nucleoids in sucrose gradients containing different concentrations of ethidium is expressed as a ratio relative to nucleoids sedimenting under standard conditions (unirradiated nucleoids in the absence of ethidium). The difference between the sedimentation properties of irradiated nucleoids and nucleoids at the minimum caused by ethidium (5 μg/ml) may reflect an average superhelical conformation with significant differences in superhelical density in different domains. (From Cook and Brazell, 1975, reproduced by permission.)

lation of ethidium first relaxes the negative superhelical turns in the nucleoid causing a decrease in the sedimentation rate. Ethidium at higher concentrations introduces positive superhelical turns and causes an increase in sedimentation rate. When nicks are present in the DNA after X irradiation, ethidium intercalation no longer influences superhelix density and only causes a slight decrease in sedimentation rate. That is, the DNA behaves as if the topological constraint restricting DNA strand rotation has been removed.

Analysis of the radiation dose response leads to an estimate of about 10^9 daltons of DNA in each topologically closed domain. There would be about 2000 such domains per haploid DNA content if all the DNA is in this conformation. Actually, 10^9 daltons may be an overestimate of the average if there is considerable variability in size. Large domains (large X-ray targets) would have an inordinately large effect in sedimentation rate relative to small domains, so that these smaller domains may be missed in the analysis.

Cook and co-workers interpret their data as a demonstration that isolation of nuclei and removal of proteins produces a structure in which DNA is found in a high-energy supercoiled state. This configuration is unstable, and a relaxed low-energy state is produced by the introduction of single-strand breaks. From the previous discussion of the influence of nucleosomal histones on the superhelical density of DNA in SV40, it is easy to see how this situation may occur. One need only consider the native chromosome fiber containing assemblies of nucleosomes to be folded so that it forms topologically closed domains for the DNA. Multiple attachments to the nuclear periphery as described above may supply the necessary constraints. The folding of DNA into nucleosomes then provides the negative supertwists that are observed after removal of the histones. To strengthen this view further, Cook and Brazell (1977) have determined the number of superhelical twists in nucleoids to be about one in 200 bp of DNA, just as one would expect from the results with the SV40 minichromosome (Germond et al., 1975). To what extent the second level of fiber folding involving histone H1 may affect supercoiling is not known. The coiling of nucleosomes in the solenoidal model of Finch and Klug (1976) might be expected to affect the superhelix density if there were a preference for one supercoiling sense over the other. For the superbead model of fiber organization the situation is difficult to predict, since little is known about the internal organization of a superbead. In this general regard, it is interesting that histone H1 preferentially binds to supercoiled SV40 DNA over uncoiled DNA (Vogel and Singer, 1975). It may simply be that H1 is able to cross-link supercoiled DNA because of an increase in the number of cross-over points, since both negative and positive superhelical molecules are favored to the same degree over relaxed molecules.

In any case, the topological constraints that result in the super-coiled configuration of DNA after removal of nucleosomal proteins from nuclei may be responsible for the third level of chromosome structure, which sets the morphological pattern for interphase nuclei and for mitotic and meiotic chromosomes.

2. *Polytene Chromosomes of Dipterans*

In contrast to the near invisibility of fine structural features of chromosomes of interphase mammalian cells, even at the electron microscopic level, the chromosomes in certain tissues of dipteran species show a high degree of structural order at both the light and electron microscopic levels (Heitz and Bauer, 1933; Sorsa, 1973). For example, in salivary gland cells of *Drosophila* species, interphase chromatids undergo many cycles of replication with neither separation nor alternating condensation and decondensation at mitosis. All chromatids (of both homologous chromosomes) remain tightly bound so that many hundreds of chromatids lie side by side in perfect register. Thus the chromatids, in their extended interphase state, display their morphological features amplified more than a thousandfold. This feature gives them great value in studies of the morphology of interphase chromosomes.

The chromosomes appear as a linear array of dark "bands" and light "interbands." In *Drosophila melanogaster* the total number of bands is about 5000, each containing on the average 30,000 bp of DNA. There is much more DNA in the bands than in the interbands (Swift, 1962), but the total length of polytene chromosomes is distributed about evenly between them. There is strong genetic evidence that each band–interband region corresponds to a functional unit specifying one or only a few structural genes [for a review, see Beerman (1972)].

Until recently, however, it has not been possible to reconcile the striking morphological features of the interphase polytene nucleus with the diploid nuclear features in the same or other species at interphase. The connection may be provided by the work of Cook and Brazell (1976) and Benyajati and Worcel (1976) on the conformational constraints of nuclear DNA in diploid cell lines of *Drosophila*. These experiments parallel those of Cook and Brazell (1975) on human interphase nuclei. They demonstrate with the same techniques that DNA in *Drosophila* diploid nuclei contains negative superhelical turns that are due to the presence of nucleosomes in the chromosome fibers. Analysis of domain size by digestion with pancreatic DNase yields a value of about 85,000 bp, which is in the range of the amount of DNA per chromatid in the larger polytene bands (Benyajati and Worcel, 1976). As in the studies on mammalian nucleoid domain size, this method is likely to produce an overestimate of the average size. Therefore, the counterpart of the bands may have been found in the diploid state of *Drosophila* and by extension in diploid interphase nuclei of mammalian cells. The DNA superhelix density for *Drosophila* cells as measured by the interaction with ethidium bromide is also similar to that of mammalian cells, or about one negative supertwist per 200 bp, which again can be accounted for in terms of nucleosomal winding of DNA. In addition, Cook *et al.* (1976) using 2 *M* NaCl, detergent, and EDTA have observed the banding structure of polytene chromosomes complexed with ethidium bromide by fluorescence microscopy. The band–interband pattern is seen as bright and weakly fluorescing regions under conditions that simulate those used in their sedimentation studies. The analogous behavior of *Drosophila* and mammalian nucleoids in ethidium bromide–sucrose gradients and the maintenance of polytene chromosome morphology under similar conditions strongly suggest that polytene bands are the morphological equivalent of the topologically closed domains in diploid nuclei.

B. MITOTIC CHROMOSOMES

It has already been mentioned that mitotic chromosomes contain nucleosomes and that the native chromosome fiber at mitosis has the same overall morphology as in interphase. However, there are some aspects of mitotic chromosome morphology at the light microscopic level that are difficult to interpret in terms of fiber structure and organization. Of particular interest is the banding pattern that appears when mitotic chromosomes are fixed with methanol and acetic acid and stained with quinacrine or giemsa after appropriate pretreatment [for a review, see Evans (1977)]. The bands are alternating regions of light and dark staining material which are highly reproducible and characteristic for each chromosome. They are identically placed on sister chromatids. In humans, the total number of dark and light bands is about 320 in metaphase, but the number is higher in earlier stages of chromosome condensation. There are approximately 1260 bands per haploid set at late prophase (Yunis and Sanchez, 1975; Yunis, 1976).

The structural basis for mitotic chromosome banding is a matter of intense current research. Among the possibilities are that the various dyes used in staining are specific for certain regions along the chromosome, that proteins or DNA may be differentially extracted during processing, or that banding reflects the inherent distribution of chromatin in condensed chromosomes. In any case, it should be recognized that one is seeing only the remnants of the original native structure. Most of the histones are removed during fixation (Sumner *et al.*, 1973; Brody, 1974; Retief and Rüchel, 1977) so that the nucleosomal and native fiber levels of chromosome structure must be significantly altered. However, a large proportion of nonhistone protein is still present after fixation (Vogel *et al.*, 1974), perhaps leaving higher orders of structure preserved.

At least part of the structural basis for banding may be found in the distribution of chromosome fibers in surface spread electron microscopic studies of prophase chromosomes. Thre are characteristic discontinuities in fiber organization in the chromatids (Bahr *et al.*, 1973). Dense clusters of fiber loops, in which most of the loops run transverse to the chromatid axis, are spaced by regions where a smaller number of fibers run longitudinally (Fig. 7). Densitometric tracing of electron micrographs of human (Bahr and Larsen, 1974) and Indian muntjac (Green and Bahr, 1975) chromosomes show a general correspondence of high and low density fiber groupings with tracings of dark and light bands in giemsa- and quinacrine-banded chromosomes.

Chromosome fiber morphology does not change during chromosome condensation. This is in accordance with the persistence of nucleosomes throughout mitosis. The presence of histone H1 in mitotic chromosomes also makes it likely that the same sort of nucleosome fiber superstructure that exists in interphase chromosome fibers also persists in mitosis. Mitotic chromosomes show the same 200 Å knobby fiber as is seen in interphase. Therefore, all indications point to the stability of chromosome fiber organization at the nucleosomal level and at the level of nucleosome fiber folding by H1. Mechanisms of chromosome condensation may involve higher levels of folding. In fact, recent electron microscopic studies of partially deproteinized mitotic chromosomes show that long fibers form loops attached to central protein-rich regions along each chromatid called chromosome "scaffolds" (Paulson and Laemmli, 1977) or "cores" (Wray *et al.*, 1978). These structures may play some role in reorganizing chromosome fibers at mitosis.

C. MEIOSIS AND EMBRYOGENESIS

The germ line cells of eukaryotes undergo a special form of nuclear division in which the number of chromosomes is halved to produce a haploid set. The basic mechanism involves two successive divisions without intervening chromosome replication. The first meiotic division includes a particularly long and complex prophase in which the nuclear membrane persists and homologous chromosomes are paired along a highly organized structure called the synaptonemal complex (Moses, 1968). Genetic information is exchanged between homologues at this time. The following metaphase and anaphase lead, not to the separation of sister chromatids, but to the separation of homologous chromosomes. The second meiotic division resembles mitosis in that sister chromatids are separated. The net effect is four daughter cells, each with a haploid chromosome content. In males, all four go on to become sperm, while in females, only one of the four forms an egg while the other three form small polar bodies. Embryos are formed by the fusion of egg and sperm to form a new diploid cell, which undergoes rapid mitotic division. In some organisms the first several rounds of replication are very rapid and nuclear division may occur in the absence of cell division.

First meiotic prophase has been studied extensively by electron microscopy, but the emphasis has been on the structure and organization of the synaptonemal complex. The morphology of meiotic chromosome fibers has not been studied as extensively as mitotic fibers. Nevertheless, surface-spread meiotic chromosomes present the same sort of chromosome fiber morphology as in mitosis and interphase (Burkholder et al., 1972; Comings and Okada, 1975). Fibers are about 200 Å in diameter and have a knobby appearance. They can be seen attached to the synaptonemal complex and to the nuclear envelope during the early stages of meiotic prophase.

Nucleosomes have not yet been reported in meiotic nuclei, and the chemical makeup of meiotic fibers has not been extensively studied. Nevertheless, developmental stages of germ line cells before and after meiosis are known to contain nucleosomal histones. In fact, Adamson and Woodland (1974) have suggested that a pool of excess histone (in excess of the 1:1 weight ratio with DNA in chromatin) is built up during *Xenopus laevis* oogenesis and that the excess is used during DNA synthesis in early embryos. Estimates of the amount of histone in unfertilized oocytes range from 40 μg (Laskey et al., 1977) to 100 μg (Adamson and Woodland, 1974), while the eggs contain only about 6 pg of DNA. Thus there is a sufficient pool of histone to supply several rounds of DNA replication following fertilization. These histones would presumably be used to form nucleosomes with the new DNA.

It has recently been shown by McKnight and Miller (1976) that early cleavage embryos of *Drosophila* do indeed contain nucleosomes. The nucleosomes appear in the electron microscope to be similar to those in later stages of development (Woodcock *et al.*, 1976a). Biochemical studies (Spadafora *et al.*, 1976; Spadafora and Geraci, 1976) on sea urchin embryos involving micrococcal nuclease digestion of nuclei reveal the typical pattern of protected monomer and higher-order multimer DNA characteristic of nucleosomes.

For all the cases mentioned so far, nucleosomal organization appears to be the most likely first level structural arrangement. The evidence comes from nuclease digestion and electron microscopy, or at least from chemical composition studies showing the presence of histones. Unfortunately, even less is known about higher levels of fiber organization in gametes and embryos. Spermatozoa from some species (carp, goldfish, and sea cucumber) are known to possess a full set of histones, including H1, and the X-ray diffraction patterns from isolated nuclei are very similar to that of typical interphase chromatin (Subirana *et al.*, 1975). It is possible that the usual interphase fiber organization is retained. The histones may persist even in mature sperm of these species (DuPraw, 1970) so that nucleosome structure is never lost. However, order at the level of fiber folding may be very different in sperm in order to achieve a very high degree of chromatin packing.

There is at least one situation in which the nucleosomal level of organization may be completely absent. In some species all of the histones are removed from the DNA during spermatogenesis. The most thoroughly studied examples are trout and salmon, in which all histones are replaced by a special class of proteins known as protamines. These are small polypeptides which have an average molecular weight of about 4000 and are very rich in arginine residues. X-Ray diffraction patterns of intact sperm heads and isolated DNA–protamine show a prominent ring at about 29 Å (Feughelman *et al.*, 1955; Luzzati and Nicolaieff, 1963) over a wide concentration range, instead of the usual concentration-dependent series of diffraction rings from nucleohistone. Therefore, if one accepts the interpretation that the nucleohistone X-ray diffraction pattern is due to nucleosomal organization, it seems unlikely that protamines are capable of folding DNA into nucleosomes.

VI. Functional Aspects of Chromosome Fiber Organization

A. Fiber Structure During DNA Replication and Transcription

The nucleosomal and higher levels of fiber folding by nature affect the accessibility of DNA for interaction with regulatory elements and with enzymes. Of particular interest is the state of the chromosome fiber during replication of the chromosome and during transcription.

1. DNA Replication

DNA replication in eukaryotic cells takes place simultaneously at many growing points (replication forks). It seems likely that the chromosome fiber unfolds at the replicating fork so that the new complementary DNA strands may be synthesized and new chromosomal proteins added to the fiber. The biochemistry of DNA replication has been studied in detail, but until recently the assembly of new chromosome fibers has not been given much attention. Prescott and Bender (1963) had previously shown autoradiographically that tritium-labeled chromosomal proteins in ameba are randomly distributed between sister chromatids in subsequent generations. A more recent study using density labeling techniques (Jackson et al., 1975) reaches the same conclusion. However, neither approach yields information about the nucleosome level of chromosome assembly at the growing point.

Weintraub (1973), using very short pulses with [³H]thymidine to label the DNA in the vicinity of the replication forks in chick erythroblasts, has studied the sensitivity of this newly replicated DNA to pancreatic DNase. He found that very short pulses (less than 1 minute) labeled DNA that was more resistant to DNase than longer pulses. Based on an estimated rate of fork movement of 75 bp of DNA per second, approximately 500 Å of newly synthesized DNA seemed to be protected above the level of DNA in longer term pulses. It is not known if this protection is due to the presence of a large replication complex or some other special arrangement of chromatin at the growing fork.

After 1 minute the sensitivity of newly replicated DNA remains constant, indicating that newly made histone rapidly associates with the new DNA (Weintraub, 1973). When new histone synthesis is blocked with cycloheximide, DNA synthesis will continue for at least 30 minutes, although at a reduced rate (Weintraub and Holtzer, 1972).

Weintraub (1973) showed that after the initial period of high resistance to DNase I, the level of resistance in cycloheximide-treated cells dropped to about one-half the level of untreated cells, indicating that newly made DNA is packaged by fewer proteins when their synthesis is inhibited. Apparently, old histones do not redistribute to cover the new DNA.

The nucleosomal concept lead to refinements in these experiments by which one could distinguish between two possible modes of association of nucleosomal histones with newly replicated DNA. Weintraub (1976) showed that while 50% of the DNA is resistant to micrococcal nuclease digestion in isolated chick erythroblast nuclei, only 25% is resistant after replication in the presence of cycloheximide. Yet cycloheximide DNA displays a typical pattern of nucleosomal monomer and multimer species. This result appears to distinguish between cooperative and randomly dispersive distribution of old nucleosome histones of new DNA (Fig. 20). A dispersive alignment of parental nucleosomes would give normal nuclease resistance at the monomer level, but the multimer nucleosomal DNA species would not be produced in the usual form of discrete DNA sizes. However, a cooperative alignment would give the observed normal pattern of nucleosomal multimer DNA species. Furthermore, there is another interesting aspect of the alignment of nucleosomes in the absence of histone synthesis. The extent of nucleosome cooperativity appears to extend only to the octamer level, above which few DNA bands are seen in polyacrylamide gels (Weintraub, 1976). This may be related to

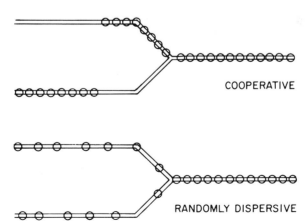

FIG. 20. Models explaining the alignment of old (parental) nucleosomes (open circles) on DNA when synthesis of new histones is blocked with cycloheximide. (From H. Weintraub, 1976, *Cell* **9**, 419–422. Copyright © 1976 MIT Press.)

the cooperative folding of an average octamer chain of nucleosomes by histone H1 as proposed in the superbead model of fiber structure (Hozier *et al.*, 1977; Renz *et al.*, 1978). Presumably, eight old nucleosomes are aligned on new DNA on one side of the replication fork and folded by H1 and the next eight on the other side, leaving wide gaps of free DNA (see Fig. 20). The average octamer would then be very prominent in gels after nuclease digestion. The gaps between new superbeads would account for the extra 25% digestion of DNA in cycloheximide-treated cells.

Using a different approach in which the cleavage kinetics of pulse labeled chromatin DNA was compared with that of parental DNA, Seale (1976) found that the nucleosome spacings on new chromatin closely approximate those on old chromatin after digestion with micrococcal nuclease. These results also indicate a nondispersive distribution of nucleosomes on new DNA, with the technical advantage that inhibition of protein synthesis was not necessary for the analysis. For very short digestion times the new chromatin DNA showed a broad but distinct peak at the octamer position similar to that in cycloheximide blocking, again indicating the cooperative association of average octamer chains of nucleosomes.

The cooperative nature of nucleosome alignment on newly replicated DNA strands may require other factors in addition to nucleosomal histones and DNA, for Oudet *et al.* (1975) failed to detect cooperative interactions among nucleosomes in reconstitution experiments. From previous arguments, it seems likely that histone H1 provides this cooperativity in folding nucleosomes into higher-order structures (Renz, 1975; Renz *et al.*, 1977; Hozier *et al.*, 1977). Consistent with this possibility are the DNase I digestion experiments of Burgoyne *et al.* (1976). Under their digestion conditions, nuclear DNA is resistant to DNase I. However, if nuclei are predigested with trypsin under conditions where H1 is preferentially degraded, the nuclease resistance is abolished. Newly replicated chromatin DNA also lacks resistance to DNase I attack, but the resistance is quickly regained after replication. Burgoyne *et al.* (1976) interpret these results in terms of a higher level folding of nucleosome chains, a function proposed here for histone H1.

2. Transcription and Fiber Structure

A persistent problem in analyzing the sites of active transcription is the difficulty in isolating them from the rest of the inactive DNA in chromosome fibers. The isolation techniques invariably use shear to break up the fibers. Afterward, the fragments are separated according to various criteria which are believed to be related to transcriptional activity [for a review, see Gottesfeld (1977)]. Of the various techniques, differential solubility of DNase II-sheared chromatin in 2 mM $MgCl_2$ has been characterized most thoroughly in terms of structure of the transcriptionally active fraction (Gottesfeld et al., 1974, 1975). Isolated chromatin is first digested with DNase II, followed by centrifugation to remove about 85% of chromatin DNA in the pellet. The supernatant is then made 2 mM in $MgCl_2$ and recentrifuged. The final supernatant, containing about 11% of the starting material, has a higher transcriptional activity, a higher RNA and nonhistone protein content, and a lower DNA sequence complexity and histone content than bulk chromatin. These are properties one might expect of transcriptionally active chromatin. Further DNase digestion of the first pellet material (presumably inactive chromatin) yields an 11 S peak in sucrose gradients. Chromatin in this fraction has the properties of nucleosomes, including a resemblance to nucleosome monomers in the electron microscope. However, the "active" soluble fraction displays peaks at 3–5 S, 14 S, and 19 S. There is no indication of a precursor–product relationship among these peaks. Therefore, chromatin in the soluble "active" fraction differs significantly in structure from the "inactive" fraction. The soluble fraction does not appear to be made up of typical nucleosome chains.

Another approach to the study of total transcriptionally active sites involves the isolation of total cytoplasmic mRNA and the use of DNA synthesized from it by RNA-dependent DNA polymerase (cDNA) as a probe for the structure of messenger-specifying chromatin. Kuo et al. (1976) have studied the kinetics of hybridization of cDNA made from total human leukemic leukocyte mRNA to total isolated mononucleosome DNA and showed that most of the sequences in mRNA are present in nucleosomes. The nucleosomes were produced by micrococcal nuclease digestion of isolated chromatin, and the authors have not ruled out the possibility that some redistribution of nucleosomal histones to cover transcribing regions may occur during digestion. However, their nucleosome fraction could support transcription in vitro with E. coli RNA polymerase, suggesting that nuclease digestion does not completely suppress template capacity.

These results are difficult to compare with those of Gottesfeld et al. (1975), since mRNA-specifying sequences make up only a small proportion of the total extent of transcription in the nucleus (Darnell, 1975). The presence of nucleosomes in messenger-specifying sequences is not incompatible with their apparent scarcity in a total active chromatin fraction.

In contrast to these approaches, in which total active fractions or total mRNA specifying sequences are studied, others have focused on the ultrastructure of individual genes. Both the globin gene in chick red blood cells (Weintraub and Groudine, 1976) and the ovalbumin gene in hen oviduct (Garel and Axel, 1976) appear to be preferentially sensitive to DNase I digestion. When 10% of the DNA of chick red cell nuclei is digested, hybridization experiments between the surviving DNA and cDNA prepared from globin mRNA shows that the globin-specifying sequences are removed. The globin sequences are not preferentially digested from other chick tissues which do not actively transcribing globin messenger. Micrococcal nuclease does not seem to recognize the same structural features of active globin genes as DNase I, since it does not show preferential digestion. Using the same approach with ovalbumin genes Garel and Axel (1976) have shown DNase I sensitivity of active genes in oviduct tissue and lack of sensitivity in nontranscribing tissues. Digestion of oviduct cell nuclei with micrococcal nuclease does not produce differential sensitivity of ovalbumin genes.

The two studies are in general agreement on the differential sensitivity of actively transcribed genes to DNase I and lack of sensitivity to micrococcal nuclease, and on the interpretation that nucleosomes are present in active genes in some altered configuration. However, there is a major difference in the possible nature of this altered configuration. For the globin gene, monomer nucleosomes prepared by digestion with micrococcal nuclease retain their differential sensitivity to DNase I (Weintraub and Groudine, 1976), while for ovalbumin genes, isolated nucleosomes are no longer differentially sensitive to DNase I (Garel and Axel, 1976). Based on their results, Weintraub and Groudine (1976) suggest that differential sensitivity in active genes is due to an altered nucleosome substructure, while Garel and Axel (1976) suggest that it is due to an alteration in the higher level folding of nucleosome chains. Again, the possibility of protein redistribution during digestion complicates these interpretations. Possibly *both* levels of fiber structure must be altered to allow transcriptional activity.

Electron microscopic studies of active gene transcription are presented in Chapter IV, so that only a few observations bearing on fiber structure will be mentioned here. For ribosomal genes, both the appearance of the fibers and the very low DNA packing ratio (McKnight and Miller, 1976; Laird *et al.*, 1976; Woodcock *et al.*, 1976c) seem to preclude the packing of active sequences into nucleosomes. However, spacer regions and inactive ribosomal genes do have the appearance of nucleosome chains (McKnight and Miller, 1976; Woodcock *et al.*, 1976c). Nucleosomes in robosomal DNA have also been detected by micrococcal nuclease digestion (Mathis and Gorovsky, 1976). These may correspond to the spacer regions or inactive genes. Alternatively, the transcriptional complex may offer some protection to the DNA being actively transcribed mimicking that usually provided by nucleosomes.

Transcriptional units considered to be nonribosomal generally have a lower density of RNA transcripts per unit length and a higher packing ratio than ribosomal genes (Laird *et al.*, 1976; Foe *et al.*, 1976). It seems most likely that they represent sites of heterogeneous nuclear RNA transcription. Nucleosomes are seen within and between these transcriptional units. It should be noted that the techniques used to visualize active transcription probably do not preserve higher orders of fiber structure.

The question of nucleosome arrangement in actively transcribed regions of the chromosome seems quite unsettled at present. The biochemical data indicate that nucleosomes are present on active fibers in an altered form. This, at least, seems consistent with the electron microscopic data, for even where nucleosomes are seen within transcribed regions, the calculated packing ratio is considerably lower than for inactive chromatin (Laird *et al.*, 1976).

B. HIGHER LEVELS OF FIBER ORGANIZATION AND THE SITES OF REPLICATION AND TRANSCRIPTION

1. *Electron Microscopic Studies of Diploid Nuclei*

In diploid nuclei, the sites of transcription and replication have been studied by cytochemical and autoradiographic techniques at the electron microscopic level. These studies have roughly localized the major concentrations of DNA, RNA, and protein within the various nuclear structures and are beginning to give valuable information concerning their metabolism [for a review, see Bouteille *et al.* (1974)]. Recent results localizing the sites of DNA and RNA synthesis in chromatin will be discussed briefly here.

As stated above, thin section electron microscopy reveals a rather sharp partition of chromatin into condensed and diffuse regions. The condensed chromatin is usually found in close association with the nuclear envelope and surrounding the nucleolus. Diffuse chromatin fills most of the remaining nuclear space.

The sites of DNA replication have been investigated by autoradiography, using tritium-labeled thymidine. Differences in pulse label length and cell synchronization techniques have probably been responsible for the variety of nuclear sites reported for DNA synthesis, including both condensed and diffuse chromatin (Bernier and Jensen, 1966; Blondel, 1968). Some studies have used radioactive pulses followed by nonradioactive chases in phytohemagglutinin-stimulated lymphocytes and other differentiating systems which do not require artificial synchronization (Milner and Hayhoe, 1968; Milner, 1969). Here a decondensation–recondensation cycle appears to be important in replication, and a large proportion of pulse label radioactivity is localized in the perichromatin area, the junction between diffuse and condensed chromatin. This is also the site of the most intense labeling with amino acid precursors, indicating that new chromosome fibers may be assembled here (Bouteille, 1972).

Autoradiographic studies of transcription are subject to the same sort of problems as in replication, but have led to localizations in diffuse chromatin and in the perichromatin region (Littau et al., 1964; Karasaki, 1968). Kinetic studies suggest that the perichromatic region is the site of initiation, followed by migration of transcripts to diffuse chromatin (Milner and Hayhoe, 1968; Fakan and Bernhard, 1971). In this regard, it is interesting that perichromatin granules, which are located at the condensed chromatin–diffuse chromatin junction, have been shown by cytochemical techniques to contain RNA (Bernhard, 1969). These particles are very similar in morphology to Balbiani granules in polytene chromosomes, and they respond to certain drugs in a manner suggestive of structures associated with nonnucleolar RNA synthesis (Heine et al., 1971).

These studies implicate diffuse chromatin and perichromatin in both DNA and RNA synthesis, and suggest that decondensation, or unfolding, of chromosome fibers is necessary for these functions. They are consistent with biochemical studies on isolated active chromatin fractions and on the structure of individual active genes, in that changes in fiber structure are necessary for activity.

2. Transcription in Polytene Chromosomes

Polytene chromosomes remain the best system for studying the structure–function relationship in transcription because of their banding pattern and the changes in the bands which are associated with transcription. It has long been known that certain bands become diffuse and swollen during development. This phenomenon, known as puffing, is specific for different developmental stages and usually occurs in about 10% of the total number of bands. Various lines of evidence have demonstrated that puffing is the morphological expression of active RNA synthesis. Of the three major classes of RNA, it appears that heterogeneous nuclear RNA (hnRNA) is the primary product of puff transcription, since the other two classes, low molecular weight RNA and ribosomal RNA, have been localized to a few bands and to the nucleolar organizer, respectively.

The banding pattern represents a differential compaction of chromatin along the chromosome axis. A quantitative look at the distribution of chromosomal components may help in understanding the structure of bands and their function in transcription. Although the 5000 bands and interbands in *Drosophila* polytene chromosomes are approximately equal in average length, the bands contain about 95% of the DNA (Beerman, 1972). For instance, in the X chromosome, which contains about 2.6×10^7 bp of DNA, there are about 800 bands and interbands with an average thickness of 0.12 μm. Therefore, there are about 30,000 bp of DNA in the average band and 1500 bp in the average interband on a single chromatid basis. The calculated packing ratios for bands and interbands are about 130:1 and 5:1, respectively. The packing ratio for interbands is that expected for an open chain of nucleosomes without histone H1, while the packing ratio in bands is greater than the 30:1 value expected from a native chromosome fiber containing histone H1. Indirect immunofluorescence studies (Plagens et al., 1976; Jamrich et al., 1977) using anti-H1 antibodies and aldehyde-fixed *Drosophila* polytene chromosomes are in good agreement with this view, since anti-H1 binds preferentially to bands (Fig. 21). It is then possible that interbands correspond to open chains of nucleosomes, while the higher packing of bands is the result of the crosslinking of nucleosomes by H1, plus additional foreshortening of the fiber by folding into topologically closed structures (see Section V).

Interbands and puffs react strongly with antibodies against RNA polymerase B, which is involved in hnRNA synthesis (Plagens et al., 1976; Jamrich et al., 1977). In contrast, newly induced puffs do not react with anti-H1 serum. This is consistent with a decondensation of the H1-folded native fiber structure during transcription. Jamrich et al. (1977) have interpreted the binding of anti-polymerase B at interbands as an indication of RNA synthesis at these sites. The presence of H1 in bands and its absence in newly induced puffs suggest that unpacking at the superbead level of fiber structure is necessary for active transcription.

C. CONDENSATION OF INTERPHASE CHROMOSOME FIBERS

At the molecular level very little is known about the events that lead to chromosome condensation at mitosis. However, as previously

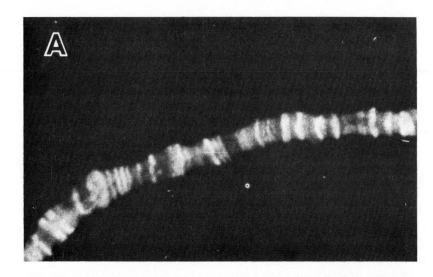

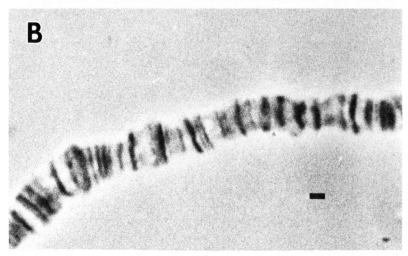

FIG. 21. Distribution of chromatin and histone H1 in the *Drosophila* X chromosome (from region 4A to 7E). (A) Indirect immunofluorescent visualization of histone H1 demonstrating primary localization in bands. (B) Phase contrast pattern of the same specimen after orcein staining, showing chromatin concentrated in bands. Scale line represents 1 μm. (From Jamrich *et al.*, 1977, reproduced by permission.)

mentioned, there are some features of chromosome fiber structure which appear to be constant throughout condensation. Nucleosomes can be detected in mitotic chromosomes by electron microscopic and biochemical techniques. Native chromosome fiber morphology in higher eukaryotes does not seem to change. In the acellular slime mold *Physarum polycephalum* the kinetics of fiber digestion is the same in interphase and mitotic nuclei (Hozier and Kaus, 1976) and a superbeadlike structure can be seen in interphase and mitotic *Physarum* chromosomes (Fig. 22). The kinetics of digestion and the superbead morphology, both dependent on histone H1, suggest that H1 plays a similar role in fiber structure at interphase and mitosis.

However, studies on the enzymatic phosphorylation of H1 suggest other roles for H1 in chromosome condensation. During the cell cycle in many species, histone H1 is phosphorylated at specific amino acid residues. In proceeding from G_1 through S phase and G_2 the level of phosphorylation increases and reaches a maximum level just before mitosis. This phenomenon has been most extensively studied in Chinese hamster cells (Gurley *et al.*, 1974; Hohman *et al.*, 1976) and in *Physarum polycephalum* (Bradbury *et al.*, 1974), and has implicated H1 in the condensation of chromosomes at mitosis.

In highly synchronized Chinese hamster cells, phosphorylation is absent in early G_1 cells or G_1-arrested cells. The first phosphorylation events precede the S phase. However, the most rapid phosphorylation occurs as the cells move from G_2 into mitosis. During the transition from mitosis to early G_1 the phosphate groups added before mitosis are lost. The mitotic phosphorylation involves at least four specific sites on histone H1. Langan and Hohman (1974) have isolated a chromatin-bound histone kinase which may be responsible for mitotic H1 phosphorylation.

Physarum facilitates studies of the cell cycle and the control of mitosis since it can be grown as a macroplasmodium in which up to 10^7 nuclei, contained in a common cytoplasm, traverse the cell cycle in nearly perfect synchrony. In addition, *Physarum* mitosis is intranuclear, so that mitotic nuclei can be isolated and compared with interphase nuclei. Studies on *Physarum* have shown that a peak in nuclear phosphokinase activity occurs in G_2, just before chromosome condensation begins. Associated with the increased phosphokinase activity is a peak in H1 phosphate content. Based on the temporal correlation of kinase activity and H1 phosphorylation with the onset of chromosome condensation, Bradbury *et al.* (1974) have proposed that H1 phosphorylation acts as a "mitotic trigger" to initiate chromosome condensation. In their model of condensation, the specific phosphorylation of H1 in late G_2 causes the phosphorylated regions to become more loosely associated with the chromosome fiber on account of repulsive interactions with DNA phosphates. The loosened segments of H1 would then be free to interact with each other, forming cross-links and a high degree of chromosome fiber aggregation.

Despite the temporal correlation between histone H1 phosphory-

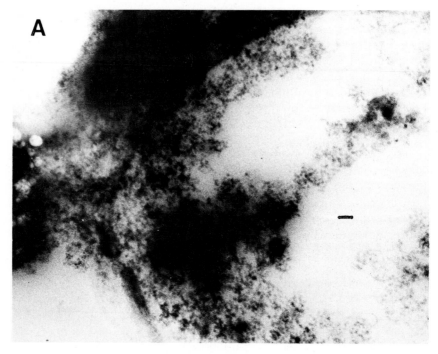

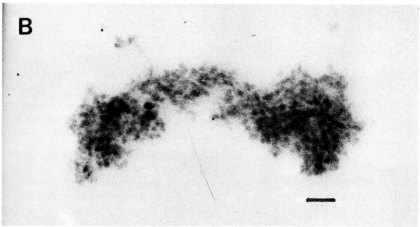

FIG. 22. Electron micrographs of surface spread chromosome fibers from *Physarum polycephalum* interphase (A) and mitotic (B) nuclei prepared by the Ficoll drying technique (Moses and Solari, 1976). Globular "superbead" structures have an average diameter of about 200 Å in both interphase and mitotic chromosome fibers. Scale lines represent 1000 Å.

lation and chromosome condensation, other factors must have a role in the formation and maintenance of mitotic chromosomes. This can be shown in the vegetative cells of the ciliated protozoan *Tetrahymena pyriformis*, which contain two different kinds of nuclei. The macronucleus divides amitotically and is responsible for RNA synthesis, while the micronuclei divide mitotically and are transcriptionally inert. Gorovsky and Keevert (1975) have reported that micronuclei contain no histone H1, even though they show distinct changes in the degree of chromosome condensation during the cell cycle. Furthermore, the presence of H1 in macronuclei and the correlation of its level of phosphorylation with transcriptional activity suggest a role in control of gene expression rather than chromosome condensation. In studies on mammalian cells, Tanphaichitr *et al.* (1976) have shown that a G_1 phosphatase activity can be inhibited with $ZnCl_2$ so that histone H1 is maintained in its highly phosphorylated mitotic form. Yet synchronized mitotic cells grown in the presence of $ZnCl_2$ pass into G_1 with normal chromosome decondensation and nuclear membrane formation in spite of the maintenance of high levels of H1 phosphorylation.

These points argue against a direct role for histone H1 or its attendant phosphorylation in chromosome condensation, although H1 phosphorylation may be a "trigger" in the sense that it is a requirement that must be fulfilled before mitotic condensation (in nuclei containing H1) can proceed. In their model Bradbury *et al.* (1974) state that other factors are indeed responsible for the completion of mitotic chromosome condensation.

Taylor (1969) and Comings and Riggs (1971) have proposed models for chromosome condensation in which long-range interactions along the chromosome fiber cause it to fold into a compact structure. In these models very little need happen at the level of the folding of nucleosomes into the native fiber. Instead, specific sites are envisioned to be present at rather long intervals along the fiber that are able to pair in some fashion to form fiber loops.

It has already been mentioned that the results of sucrose gradient–ethidium bromide studies of nucleoids are consistent with loop structures at interphase. The retention of distinct metaphase chromosome morphology under conditions of nucleoid isolation has also been demonstrated (Cook *et al.*, 1976). It would be interesting to study the sedimentation properties of metaphase chromsomes in the presence of ethidium bromide to see if their behavior is similar to interphase nucleoids. If this is the case, then perhaps chromosome condensation is caused by a rearrangement of fiber loops rather than by any important structural change within the fiber. Such a rearrangement of interphase chromosome fiber loops to form mitotic chromosomes could help explain the types of chromosome aberrations caused by X rays and other mutagens [for a review, see Comings (1974b)].

Just what factors may be involved in the rearrangement of chromosome fibers are not known, but some studies indicate that both nonhistone proteins and DNA are important. In premature condensation of interphase chromosomes caused by Sendai virus-mediated fusion with mitotic cells, tritium-labeled proteins in the cytoplasm of the mitotic cells appear to migrate specifically to the prematurely condensed interphase chromosomes (Rao and Johnson, 1974). In these autoradiography experiments, the migrating proteins are probably nonhistones, since the ratio of grains on interphase chromosomes relative to mitotic chromosomes is higher when labeled tryptophan is used than when labeled lysine or arginine, which are abundant in histones, are used.

In addition, chromosomes are often observed condensing onto the nuclear envelope at early prophase, before the dissolution of the nuclear membrane (Comings and Okada, 1971), and nuclear pore complexes appear to be attached to the fibers of metaphase chromosomes (Lampert, 1971; Maul, 1977). Since nuclear pore complexes are attached to interphase chromosome fibers (Comings and Okada, 1970) they may take part in the fiber rearrangement process at mitosis.

The possibility that the DNA backbone is at least indirectly involved in condensation comes from studies of 5-bromodeoxyuridine (BrdUrd) and 5-bromodeoxycytidine (BrdCyd) substitution on chromosome morphology at both the light and electron microscopic levels. BrdUrd and BrdCyd substitution leads to elongated chromosome forms at metaphase (Zakharov *et al.*, 1974), which appear not to be able to condense fully. Banded studies reveal extensive changes in chromosome morphology at the light microscopic level. Korenberg and Ris (1975) studied the morphology of BrdUrd-substituted chromosomes at the electron microscopic level and found that the elongation of chromatids is due to differences in the manner of fiber folding rather than in any change in the fiber itself. Therefore, the effects of DNA base analogue substitution appears to be at the level of fiber organization rather than fiber structure.

More detailed models of chromosome condensation can be constructed with these factors in mind. One might imagine that there are repetitive DNA sequences spaced by the length of DNA in each loop of chromosome fiber. Condensation would occur when these repetitive sequences are brought together by some recognition factor, such as the allosteric protein suggested by Comings and Riggs (1971) (Fig. 23). To keep condensation orderly, different chromosomes should have different sets of repetitive "condensation sequences" to prevent

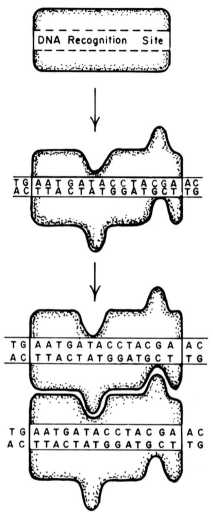

FIG. 23. Possible mechanism for the formation of multiple fiber loops by interactions of pairing proteins with specific DNA base sequences. After binding, allosteric changes occur which allow the proteins to bind to each other, forming loops of fiber. Multiple interactions of this type would cause chromosome condensation. (From Comings and Riggs, 1971, reproduced by permission.)

cross-condensation of chromosomes. Evidence for chromosome-specific intermediate repetitive DNA sequences has recently been presented for the human Y chromosome (Kunkel *et al.*, 1976) and for Chinese hamster chromosomes (Benz and Burki, 1978).

This level of chromosomal organization may prove to be important with regard to the mechanisms involved in the production and transmittal of chromosome aberrations. It almost certainly forms the basis of chromosome banding, so that a thorough understanding of the organization of chromosome fibers is crucial to further progress in the analysis of genetic defects and in phenotypic mapping. A combined biochemical and electron microscopic approach in studying the organization of chromosome fibers may present the best hope for understanding these structures at the molecular level.

REFERENCES

Aaronson, R. P., and Blobel, G. (1975). *Proc. Natl. Acad. Sci. U.S.A.* **72**, 1007–1011.
Adamson, E. D., and Woodland, H. R. (1974). *J. Mol. Biol.* **88**, 263–285.
Alberts, B., Worcel, A., and Weintraub, H. (1977). *In* "The Organization and Expression of the Eucaryotic Genome" (E. M. Bradbury and K. Javaherian, eds.), pp. 165–180. Academic Press, New York.
Bahr, G. F. (1970). *Exp. Cell Res.* **62**, 39–49.
Bahr, G. F. (1975). *Fed. Proc., Fed. Am. Soc. Exp. Biol.* **34**, 2209–2217.
Bahr, G. F. (1977). *In* "Molecular Structure of Human Chromosomes" (J. Yunis, ed.), pp. 143–203. Academic Press, New York.
Bahr, G. F., and Golomb, H. M. (1974). *Chromosoma* **46**, 247–254.
Bahr, G. F., and Larsen, P. M. (1974). *Adv. Cell Mol. Biol.* **3**, 191–212.
Bahr, G. F., Mikel, U., and Engler, W. F. (1973). *In* "Chromosome Identification: Technique and Applications in Biology and Medicine" (T. Caspersson and L. Zech, eds.), pp. 280–288. Academic Press, New York.
Baldwin, J. P., Boseley, P. G., Bradbury, E. M., and Ibel, K. (1975). *Nature (London)* **253**, 245–249.
Beerman, W. (1972). *In* "Developmental Studies on Giant Chromosomes" (W. Beerman, ed.), pp. 1–33. Springer-Verlag, Berlin and New York.
Benyajati, C., and Worcel, H. (1976). *Cell* **9**, 393–407.
Benz, R. D., and Burki, H. J. (1978). *Exp. Cell Res.* **112**, 155–165.
Berezney, R., and Coffey, D. S. (1974). *Biochem. Biophys. Res. Commun.* **60**, 1410–1417.
Bernhard, W. (1969). *J. Ultrastruct. Res.* **27**, 250–265.
Bernier, G., and Jensen, W. A. (1966). *Histochemie* **6**, 85–92.
Blondel, B. (1968). *Exp. Cell Res.* **53**, 348–356.
Bouteille, M. (1972). *Exp. Cell Res.* **74**, 343–354.
Bouteille, M., Laval, M., and Dupuy-Coin, A. M. (1974). *In* "The Cell Nucleus" (H. Busch, ed.), Vol. 1, pp. 3–71. Academic Press, New York.
Bradbury, E. M. (1975). *FEBS, Proc. FEBS Meet., 10th, 1975,* pp. 81–92.
Bradbury, E. M., Molgaard, H. V., Stephens, R. M., Bolund, L. A., and Johns, E. W. (1972). *Eur. J. Biochem.* **31**, 474–482.
Bradbury, E. M., Inglis, R. J., Matthews, H., and Langan, T. A. (1974). *Nature (London)* **249**, 553–556.
Bradbury, E. M., Danby, S. E., Rattle, H. W. E., and Giancotti, V. (1975). *Eur. J. Biochem.* **57**, 97–105.
Brasch, K., Seligy, V. L., and Setterfield, G. (1971). *Exp. Cell Res.* **65**, 61–72.
Brody, T. (1974). *Exp. Cell Res.* **85**, 255–263.

Burgoyne, L. A., Mobbs, J. D., and Marshall, A. J. (1976). *Nucleic Acids Res.* **3**, 3293–3304.

Burkholder, G. D., Okada, T. A., and Comings, D. E. (1972). *Exp. Cell Res.* **75**, 497–511.

Bustin, M., and Cole, R. D. (1968). *J. Biol. Chem.* **243**, 4500–4505.

Cairns, J. (1966). *J. Mol. Biol.* **15**, 372–373.

Callan, H. G., and MacGregor, H. C. (1958). *Nature (London)* **181**, 1479–1480.

Carlson, R. D., and Olins, D. E. (1976). *Nucleic Acids Res.* **1**, 89–100.

Carpenter, B. G., Baldwin, J. P., Bradbury, E. M., and Ibel, K. (1976). *Nucleic Acids Res.* **3**, 1739–1745.

Champoux, J. J., and Dulbecco, R. (1972). *Proc. Natl. Acad. Sci. U.S.A.* **69**, 143–146.

Chapman, G. E., Hartman, P. G., and Bradbury, E. M. (1976). *Eur. J. Biochem.* **61**, 69–75.

Cleaver, J. E. (1974). *Biochem. Biophys. Res. Commun.* **59**, 92–99.

Comings, D. E. (1968). *Am. J. Hum. Genet.* **20**, 440–460.

Comings, D. E. (1972). *Adv. Hum. Genet.* **3**, 237–431.

Comings, D. E. (1974a). *Cold Spring Harbor Symp. Quant. Biol.* **38**, 145–153.

Comings, D. E. (1974b). *In* "Chromosomes and Cancer" (J. German, ed.), pp. 95–133. Wiley, New York.

Comings, D. E., and Okada, T. A. (1970). *Exp. Cell Res.* **62**, 293–302.

Comings, D. E., and Okada, T. A. (1971). *Exp. Cell Res.* **63**, 471–473.

Comings, D. E., and Okada, T. A. (1975). *Exp. Cell Res.* **93**, 267–274.

Comings, D. E., and Riggs, A. D. (1971). *Nature (London)* **233**, 48–50.

Compton, J. L., Bellard, M., and Chambon, P. (1976a). *Proc. Natl. Acad. Sci. U.S.A.* **73**, 4382–4386.

Compton, J. L., Hancock, R., Oudet, P., and Chambon, P. (1976b). *Eur. J. Biochem.* **70**, 555–568.

Cook, P. R., and Brazell, I. A. (1975). *J. Cell Sci.* **19**, 261–279.

Cook, P. R., and Brazell, I. A. (1976). *J. Cell Sci.* **22**, 287–302.

Cook, P. R., and Brazell, I. A. (1977). *Eur. J. Biochem.* **74**, 527–531.

Cook, P. R., Brazell, K. A., and Jost, E. (1976). *J. Cell Sci.* **22**, 303–324.

Darnell, J. E. (1975). *In* "The Eukaryotic Chromosome" (W. J. Peacock and R. D. Brock, eds.), pp. 185–198. Aust. Natl. Univ. Press, Canberra.

Davies, H. G. (1968). *J. Cell Sci.* **3**, 129–150.

Davies, H. G., and Haynes, M. E. (1976). *J. Cell Sci.* **21**, 315–327.

Davies, H. G., and Small, J. V. (1968). *Nature (London)* **217**, 1122–1125.

Davison, P. F. (1959). *Proc. Natl. Acad. Sci. U.S.A.* **45**, 1560–1568.

Diacumakos, E. G., Holland, S., and Pecora, P. (1971). *Nature (London)* **232**, 33–36.

DuPraw, E. J. (1965). *Proc. Natl. Acad. Sci. U.S.A.* **53**, 161–168.

DuPraw, E. J. (1970). "DNA and Chromosomes." Holt, New York.

DuPraw, E. J. (1974). *Cold Spring Harbor Symp. Quant. Biol.* **38**, 87–98.

DuPraw, E. J., and Bahr, G. F. (1969). *Acta Cytol.* **13**, 188–205.

Elgin, S. C. R., and Bonner, J. (1972). *Biochemistry* **11**, 772–781.

Elgin, S. C. R., and Weintraub, H. (1975). *Annu. Rev. Biochem.* **44**, 725–774.

Elkind, M., and Kamper, C. (1970). *Biophys. J.* **10**, 237–245.

Evans, H. J. (1977). *Adv. Hum. Genet.* **8**, 347–438.

Evenson, D. P., Mego, W. A., and Taylor, J. H. (1972). *Chromosoma* **39**, 225–235.

Fakan, S., and Bernhard, W. (1971). *Exp. Cell Res.* **67**, 129–141.

Feughelman, M., Langridge, R., Seeds, W. E., Stokes, A. R., Wilson, H. R., Hooper, C. N., Wilkins, M. H. F., Barclay, R. K., and Hamilton, L. D. (1955). *Nature (London)* **175**, 834–838.

Finch, J. T., and Klug, A. (1976). *Proc. Natl. Acad. Sci. U.S.A.* **73**, 1897–1901.

Finch, J. T., Noll, M., and Kornberg, R. D. (1975). *Proc. Natl. Acad. Sci. U.S.A.* **72**, 3220–3222.

Foe, V. E., Wilkinson, L. E., and Laird, C. D. (1976). *Cell* **9**, 131–146.

Franke, W. W., and Scheer, U. (1974). *In* "The Cell Nucleus" (H. Busch, ed.), Vol. 1, pp. 220–238. Academic Press, New York.

Franke, W. W., Scheer, M. F., Trendelenberg, H., and Zentgraf, H. (1976). *Cytobiologie* **13**, 401–434.

Frenster, J. H. (1974). *In* "The Cell Nucleus" (H. Busch, ed.), Vol. 1, pp. 565–580. Academic Press, New York.

Fujiwara, Y. (1975). *Cancer Res.* **35**, 2780–2789.

Gall, J. G. (1963). *Nature (London)* **198**, 36–38.

Garel, A., and Axel, R. (1976). *Proc. Natl. Acad. Sci. U.S.A.* **73**, 3966–3970.

Georgiev, G. P. (1974). *In* "The Cell Nucleus" (H. Busch, ed.), Vol. 3, pp. 67–108. Academic Press, New York.

Germond, J. E., Hirt, B., Oudet, P., Gross-Bellard, M., and Chambon, P. (1975). *Proc. Natl. Acad. Sci. U.S.A.* **72**, 1843–1847.

Gorovsky, M. A., and Keevert, J. B. (1975). *Proc. Natl. Acad. Sci. U.S.A.* **72**, 2672–2676.

Gottesfeld, J. M. (1977). *Methods Cell Biol.* **16**, 421–436.

Gottesfeld, J. M., Garrard, W. T., Bagi, G., Wilson, R. F., and Bonner, J. (1974). *Proc. Natl. Acad. Sci. U.S.A.* **71**, 2193–2197.

Gottesfeld, J. M., Murphy, R. F., and Bonner, J. (1975). *Proc. Natl. Acad. Sci. U.S.A.* **72**, 4404–4408.

Green, R. J., and Bahr, G. F. (1975). *Chromosoma* **50**, 53–67.

Griffith, J. D. (1975a). *ICN-UCLA Symp. Mol. & Cell. Biol.* **3**, 201–208.

Griffith, J. D. (1975b). *Science* **187**, 1202–1203.

Gurley, L. R., Walters, R. A., and Tobey, R. A. (1974). *J. Cell Biol.* **60**, 356–364.

Hartman, P. G., Chapman, G. E., Moss, T., and Bradbury, E. M. (1977). *Eur. J. Biochem.* **77**, 45–51.

Heine, V., Sverak, L., Kondratick, J., and Bonar, R. A. (1971). *J. Ultrastruct. Res.* **34**, 375–396.

Heitz, E., and Bauer, H. (1933). *Z. Zellforsch. Mikrosk. Anat.* **17**, 68–82.

Henley, C. (1973). *Chromosoma* **42**, 163–174.

Hjelm, R. P., Baldwin, J. P., and Bradbury, E. M. (1978). *Methods Cell Biol.* **18**, 295–324.

Hohman, P., Tobey, R. A., and Gurley, L. H. (1976). *J. Biol. Chem.* **251**, 3685–3692.

Hoskins, G. C. (1968). *Nature (London)* **217**, 748–750.

Howze, G. B., Hsie, A. W., and Olins, A. L. (1976). *Exp. Cell Res.* **100**, 424–428.

Hozier, J. C., and Kaus, R. (1976). *Chromosoma* **57**, 95–102.

Hozier, J. C., and Taylor, J. H. (1975). *J. Mol. Biol.* **93**, 181–201.

Hozier, J. C., Nehls, P., and Renz, M. (1977). *Chromosoma* **61**, 301–317.

Huberman, J. A., and Riggs, A. D. (1966). *Proc. Natl. Acad. Sci. U.S.A.* **55**, 599–606.

Jackson, V., Granner, D. K., and Chalkley, R. (1975). *Proc. Natl. Acad. Sci. U.S.A.* **72**, 4440–4444.

Jamrich, M., Greenleaf, A. L., and Bautz, E. K. F. (1977). *Proc. Natl. Acad. Sci. U.S.A.* **74**, 2079–2083.

Johnson, J. D., Douvas, A. S., and Bonner, J. (1974). *Int. Rev. Cytol., Suppl.* **4**, 273–361.

Jolley, G. M., and Ormerod, M. G. (1974). *Biochim. Biophys. Acta* **353**, 200–214.

Karasaki, S. (1968). *Exp. Cell Res.* **52**, 13–26.

Kavenoff, R. (1972). *J. Mol. Biol.* **72**, 801–806.

Kavenoff, R., and Zimm, B. H. (1973). *Chromosoma* **41**, 1–27.
Kavenoff, R., Klotz, L. C., and Zimm, B. H. (1974). *Cold Spring Harbor Symp. Quant. Biol.* **38**, 1–8.
Kiryanov, G. I., Manamshjan, T. A., Polyakov, V. Ya., Fais, D., and Chentsov, Ju. S. (1976). *FEBS Lett.* **67**, 323–327.
Klotz, L. C., and Zimm, B. H. (1972). *J. Mol. Biol.* **72**, 779–800.
Korenberg, J. R., and Ris, H. (1975). *J. Cell Biol.* **67**, 221a.
Kornberg, R. D. (1974). *Science* **184**, 868–871.
Kornberg, R. D., and Thomas, J. O. (1974). *Science* **184**, 865–868.
Kraemer, P. M., Deaven, L. L., Crissman, H. A., Steinkamp, J. A., and Petersen, D. F. (1974). *Cold Spring Harbor Symp. Quant. Biol.* **38**, 133–144.
Kunkel, L. M., Smith, K. D., and Boyer, S. H. (1976). *Science* **191**, 1189–1190.
Kuo, M. T., Sahasrabudde, C. G., and Saunders, G. F. (1976). *Proc. Natl. Acad. Sci. U.S.A.* **73**, 1572–1575.
Laird, C. D., Wilkinson, L. E., Foe, V. E., and Chooi, W. Y. (1976). *Chromosoma* **58**, 169–192.
Lampert, F. (1971). *Humangenetik* **13**, 285–295.
Langan, T. A., and Hohman, P. (1974). *Fed. Proc., Fed. Am. Soc. Exp. Biol.* **33**, 1597.
Lange, C. S. (1975). *Biophys. Soc. Abstr.* p. 205a.
Laskey, R. A., Mills, A. D., and Morris, N. R. (1977). *Cell* **10**, 237–243.
Lauer, G. D., and Klotz, L. C. (1975). *J. Mol. Biol.* **95**, 309–326.
Lengyel, J., and Penman, S. (1975). *Cell* **5**, 281–290.
Lett, J. T., Klucis, E. S., and Sun, C. (1970). *Biophys. J.* **10**, 277–292.
Lipchitz, L., and Axel, R. (1976). *Cell* **9**, 355–364.
Littau, V. C., Allfrey, V. G., Frenster, J. H., and Mirsky, A. E. (1964). *Proc. Natl. Acad. Sci. U.S.A.* **52**, 93–100.
Luzzati, V., and Nicolaieff, A. (1959). *J. Mol. Biol.* **1**, 127–133.
Luzzati, V., and Nicolaieff, A. (1963). *J. Mol. Biol.* **7**, 142–163.
McBurney, M., Graham, F., and Whitmore, G. (1971). *Biochem. Biophys. Res. Commun.* **44**, 171–176.
McKnight, S. L., and Miller, O. L., Jr. (1976). *Cell* **8**, 305–319.
Mathis, D. J., and Gorovsky, M. A. (1976). *Biochemistry* **15**, 750–755.
Maul, G. G. (1977). *J. Cell Biol.* **74**, 492–500.
Milner, G. R. (1969). *Nature (London)* **221**, 71–72.
Milner, G. R., and Hayhoe, F. G. J. (1968). *Nature (London)* **218**, 785–787.
Mohberg, J., and Rusch, H. P. (1970). *Arch. Biochem. Biophys.* **138**, 418–432.
Mortimer, R. K., and Hawthorne, D. C. (1973). *Genetics* **74**, 33–54.
Moses, M. J. (1968). *Annu. Rev. Genet.* **2**, 363–412.
Moses, M. J., and Solari, A. J. (1976). *J. Ultrastruct. Res.* **54**, 109–114.
Müller, U., Zentgraf, H., Eiken, I., and Keller, W. (1978). *Science* **201**, 406–415.
Musich, P. R., Brown, F. L., and Maio, J. J. (1977). *Proc. Natl. Acad. Sci. U.S.A.* **74**, 3297–3301.
Noll, M. (1974). *Nature (London)* **251**, 249–251.
Noll, M., and Kornberg, R. D. (1977). *J. Mol. Biol.* **109**, 393–404.
Olins, A. L., and Olins, D. E. (1974). *Science* **183**, 330–332.
Olins, A. L., Carlson, R. D., and Olins, D. E. (1975). *J. Cell Biol.* **64**, 528–537.
Olins, A. L., Carlson, R. D., Wright, E. B., and Olins, D. E. (1976). *Nucleic Acids Res.* **3**, 3271–3291.
Oosterhof, D. K., Hozier, J. C., and Rill, R. L. (1975). *Proc. Natl. Acad. Sci. U.S.A.* **72**, 633–637.

Ormerod, M. G., and Lehmann, A. R. (1971). *Biochim. Biophys. Acta* **247**, 369–372.
Oudet, P., Gross-Bellard, M., and Chambon, P. (1975). *Cell* **4**, 281–300.
Pardon, J. F., and Wilkins, M. H. F. (1972). *J. Mol. Biol.* **68**, 115–124.
Pardon, J. F., Worchester, D. L., Wooley, J. C., Tatchell, K., Van Holde, K. E., and Richards, B. M. (1975). *Nucleic Acids Res.* **2**, 2163–2176.
Paulson, J. R., and Laemmli, U. K. (1977). *Cell* **12**, 817–828.
Plagens, U., Greenleaf, A. L., and Bautz, E. K. F. (1976). *Chromosoma* **59**, 157–165.
Prescott, D. M., and Bender, M. A. (1963). *J. Cell. Comp. Physiol.* **62**, Suppl. 1, 175–194.
Rao, P. N., and Johnson, R. T. (1974). *In* "Control of Proliferation in Animal Cells" (B. Clarkson and R. Baserga, eds.), pp. 785–800. Cold Spring Harbor Lab., Cold Spring Harbor, New York.
Rattner, J. B., Branch, A., and Hamkalo, B. A. (1975). *Chromosoma* **52**, 329–338.
Renz, M. (1975). *Proc. Natl. Acad. Sci. U.S.A.* **72**, 733–736.
Renz, M., and Day, L. A. (1976). *Biochemistry* **15**, 3220–3228.
Renz, M., Nehls, P., and Hozier, J. (1977). *Proc. Natl. Acad. Sci. U.S.A.* **74**, 1879–1883.
Renz, M., Nehls, P., and Hozier, J. (1978). *Cold Spring Harbor Symp. Quant. Biol.* **42**, 245–252.
Retief, A. E., and Rüchel, R. (1977). *Exp. Cell Res.* **106**, 233–237.
Riley, D. E., Keller, J. M., and Byers, B. (1975). *Biochemisty* **14**, 3005–3113.
Rill, R. L., and Van Holde, K. E. (1973). *J. Biol. Chem.* **248**, 1080–1083.
Rill, R. L., Nelson, D. A., Oosterhof, D. K., and Hozier, J. C. (1977). *Nucleic Acids Res.* **4**, 771–789.
Ris, H. (1969). *In* "Handbook of Molecular Cytology" (A. Lima-de-Faria, ed.), pp. 221–250. Am. Elsevier, New York.
Ris, H. (1975). *Struct. Funct. Chromatin, Ciba Found. Symp., 1975* No. 28, pp. 7–23.
Ris, H. (1978). *J. Cell Biol.* **79**, 107a.
Rudkin, G. T. (1965). *In Vitro* **1**, 12–20.
Sadgopal, A., and Bonner, J. (1970). *Biochim. Biophys. Acta* **207**, 227–239.
Seale, R. L. (1975). *Nature (London)* **255**, 247–249.
Seale, R. L. (1976). *Cell* **9**, 423–429.
Senior, M. B., Olins, A. D., and Olins, D. E. (1975). *Science* **187**, 173–175.
Simpson, J. R., Nagle, W. A., Bick, M. D., and Belli, J. A. (1973). *Proc. Natl. Acad. Sci. U.S.A.* **70**, 3660–3664.
Skaer, R. J., Whytock, S., and Emmines, J. P. (1976). *J. Cell Sci.* **21**, 479–496.
Solari, A. J. (1971). *Exp. Cell Res.* **67**, 161–170.
Solari, A. J. (1974). *In* "The Cell Nucleus" (H. Busch, ed.), Vol. 1, pp. 493–535. Academic Press, New York.
Sorsa, V. (1973). *Hereditas* **73**, 147–151.
Spadafora, C., and Geraci, G. (1976). *Biochem. Biophys. Res. Commun.* **69**, 291–295.
Spadafora, C., Noviello, L., and Geraci, G. (1976). *Cell Differ.* **5**, 225–231.
Sponar, J., and Sormova, Z. (1972). *Eur. J. Biochem.* **29**, 99–103.
Stein, G. S., and Kleinsmith, L. J., eds. (1975). "Chromosomal Proteins and Their Role in the Regulation of Gene Expression." Academic Press, New York.
Subirana, J. A., Puigjaner, L. C., Roca, J., Remedios, L., and Suau, P. (1975). *Struct. Funct. Chromatin, Ciba Found. Symp., 1975* No. 28, pp. 157–174.
Sumner, A. T., Evans, H. J., and Buckland, R. A. (1973). *Exp. Cell Res.* **81**, 214–222.
Swift, H. (1962). *In* "The Molecular Control of Cellular Activity" (J. Allen, ed.), pp. 73–125. McGraw-Hill, New York.
Tanphaichitr, N., Moore, K. C., Granner, D. K., and Chalkley, R. (1976). *J. Cell Biol.* **69**, 43–50.

Taylor, J. H. (1963). *In* "Molecular Genetics" (J. H. Taylor, ed.), Part I, pp. 65–113. Academic Press, New York.

Taylor, J. H. (1969). *In* "Genetic Organization" (E. W. Caspari and A. W. Ravin, eds.), Vol. 1, pp. 163–221. Academic Press, New York.

Taylor, J. H., Woods, P., and Hughes, W. (1957). *Proc. Natl. Acad. Sci. U.S.A.* **43**, 122–128.

Van Holde, K. E., Sahasrabudde, C. G., and Shaw, B. (1974). *Nucleic Acids Res.* **1**, 1579–1586.

Varshavsky, A. J., Bakayev, V. V., Chumackov, P. M., and Georgiev, G. P. (1976). *Nucleic Acids Res.* **3**, 2101–2113.

Vogel, F., and Schroeder, T. M. (1974). *Humangenetik* **25**, 265–297.

Vogel, T., and Singer, M. F. (1975). *Proc. Natl. Acad. Sci. U.S.A.* **72**, 2597–2600.

Vogel, W., Faust, J., Schmid, M., and Siebers, J. W. (1974). *Humangenetik* **21**, 227–236.

Wang, J. C. (1974). *J. Mol. Biol.* **89**, 783–801.

Weintraub, H. (1973). *Cold Spring Harbor Symp. Quant. Biol.* **38**, 247–256.

Weintraub, H. (1976). *Cell* **9**, 419–422.

Weintraub, H., and Groudine, M. (1976). *Science* **193**, 848–856.

Weintraub, H., and Holtzer, H. (1972). *J. Mol. Biol.* **66**, 13–35.

Wilkins, M. H. F., Zubay, G., and Wilson, H. R. (1959). *J. Mol. Biol.* **1**, 179–185.

Wolfe, S. L. (1968). *J. Cell Biol.* **37**, 610–620.

Woodcock, C. L. F., Safer, J. P., and Stanchfield, J. E. (1976a). *Exp. Cell Res.* **97**, 101–110.

Woodcock, C. L. F., Sweetman, H. E., and Frado, L. (1976b). *Exp. Cell Res.* **97**, 111–119.

Woodcock, C. L. F., Frado, L. L. Y., Hatch, C. L., and Ricciardiello, L. (1976c). *Chromosoma* **58**, 33–39.

Worcel, A. (1978). *Cold Spring Harbor Symp. Quant. Biol.* **42**, 313–324.

Wray, W., Mace, M., Daskal, Y., and Stubblefield, E. (1978). *Cold Spring Harber Symp. Quant. Biol.* **42**, 361–365.

Yunis, J. J. (1976). *Science* **191**, 1268–1270.

Yunis, J. J., and Sanchez, O. (1975). *Humangenetik* **27**, 167–172.

Zakharov, A. F., Baranovskaya, L. I., Ibriamov, A. I., Benjusch, V. A., Dementseva, V. S., and Oblapenko, N. G. (1974). *Chromosoma* **44**, 343–359.

Zentgraf, H., Falk, H., and Franke, W. W. (1975). *Cytobiologie* **11**, 10–29.

Zirkin, B. R., and Wolfe, S. L. (1972). *J. Ultrastruct. Res.* **39**, 496–508.

INDEX

A

5-Adenosylmethionine, 5, 89, 98–99
Autoradiography, of chromosomes, 318

B

Bromodeoxyuridine (BrdUrd), effects on
  methylation of DNA, 99–101
Buoyant density, of DNA, 142

C

CG doublets, 221
Chromatin
  core particles of, 96
  digestion, with micrococcal nuclease,
    270–277
Chromatography
  reverse-phase, 25
  RPC-5 column, 25
Chromosome
  components, 332–333
  condensation, 373–380
  continuity of DNA, 317–321
  DNA packing ratio, 333–334
  electron microscope analysis, 324–329
  interphase, 325–331
  mitotic, 331–332
  morphology, 316–317
  organization, 322–324
  polytene, 143, 165
  puffs, 165
Chromosome fibers, 346–357
  chain of nucleosomes, 351–357
  components, 334–346
  in DNA replication and transcription,
    366–371
  in early embryos, 364–365
  folding, 349–351
  helix of nucleosomes, 353–356
  in interphase nuclei of mammals, 358–
    361
  in mitotic chromosomes, 362–363
  in polytene chromosomes, 361–362
  organization, 358–380
  in meiosis, 364–365
  superbead model, 356–357

Circular dichroism, of histones, 264–265
Codon, bias in utilization, 226
  selection, 225
  synonym, 224
Core of nucleosomes, see Nucleosomes,
  core

D

Dideoxynucleotide, extension technique
  for DNA sequencing, 235
DNA
  in chromosomes, loop structure at inter-
    phase, 376–378
  coding, 5 S RNA, 137
  fragments, end labeling procedures, 31
  gyrase, 80, 82
  initiation of replication, 179
  initiation of transcription, 179, 183–
    185
  molecular cloning, 43–44
  negative superhelical turns, 68, 76–77
  noncoding, 158
  nonrepetitive, 134
  nontranscribed spacer, 129–131
  origins for replication, 86
  origins of replication, 213–220
    "A" protein, 212
    Bellet protein, 214
    sequences, 217–220
  protein complexes, binding assays, 41
  recombinant technology, 42–44
  repeated sequences, 230
  repetitive, 134
  replication initiation, 134
  replication, rolling circle model, 133
  sequence analysis, 178
  sequence change mechanisms, 154–155
  sequence
    genomic, 166
    intervening, 166
    repetitive, 118
    spacer, 155
  sequencing methods, 232–235
  supercoiling, 79, 84
  termination of transcription, 185–186

387

Molecular Biology

An International Series of Monographs and Textbooks

Editors

BERNARD HORECKER

Roche Institute of Molecular Biology
Nutley, New Jersey

NATHAN O. KAPLAN

Department of Chemistry
University of California
At San Diego
La Jolla, California

JULIUS MARMUR

Department of Biochemistry
Albert Einstein College of Medicine
Yeshiva University
Bronx, New York

HAROLD A. SCHERAGA

Department of Chemistry
Cornell University
Ithaca, New York

WALTER W. WAINIO. The Mammalian Mitochondrial Respiratory Chain. 1970

LAWRENCE I. ROTHFIELD (Editor). Structure and Function of Biological Membranes. 1971

ALAN G. WALTON AND JOHN BLACKWELL. Biopolymers. 1973

WALTER LOVENBERG (Editor). Iron-Sulfur Proteins. Volume I, Biological Properties—1973. Volume II, Molecular Properties—1973. Volume III, Structure and Metabolic Mechanisms—1977

A. J. HOPFINGER. Conformational Properties of Macromolecules. 1973

R. D. B. FRASER AND T. P. MACRAE. Conformation in Fibrous Proteins. 1973

OSAMU HAYAISHI (Editor). Molecular Mechanisms of Oxygen Activation. 1974

FUMIO OOSAWA AND SHO ASAKURA. Thermodynamics of the Polymerization of Protein. 1975

LAWRENCE J. BERLINER (Editor). Spin Labeling: Theory and Applications. Volume I, 1976. Volume II, 1978

T. BLUNDELL AND L. JOHNSON. Protein Crystallography. 1976

HERBERT WEISSBACH AND SIDNEY PESTKA (Editors). Molecular Mechanisms of Protein Biosynthesis. 1977

J. HERBERT TAYLOR (Editor). Molecular Genetics, Part III, Chromosome Structure